BASIC
MATHEMATICS

For Grade **9**

BASIC
MATHEMATICS

For Grade 9

ALGEBRA AND GEOMETRY

*GRAPHS OF BASIC POWER
AND RATIONAL FUNCTIONS*

TESFAYE LEMA BEDANE

Order this book online at www.trafford.com
or email orders@trafford.com

Most Trafford titles are also available at major online book retailers.

Printed in the United States of America.

ISBN: 978-1-4269-9764-8 (sc)
ISBN: 978-1-4269-9766-2 (hc)
ISBN: 978-1-4269-9765-5 (e)

Library of Congress Control Number: 2011962914

Trafford rev. 08/09/2012

 www.trafford.com

North America & international
toll-free: 1 888 232 4444 (USA & Canada)
phone: 250 383 6864 ♦ fax: 812 355 4082

CONTENTS

CHAPTER-1: LINEAR FUNCTION

CHAPTER-2: POLYNOMIAL AND QUADRATIC FUNCTIONS

CHAPTER-3: CUBIC, SQUARE ROOT AND RATIONAL FUNCTIONS

CHAPTER-4: GEOMETRY; REASONING AND PROOF, LINES AND CONGRUENT TRIANGLES

CHAPTER-5: RELATIONSHIPS IN TRIANGLES AND QUADRILATERALS

CHAPTER-6: STATISTICS

PREFACE

This book contains six chapters, each chapter contains notes with explanation, examples of solved problems and at the end of each chapter review problems with answers at the back of the book. The topics in this book are designed based on with the current level of high school grade nine text book. But I have added some extra topics in chapter four about logics just to help the users of this book in gaining wide concepts about logics.

My experience of teaching mathematics and science in my home town, Ethiopia for more than ten years in different middle and high school and giving tutor maths and science for middle and high school students in Atlanta helped me a lot to prepare this book. Finally, the goal of this book is to help the users assess their understanding of the concepts of high school mathematics and students self preparing for examination just to use as a reference.

TESFAYE LEMA BEDANE

ACKNOWLEDGMENTS

It is a pleasure to thankful to those who made this little book possible, such as my friends Solomon Nerisho, Zerihun Maruta and my nephew Solomon Heyi, they gave me the moral support I required. I also would like to give special thanks to my best friend Dessalegn Nemera, who is currently studying Bio-chemistry in University of Cincinnati for motivating me from the initial to the final steps of writing this little book.

Lastly, I offer my regards and blessings to all of those who supported me in any respect during the completion of this book.

Tesfaye Bedane

CHAPTER ONE

1-Linear Function

1.1   Definition of Linear Function
*   A function that can be graphically represented in the Cartesian coordinate plane by a straight line is called a Linear Function.

More about Linear Function
*   A Linear function is a first degree polynomial of the form, F(x) = mx + b, where m and b are constants and x is a real variable.
*   The constant m is called slope and b is called y-intercept.

Example-1: $y = 3x + 2$
               $y = 4x - 6$ $\Big\}$ are linear functions
               $y = -x + 1$

*   The graph of a linear function can be determined by two points.

1.2   Slope of a Line
        In a Cartesian coordinate plane or x, y coordinate plane the slope of a line is the ratio of rise to run or ratio of vertical increase to horizontal increase. If two distinct points $P_1(X_1, Y_1)$ and $P_2(X_2, Y_2)$ are points on a non-vertical line l, then the slope of the line l is given by:-

$$\text{Slope } m = \frac{\text{rise}}{\text{run}} = \frac{\text{vertical increase}}{\text{horizontal increase}} = \frac{y_2 - y_1}{x_2 - x_1}$$

Example 2:   What is the slope of the line passing through the points $P_1(3, 4)$ and $P_2(11, 8)$?

Solution: Slope $M = \dfrac{y_2 - y_1}{x_2 - x_1} = \dfrac{8 - 4}{11 - 3} = \dfrac{4}{8} = \dfrac{1}{2}$

the slope of the line is ½. If we start from any point P on the line and move 1 unit up and 2 units to the right, the point obtained is also on the line, this process can be performed indefinitely by starting from different points.

Example 3) What is the slope of the line passing through the points (2, -3) and (-4, 6)?

Solution: Slope $M = \dfrac{y_2 - y_1}{x_2 - x_1} = \dfrac{6 - (-3)}{-4 - 2} = \dfrac{-9}{6} = \dfrac{-3}{2}$

Note: You can also interchange P_1 and P_2, the result still remains the same.

The above slope can also be calculated as:-

$$M = \frac{-3 - 6}{2 - (-4)} = \frac{-9}{6} = \frac{-9}{6} = \frac{-3}{2}$$

Example 4) What is the value of C such that the line passes through the points (4, 6) and (-3, c) having the slope 4/3?

Solution:- $M = \dfrac{y_2 - y_1}{x_2 - x_1} = \dfrac{C - 6}{-3 - 4} = \dfrac{4}{3}$

$$\frac{C - 6}{-7} = \frac{4}{3}$$

$$3(C - 6) = -7(4)$$

$$3C - 18 = -28$$

$$3C - 18 + 18 = -28 + 18$$

$$3C = -10$$

$$C = \frac{-10}{3}$$

Example 5) What is the slope of the line passing through the origin and the point (2, 6)?

Solution:- If the line passes through the origin, then the value of the

$$\text{point is (0, 0), thus;- } M = \frac{y_2 - y_1}{x_2 - x_1} = \frac{6-0}{2-0} = \frac{6}{2} = 3$$

Example 6) What is the slope of the line passing through the point (-1, -1) and (1, 1)?

$$\text{Solution:- } M = \frac{y_2 - y_1}{x_2 - x_1} = \frac{1-(-1)}{1-(-1)} = \frac{2}{2} = 1$$

Exercises 1.1

1) Find the slope of the line passing through the points
   a) P_1 (5, -4) and P_2 (0, 6)
   b) P_1 (0, 0) and P_2 (6, 7)

   c) $P_1\left(\frac{1}{2}, -\frac{1}{3}\right)$ and $P_2\left(\frac{-1}{5}, \frac{1}{4}\right)$

   d) P_1 (-1, -2) and (3, 5)

2) What is the value of 'b' such that the line passes through the points (9, -4) and (b, -5) having slope -1?

3) What is the slope of the line that passes through the origin and

   the point $\left(-\frac{1}{2}, 5\right)$?

4) Find the slope of each of the following line
   a) 5x - 6y = 4   c) 3y - ¼ = 6x   e) y = x
   b) 2y - 1 = ¼   d) y + 2x = 3x - 1   f) y = 6

5) Find the slope of each of the following line in terms of c and d, where c, d different from zero.
   a) 3cx + dy = 5   c) xd + cy = 1
   b) cy + dx = 2cy - 4   d) dcx + cdy = 2x + 1
                         e) dy = dcx - 3

1.3 THE NATURE OF SLOPE OF A LINE

By observing the direction of a line we can speak about the nature of the slope, whether it is positive.

negative, zero or undefined. For such case observe the following lines.

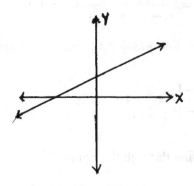

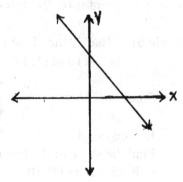

Graph of Positive slope
As we move from left to right
point is rising up

Graph of Negative slope
As we move from left to right
points is falling down.

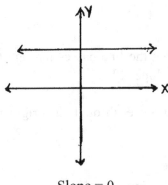

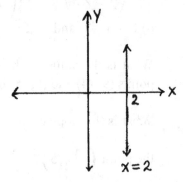

Slope = 0
Horizontal line

No slope
Vertical line

1.4 Equation of a Horizontal Line

NOTE:- A horizontal line is one that goes left-to-right, parallel to the x-axis of the coordinate plane. All points on the line will have the same y-coordinate.

The equation of a horizontal line is:-

$y = b$ where

x is the coordinate of any point on the line

b is where the line crosses the y-axis (y-intercept)

Notice that the equation of a horizontal line is independent of x. Any point on the horizontal line satisfies the equation.

1.5 Equation of a Vertical Line

- A vertical line is one that goes straight up and down parallel to the y-axis of the coordinate plane. All points on the line will have the same x-coordinate. A vertical line has no slope or put another way, for a vertical line the slope is undefined.
The equation of a vertical line is given by:-

$x = a,$ where

x - is the coordinate of any point on the line.

a is where the line crosses the x-axis

- (x-intercept).

Notice that the equation is independent of y. Any point on the vertical line satisfies the equation.

Example 7)   Draw the graph of x=5.

Solution:- First prepare a table.

x	5	5	5	5	5	5	5	5	5	5
y	-4	-3	-2	-1	0	1	2	3	4	5

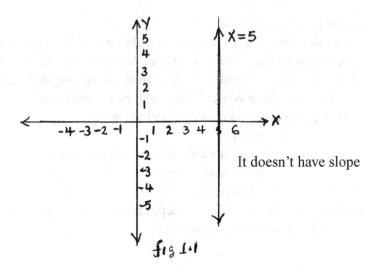

It doesn't have slope

fig 1.1

1.6 Rate of change connecting slope to real life.

The slope of a line tells us how something changes overtime. If we find the slope we can find the rate of change over that period.

Example 8)   The equation for the speed of a ball that is thrown straight up in the air is given by v = 156 - 32t, where v is the velocity (in feet/second) and t is the number of seconds after the ball is thrown.

a) What is the slope?
b) What does the slope mean in the equation?
Solution:-
a) The slope is -32, m = -32, This value tells us for an increase of 1 second, there is a decrease of 32 ft/sec
b) For every second, there is a decrease in speed by 32 feet per second. Eventually the velocity becomes zero. (When the ball reaches its maximum value), and then becomes negative

(when the acceleration due to gravity takes over and pulls the ball back down to the ground).

Example 9)   The table below shows the amount of water evaporating from a swimming pool on a hot day. Find the rate of change in gallons with respect to time.

Time (hours)	1	3	6
Gallons evaporated	2	6	12

$$\text{Rate of change} = \frac{\text{change in gallons}}{\text{change in time}} = \frac{6-2}{3-1} = \frac{4}{2} = 2$$

The rate of change is 2 gallons per hour.

1.7 Equation of a line

A linear equation or equation of a line has the following form, y = mx + b, where m is the slope and b is y-intercept which is where the line crosses the y-axis.

Example 10) Observe the slope and y-intercept of the following linear functions.
 a) y = 4x + 3 ..... (The slope is 4 and the y-intercept is 3) you can see that the graph of y = 4x + 3 is a straight line and it will crosses the y-axis at (0, 3). The slope is 4; therefore to find another point on the line and graph the function from (0, 3), move four units up and then 1 units to the right, you will find (1, 7). The graph - passes through these two points, i.e., (0, 3) and (1, 7). By drawing a line through these points you can see the graph of the linear function y = 4x + 3.

 b) $y = \frac{-1}{2}x + 6$ ... (The slope is $-\frac{1}{2}$ and the y-intercept is 6)

 c) Y = 3x .... (The slope is 3 and the y-intercept is 0)

 d) 5y + 3x - 1 = 0 ... (The slope is $-\frac{3}{5}$ and the y-intercept is $\frac{1}{5}$)

To find the slope and the y-intercept we can make some transformations to make y the subject of the equation or express y in terms of x.

$$5y + 3x - 1 = 0$$
$$5y + 3x - 3x - 1 + 1 = 0 - 3x + 1$$
$$5y = -3x + 1$$
$$y = \frac{-3}{5}x + \frac{1}{5}$$

From the final expression you can see the slope is $-\frac{3}{5}$ and the y-intercept is $\frac{1}{5}$.

Example 11) Find the equation of the line with a slope of 3 and passes through (0, -4).

Solution:- In this equation the y-intercept -4 is the point (0, -4), so to find the equation of the line, we use the slope-intercept form of a line: y = mx + b substitute values m = 3 and b = -4 into equations and you get y = 3x - 4. So that the equation of a line is y = 3x - 4.

Example 12) Find the equation of the line with slope -2 and passes through (3, 5).

Solution:- The slope is -2 and the point is (3, 5).

Note: This is a point the line passes through, but it is NOT the y-intercept so,

Step 1:- Substitute values in to equation (remember "m" represents the slope, and "b" represents the y-intercept) The y-intercept was not given in this problem. However, we were given the point, (3, 5). Thus x = 3 and y = 5

Substitute the values as follows to find the y-intercept "b".

$$y = mx + b$$
$$5 = -2(3) + b$$
$$5 = -6 + b$$
$$5 + 6 = -6 + 6 + b$$
$$11 = b$$

Thus, the equation of the line is y = -2x + 11

Example 13) Find the equation of the line passing through the points (2, 6) and (7, 5)

Solution:- First let us find the slope "m"

$$m = \frac{y_2 - y_1}{x_2 - x_1} = \frac{5-6}{7-2} = \frac{-1}{5}$$

The line is $y = \frac{-1}{5}x + b$

let us choose to use the point (2, 6)

$$y = \frac{-1}{5}x + b \quad \dots \text{ (substitute } x = 2, y = 6)$$

$$6 = -\frac{1}{5}(2) + b \Rightarrow 32 = 5b \Rightarrow b = \frac{32}{5}$$

$\therefore$ The equation of the line is $y = \frac{-1}{5}x + \frac{32}{5}$

1.7.1   Point-slope form of a line

$$y - y_1 = m (x - x_1)$$

Where m is the slope of the line and (x_1, y_1) is any point on the line. The point-slope form expresses the fact that the difference in the y coordinate between two points on a line (that is, $y - y_1$) is proportional to the difference in the x coordinate (that is, $x - x_1$). The proportionality constant is m (the slope of the line).

Example 14) What is the equation of the line passing through the point (0, 4) and slope 3?

Solution:- Given m = 3, $(x_1, y_1) = (0, 4)$

Using $y - y_1 = m (x - x_1)$

y - 4 = 3 (x - 0)

y - 4 = 3x - 0

y - 4 + 4 = 3x - 0 + 4

y = 3x + 4

Therefore, the equation of the line is y = 3x + 4

1.7.2 Two-point form of a line

$$y - y_1 = \frac{y_2 - y_1}{x_2 - x_1}(x - x_1)$$

Where (x_1, y_1) and (x_2, y_2) are two points on the line with $x_1 \neq x_2$. This is equivalent to the point-slope form above, where the slope is explicitly given as $\dfrac{y_2 - y_1}{x_2 - x_1}$

Example 15) What is the equation of the line passing through the points (2, 4) and (3, 8)?

Solution First let's find the slope m

$$M = \frac{y_2 - y_1}{x_2 - x_1} = \frac{8 - 4}{3 - 2} = \frac{4}{1} = 4$$

$$y - y_1 = \frac{y_2 - y_1}{x_2 - x_1}(x - x_1), \text{ where } \frac{y_2 - y_1}{x_2 - x_1} = 4$$

y - 4 = 4(x - 2)

y - 4 = 4x - 8

y - 4 + 4 = 4x - 8 + 4

y = 4x - 4

Therefore, the equation of the line is y = 4x - 4

1.8 Slope of Parallel and Perpendicular Lines

- Parallel lines are distinct lines lying in the same plane and they never intersect each other. Parallel lines have the same slope. In the figure below, lines AB and CD are parallel.

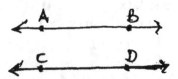

- Perpendicular lines are lines that intersect at right angles. If two lines are perpendicular to each other, then the product of their slopes is equal to -1.

In the figure shown below, the line AB and CD are perpendicular to each other.

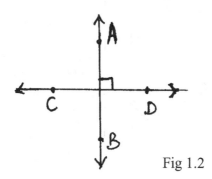

Fig 1.2

Example of two parallel lines.

As you can see from the diagram below, these lines
- Have the same slope, m = 2
- Never going to intersect.

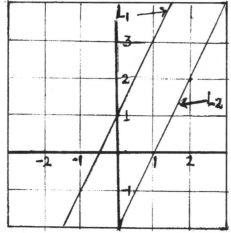

L_1: y = 2x + 1
L_2: y = 2x - 2

L_1 is parallel to L_2 and they have the same slope m = 2

Fig 1.2 shows parallel lines

Example of two perpendicular lines.

As you can see from the figure below:
- The slope of these two lines are negative reciprocals.

Thus, 2 and $-\frac{1}{2}$ are negative reciprocals.

These two lines are perpendicular and intersect at 90 degrees.

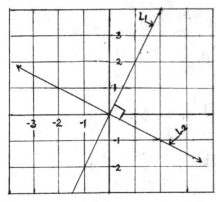

$L_1: y = 2x$

$L_2: y = -\frac{1}{2}x$

Thus, L_1 perpendicular to L_2 and their slopes are negative reciprocals.

The slope of L_1 is 2 and the slope of L_2 is $\dfrac{-1}{2}$

Fig 1.3 shows perpendicular lines

1.9 Straight Line Equations: Parallel and Perpendicular

Example 1)   Given the line 3x - 6y = 9 and the point (3, 2), find the lines through the point that are
(a) parallel to the given line and
(b) perpendicular to it.
Solution: a) 3x - 6y = 9

Step 1) Solve for "y" and you can find the reference slope

$$3x - 6y = 9$$
$$3x - 3x - 6y = 9 - 3x$$
$$-6y = -3x + 9$$
$$y = \frac{-3}{-6}x + \frac{9}{-6}$$
$$y = \frac{1}{2}x - \frac{3}{2}$$

So the reference slope from the reference line is m = ½ since the parallel lines have the same slope, then the parallel lines through the point (3, 2) will have slope m = ½. Thus, you have a point and a slope. So you can use point-slope form to find the line.

$$y - y_1 = m(x - x_1)$$

$$y - 2 = \frac{1}{2}(x - 3)$$

$$y - 2 = \frac{1}{2}x - \frac{3}{2}$$

$$y - 2 + 2 = \frac{1}{2}x - \frac{3}{2} + 2$$

$$y = \frac{1}{2}x - \frac{3 + 4}{2}$$

$$y = \frac{1}{2}x + \frac{1}{2}$$

Therefore the line is $y = \frac{1}{2}x + \frac{1}{2}$

Solution b) For the perpendicular line. You have to find the perpendicular slope. The reference slope is $M = \frac{1}{2}$, and for the perpendicular slope you will take the negative reciprocal of the reference slope, that is m = -2, so now you can do the point-slope form.

$$y - y_1 = m(x - x_1)$$
$$y - 2 = -2(x - 3)$$
$$y - 2 = -2x + 6$$
$$y - 2 + 2 = -2x + 6 + 2$$
$$y = -2x + 8$$

Therefore, the line is y = -2x + 8

Thus, the required lines are:-

Parallel: $y = \frac{1}{2}x + \frac{1}{2}$ and

Perpendicular: y = -2x + 8

Example 2) Find the equation of a line passing through the point (2, 5) and perpendicular to the line y = 3x - 1.

Solution If two straight lines are perpendicular, then the product of their slope is -1, thus

$M_1 M_2 = -1$ M_1 is slope of the line passing through
$M_1(3) = -1$ the point (2, 5) and M_2 is slope of

$M_1 = -\frac{1}{3}$ Y = 3x - 1, which is 3

Exercise 1.2

1) Find the equation of the line passing through the point (-1, -1) and parallel to the line y = -5x - 3

2) If L_1 and L_2 are two parallel lines and the slope of L_1 is -6, what is the slope of L_2?

3) If two lines L_1 and L_2 are perpendicular, and the slope of L_1 is $\dfrac{-5}{4}$, what is the slope of L_2?

4) Find the equation of the line passing through the point (-3, -2) and perpendicular to the line $y = \dfrac{-1}{5}x - 3$.

5) Given that $Y_1 = dx - 5$ and $Y_2 = 4x - 3$, If Y_1 and Y_2 are parallel, find the value of d.

6) If $L_1: y = y = -3x - 2$ and $L_2: y = bx - 1$ such that L_1 is perpendicular to L_2. What is the value of 'b'?

7) Draw the graphs of each of the following pairs of lines on the same x-y coordinate plane.
a) f(x) = 2x - 1 and g(x) = 2x + 2

b) $f(x) = \dfrac{1}{3}x + 1$ and h(x) = -3x -1

1.10 GRAPHS OF LINEAR FUNCTIONS

As mentioned before, any function of the form f(x) = MX + b, where m is not equal to zero is called a Linear Function. The domain (the value of x) of this function is the set of all real numbers. The range (the value of y) of the function is the set of all real numbers. The graph of f is a line with slope M and Y-intercept b.

Note:- A function f(x) = b, where b is a constant real number is called a constant function. Its graph is a Horizontal line at y = b.

Example) Graph the linear function f given by f(x) = 2x - 4

Solution:-   you need only two points to graph a linear function. These points may be chosen as the x and y intercepts of the graph, for example

Determine the x-intercept, f(x) = 0 and solve for x.

$f(x) = 0 \Rightarrow 2x - 4 = 0$

$2x - 4 + 4 = 0 + 4$

$2x = 4$

$x = 2$

Determine the y intercept, set x = 0 to find f(0)

$f(x) = 2x - 4$

$f(0) = 2(0) - 4$

$f(x) = -4$

The y-intercept is -4

- The graph of the above function is a line passing through the points (2, 0) and (0, -4) as shown below:-

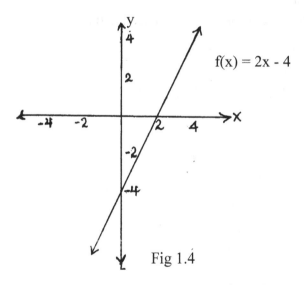

Fig 1.4

1.10.1 Using Table of Values to Graph Linear Equations.

You can draw the graph of a linear function using a table of values. A table of values is a graphic organizer or chart that helps you determine two or more points that can be used to create your graph.

Example 2) Draw the graph of a linear function y = 2x + 2 using tables of values.

$y = 2x + 2$	x	-2	-1	0	1	2
	y	-2	0	2	4	6

Why you use a Table of Values?

- In order to graph a line, you must have two points. For any given linear equation; there are an infinite number of solutions or points on that line.
- If you just find two of the solutions, then you can plot your two points and draw a line through. This will be the line that represents the equation. Every point on that line is a solution to the equation.

Example 2) Graph y = 2x + 3

Solution   1) Begin by making a table (choose convenient values for x)
2) plug the x-values into the original equation and find the values for y.
3) complete the table
4) Graph the points and draw a line by "connecting the dots."

$y = 2x + 3$	x	-2	-1	0	1	2
	y	-1	1	3	5	7

Here's what it looks like:-

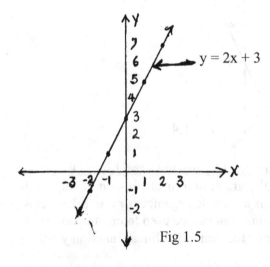

y = 2x + 3

Fig 1.5

16

Example 3)   Graph y = x - 2 or f(x) = x - 2
    To draw the graph of y = x - 2, use table

y = x - 2	x	-3	-2	-1	0	1	2	3
	y	-5	-4	-3	-2	-1	0	1

From the above table   f(-3) = -5
                        f(-2) = -4
                        f(-1) = -3
                        f(0) = -2
                        f(1) = -1
                        f(2) = 0
                        f(3) =1

Now we can observe the points (-3, -5), (-2, -4), (-1, -3), (0, -2), (1, -1), (2, 0), (3, 1) are all on the line y = x - 2, just by marking any of these two points on the plane we can draw the graph:-

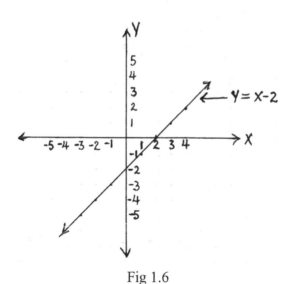

Fig 1.6

1.11   GRAPHS OF ABSOLUTE VALUE FUNCTIONS

For any real number a the absolute value or modulus of a is denoted by |a| (a vertical bar on each side of the quantity) and is defined as

$$|a| = \begin{cases} a, & \text{if } a \geq 0 \\ -a, & \text{if } a < 0 \end{cases}$$

As can be seen from the above definition, the absolute value of a is always either positive or zero, but never negative.

From an analytic geometry point of view, the absolute value of a real number is the number's distance from zero along the real number line, and more generally the absolute value of the difference of two real numbers is the distance between them.

Note:- Absolute value functions are transformations of the parent function $f(x) = |x|$.

The graph of the absolute value function $f(x) = |x|$ is similar to the graph of $f(x) = x$, except that the "negative" half of the graph is reflected over the x-axis. Here is the graph of $f(x) = |x|$:

$f(x) = |x|$	x	-3	-2	-1	0	1	2	3
	f(x)	3	2	1	0	1	2	3

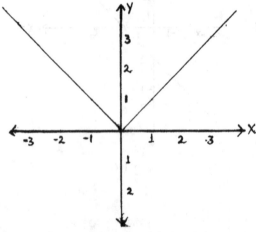

Fig 1.7 Graph of $f(x) = |x|$

The graph looks like a "V" shape, with vertex at (0, 0). Its slope is m = 1 on the right side of the vertex, and m = -1 on the left side of the vertex.

Note:- We can translate, stretch, shrink and reflect the graph.

Example) Draw the graph of y = |2x|, y = |-3x| and y = |x| on the same x-y coordinate plane.

Solution:-

y = |2x|	x	-3	-2	-1	0	1	2	3
	y	6	4	2	0	2	4	6

y = |-3x|	x	-3	-2	-1	0	1	2	3
	y	9	6	3	0	3	6	9

y = |x|	x	-3	-2	-1	0	1	2	3
	y	3	2	1	0	1	2	3

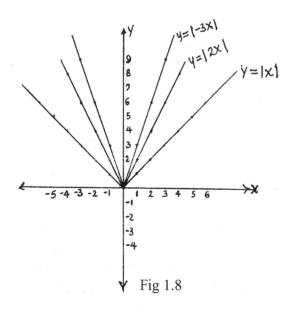

Fig 1.8

Notice that the graphs of these absolute value functions are on or above the x-axis. Absolute value always yields answers which are positive or zero.

Some observations on absolute value functions:

1) As the coefficient of x gets larger, the graph becomes thinner, closer to its line of symmetry.

2) As the coefficient of x gets smaller, the graph becomes thicker, further from its line of symmetry.

3) If the coefficient of x is negative, the graph is the same as if that coefficient of x were positive.

It is acted upon by the absolute value property.

1.12 <u>Shifting, Reflecting, Stretching and Shrinking Graphs of Absolute Value Function Across the x- and y-axes</u>

Definitions:

<u>Abscissa</u>:- The x-coordinate

<u>Ordinate</u>:- The y-coordinate

<u>Shift</u>:- A translation in which the size and shape of a graph of a function is not changed, but the location of the graph is changed.

<u>Scale</u>:- A translation in which the size and shape of the graph of a function is changed.

<u>Reflection</u>:- A translation in which the graph of a function is mirrored about an axis.

<u>Translations</u>

There are two kinds of translations that we can do to a graph of a function. They are shifting and scaling. There are three if you count reflections, but, reflections are just a special case of the second translation.

<u>Shifts</u>

A shift is a rigid translation in that it does not change the shape or size of the graph of the function. All that a shift will do is change the location of the graph. A vertical shift adds (subtracts) a constant to or from every y-coordinate while leaving the x-coordinate unchanged. A horizontal shift adds (subtracts) a constant to or from every x-coordinate while leaving the y-coordinate unchanged. Vertical and horizontal shifts can be combined into one expression.

Shifts are added (subtracted) to the x or f(x) components. If the constant is grouped with x, then it is a horizontal shift, otherwise it is a vertical shift.

1.13 Scales (Stretch/Shrink)

A Scale is a non-rigid translation in that it does alter the shape and size of the graph of the function.

A scale will multiply (divide) coordinates and this will change the appearance as well as the location. A vertical scaling multiplies (divides) every y-coordinates by a constant while leaving the x-coordinate unchanged. A horizontal scaling multiplies (divides) every x-coordinate by a constant while leaving the y-coordinate unchanged. The vertical and horizontal scalings can be combined in to one expression.

Scaling factors are multiplied (divided) by the x or f(x) components. If the constant is grouped with the x, then it is a horizontal scaling, otherwise it is a vertical scaling.

1.13.1 Reflections

A function can be reflected about an axis by multiplying by negative one. To reflect about the y-axis, multiply every x by -1 to get -x. To reflect about the x-axis, multiply f(x) by -1 to get -f(x).

1.13.2 Vertical Shifts

- If c is a positive real number, the graph of f(x) + c is the graph of y = f(x) shifted upward c units.
- If c is a positive real number, the graph of f(x) - c is the graph of y = f(x) shifted down and c units.

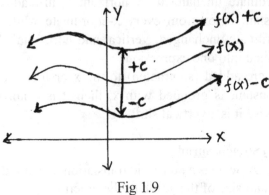

Fig 1.9

Example) Use the graph of f(x) = |x| to graph the functions g(x) = |x| + 4 and h(x) = |x| - 3

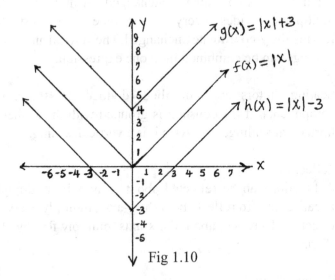

Fig 1.10

As you see from the above graphs, the graph of g(x) = |x| + 4 is the graph of f(x) = |x| shifted upward 4 units. And the graph of h(x) = |x| - 3 is the graph of f(x) = |x| shifted downward 3 units.

1.13.3 Horizontal Shifts

- If c is a positive real number, then the graph of f(x - c) is the graph of y = f(x) shifted to the right c units.

- If c is a positive real number, then the graph of f(x + c) is the graph of y = f(x) shifted to the left c units.

Example) Use the graph of f(x) = |x| to graph the function g(x) = |x - 4| and h(x) = |x + 4|

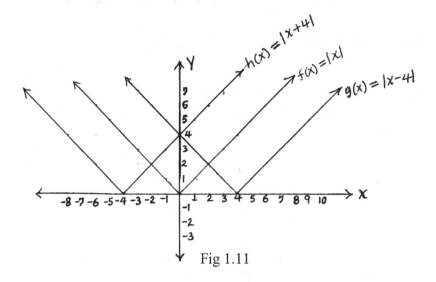

Fig 1.11

As you see from the above graphs, the graph of g(x) = |x - 4| is the graph of f(x) = |x| shifted to the right 4 units and the graph of h(x) = |x + 4| is the graph of f(x) = |x| shifted to the left 4 units.

1.13.4 Vertical Stretching and Compression (Shrinking)

- Given a function y = f(x), then the graph of y = af(x), a > 1, is the same as the graph of y = f(x) stretched with respect to the

y-axis by a factor of a. The graph of $y = \dfrac{1}{a} f(x)$, a > 1, is the

same as the graph of y = f(x) compressed with respect to the y-axis by a factor $\dfrac{1}{a}$.

1.13.5 Horizontal Stretching and Compression

Given the function y = f(x), then the graph of y = f(ax), a > 1, is the

same graph of y = f(x). Stretched with respect to the x-axis by a factor

$\frac{1}{a}$. The graph of $y = f\left(\dfrac{x}{a}\right)$, a > 1, is:- the same graph of y = f(x)

stretched with respect to the x-axis by a factor $\dfrac{1}{a}$.

Example 1) Recognizing vertical stretches

Draw the graph of f(x) = |x|, g(x) = 2|x| and h(x) = 3|x| on the same x-y coordinate plane.

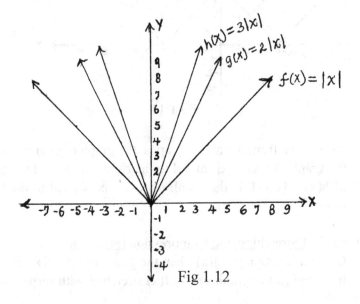

Fig 1.12

As we observe from the above graphs; if we examine closely any x-value on these graphs, we notice that the stretched function's y-value of g(x) = 2|x| is always two times further above the x-axis than the function f(x) = |x| and the stretched function's y-value of h(x) = 3|x| is always three times further above the x-axis than the function f(x) = |x|.

Example 2) Recognizing vertical shrink
Draw the graph of f(x) = |x| and g(x) = ½|x| on the same x-y coordinate plane.

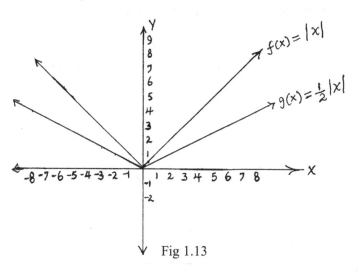

Fig 1.13

As we have seen from the above graphs, the shrinked function's y-value of g(x) = ½|x| is always half of the y-value of f(x) = |x|.

Example 3) Recognizing Horizontal Shrinks
Draw the graph of f(x) = |x| and g(x) = |2x| on the same x-y coordinate plane.

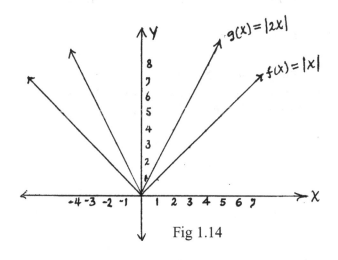

Fig 1.14

As you see from the above graphs, when x is multiplied by a number greater than 1. Just as multiplied by 2 in the above graph; the graph shrinks horizontally. This can look similar to a vertical stretch, but it is not the same thing.

Example 4), Recognizing horizontal stretches.
Draw the graphs of f(x) = |x| and g(x) = |¼x| on the same x-y coordinate plane.

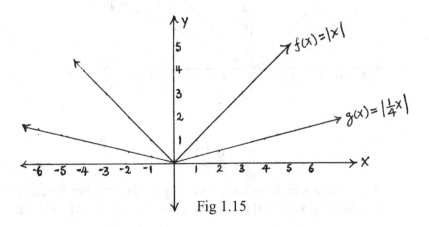

Fig 1.15

As you see from the above graphs, when x is multiplied by a number between 0 and 1, just as multiplied by ¼ in the above graph; the graph stretches horizontally. This can look similar to a vertical shrink; but it is not the same thing.

1.14   Reflections of Graphs

1.14.1 Reflection about the x-axis
Multiplying the original function by a negative causes the graph to reflect or "flip" about the x-axis.

Example) Draw the graphs of f(x) = |x| and g(x) = -|x| on the same x-y coordinate plane.

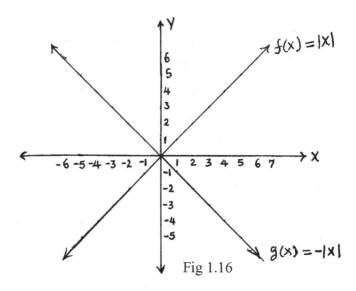

Fig 1.16

(The graph of g(x) is simply the same as the graph of f(x) folded down across the x-axis. The x-axis is the line of reflection or the mirror. (The graph of g(x) is the reflection of f(x) about the x-axis.)

Note:- If the point (a, b) is reflected about the x-axis, then it is reflected to the point (a, -b).

1.14.2 Reflection about the y-axis
Replacing every x in the equation with -x, i.e. y = f(-x), reflects the graph about the y-axis.

Note:- If a point (a, b) is reflected about the y-axis, then it is reflected to the point (-a, b).

Example) Draw the graphs of $f(x) = (x + 4)^2$ and $g(x) = (-x + 4)^2$ on the same x-y coordinate plane.

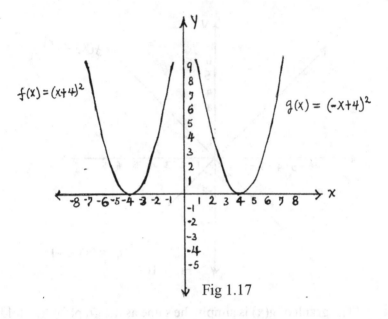

Fig 1.17

The graph of $g(x) = (-x + 4)^2$ is the reflection of the graph of $f(x) = (x + 4)^2$ about the y-axis. For every point on the right of the y-axis, there is a point equidistant to the left of the y-axis. That is for every point (x, y), there is a point (-x, y).

Note:- If the point (a, b) is reflected about the y-axis; then it is reflected to the point (-a, b).

1.14.3 Reflection about the line y = x
Reflections about the line y = x is accomplished by interchanging the x and the y-values. Thus for y = f(x) the reflection about the line y = x is accomplished by x = f(y). Thus the reflection about the line y = x for y = x^2 is the equation x = y^2. These are graphed on the following x-y coordinate plane:-

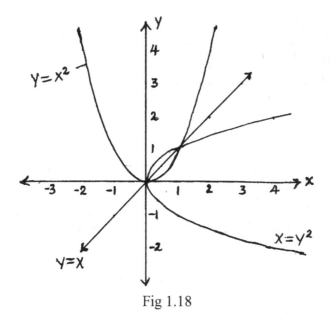

Fig 1.18

In the above graphs, the reflection of $y = x^2$ about the line $y = x$ is $x = y^2$.

Note:- If the point (a, b) is reflected about the line $y = x$, then it is reflected to the point (b, a).

- If the point (a, b) is reflected about the origin, then it is reflected to the point (-a, -b).

Exercise 1.3

1) Draw the graphs of $f(x) = |x|$ and $g(x) = -|x|$ on the same x-y coordinate plane.

2) Draw the graphs of $f(x) = |x|$, $g(x) = |2x|$ and $h(x) = \dfrac{1}{2}|x|$ on the same x-y coordinate plane.

3) Find the reflection of the point (-3, 2) through the y-axis.

4) Find the reflection of the point (1, -3) through the x-axis.

5) Find the reflection of the point (½, ⅓) through the origin.

Exercise 1.4

1 What is the slope of the line passing through the points
a) (2, 4) and (0, -4)
b) (-1, 3) and (7, 5)
c) (-2, 6) and (-1, -1)
d) (4, -5) and (3, 8)
e) (0, 0) and (4, 5)

2. Find the value of 'c' such that the line passes through the
a) points (5, 3) and (4, c); having the slope $-\frac{3}{2}$
b) points (3, -2) and (c, 6) having the slope $\frac{2}{5}$
c) origin and the point (c, $\frac{1}{4}$), having slope 2

3 Find the slope and the y-intercept of the line:-
a) $10x - 3y = 6$
b) $2y + 4x = 1$
c) $3x - 6y = 7$
d) $4x + 4y = 8$
e) $-\frac{1}{2}y - 5x = 9$

4 Determine whether these lines have positive slope, negative slope, zero slope or no slope.

figure a)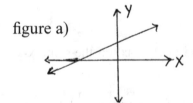

figure b)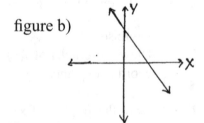

figure c)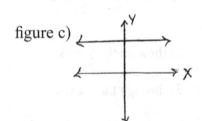

figure d)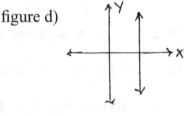

5   Find an equation of the line through the point:-
     a) (2, -3) with m = 5

     b) (4, 5) with m = -3

     c) (0, 7) with $m = \frac{1}{4}$

     d) (6, 0) with m = -1

     e) (-1, -2) with m = 2

     f) $(-3, \frac{1}{4})$ with $m = \frac{1}{3}$

6   Find an equation of the line
     a) with $m = \frac{1}{2}$ and y-intercept 5

     b) with M = 6 and y-intercept -3

     c) with $M = \frac{1}{2}$ M and y-intercept 0

     d) with M = 3 and y-intercept -6

     e) with M = 0 and y-intercept $-\frac{1}{4}$

7   Find the value of 'd' such that the line
     a) 2x + dy = 4 has slope -5

     b) dx + 3y = 1 has slope $\frac{1}{2}$

     c) 5y + 3dx = -3 has slope $-\frac{1}{3}$

     d) 4dx - 3y = -2 has slope -1

     e) $x - \frac{1}{4}dy = 8$ has slope 6

8   Identify the following as linear functions or not
     linear functions.
     a) y = 5x - 3
     b) y = x² - 1
     c) y = -6
     d) x² + y² = 4
     e) y - x = 3x - 2y
     f) y = 2ˣ
     g) x = 1

h) $y = -1$

i) $2xy - x^2 = 0$

9)   Find the equation of the horizontal line through
a) the point (4, 3)
b) the point (-2, 5)
c) the point (-1, 2)
d) ($\frac{1}{2}$, $\frac{1}{3}$)
e) (0, 6)

10   Graph the following lines:-
a) $y = 2x - 1$
b) $y = x - 3$
c) $x = 4$
d) $y = -5$

11.   Find the equation of the vertical line through:
a) the point (-3, -5)
b) the point (1, 6)
c) the point (2, -4)
d) the point (6, -5)
e) through the origin

12   Given the line $2x - 5y = 4$ and the point (-4, 3), find the lines
that are:-
a) parallel to the given line and
b) perpendicular to it.

13   If two lines, L_1 and L_2 are parallel and the slope of L_1 is $\frac{1}{3}$.
What is the slope of the line L_2?

14   If two lines L_1 and L_2 are perpendicular and the slope of L_1 is
$\frac{-3}{5}$. What is the slope of the line L_2?

15   Using table of values graph linear equation
a) $y = 2x + 1$
b) $y = x - 3$

16   Draw the graphs of y = |x|, f(x) = |3x| and g(x) = |-2x| on the same x-y coordinate plane.

17.   Draw the graphs of f(x) = |x| and g(x) = 2|x| on the same x-y coordinate plane.

18.   Draw the graphs of f(x) = |x| and g(x) = ⅓|x| on the same x-y coordinate plane.

19.   Draw the graphs of f(x) = |x| and $g(x) = \left|\dfrac{1}{5}x\right|$ on the same x-y coordinate plane.

20.   Draw the graphs of f(x) = x and g(x) = -x on the same x-y coordinate plane.

21.   Find the reflection of each of the following points through the x-axis.
a) (4, -2)
b) (6, 5)
c) (-4, 3)
d) (1, -1)

22.   Find the reflection of each of the following points through the y-axis.
a) (-5, 4)
b) (6, 0)
c) (3, -2)
d) (-1, -3)
e) (-7, 4)

23.   Find the reflection of each of the following points through the line y = x
a) (-3, 2)
b) $\left(-\frac{1}{4}, 3\right)$
c) $\left(\frac{1}{2}, -\frac{1}{4}\right)$
d) (-2, -5)
e) (-8, 7)

24. Determine whether each of the following statements is true or false.

a) If two straight lines are parallel, then they have the same slope.

b) If two straight lines are perpendicular to each other, then the product of their slope is -1.

c) The slope of a vertical line is undefined.

d) The slope of a horizontal line is zero

e) The reflection of the point (2, 3) through the x-axis is (2, -3).

CHAPTER - TWO

2- POLYNOMIALS AND QUADRATIC FUNCTIONS

2.1 Introduction to Polynomial Functions

Any function that can be described by a polynomial in one variable, such as

$$f(x) = 3x^6 - 2x^3 + x - 1,$$

is called a polynomial function. Here is a formal definition.

Definition   Polynomial Function

A polynomial function f is given by

$$f(x) = a_n x^n + a_{n-1} x^{n-1} + ... + a_2 x^2 + a_1 x + a_0,$$

where n is a nonnegative integer and $a_n, a_{n-1}, ..., a_2, a_1, a_0$ are real numbers, called the coefficients of the polynomial.

The first nonzero coefficient is assumed to be a_n and is called the leading coefficient. The degree of polynomial function is n.

Example 1) $f(x) = 3x^5 + 2x^4 - 8x^2 + 6$; is a polynomial function with degree 5 and leading coefficient 3.

Example 2) $f(x) = -4x^6 + 3x^4 - 6x + 1$, is a polynomial function with degree 6 and leading coefficient -4

Example 3) $f(x) = 4x - 6x^3 + 8x^5 - 9x^7$, is a polynomial function with degree 7 and leading coefficient -9.

Example 4) $f(x) = 6x^9 - 4x^8 + 2x^5 + x + 4$, The greatest degree is 9, so the degree of the polynomial is 9, and the leading coefficient is 6.

2.2   Monomial, Binomial, Trinomial, and Multinomial

2.2.1   Algebraic Expression:- A combination of terms obtained by the operations of "+" or "-" or both is called an algebraic expression.

Example: (i) 6x + 4y
(ii) 5m - 4n

Monomial:
- An expression containing only one term is called a monomial.
    Example: 6, x, 4x, -5x, bc, etc.
- A monomial is a number, a variable, or the product of a number and one or more variables with whole number exponents.
- A monomial is a single term which has no plus sign (+) or minus sign (-).

- The following are not monomial: $x + 8$, $\dfrac{2}{n}$, 6^x, y^{-1}, $\sqrt{x}$, $-2x^{1.5}$

Binomial
- An expression containing two terms is called a binomial.
Example: 3x + 6, 6y - c, 4x + 2y, a - b, etc.

Trinomial
- An expression containing three terms is called a trinomial.
Example: $x + y + z$, $6 - 3m + \dfrac{n}{2}$, $4 + 2x - z$, $3x - 2y + m$, etc
    Multinomial (or polynomial)
- An expression containing more than three terms is called a multinomial or a polynomial.
Example: 4x + 5y + z - 3, m + n - 4x - 2a, 9ab + 3c + 2d - 11, etc.

2.3   Degree
2.3.1   Definition of Degree
- Degree of a Monomial: The degree of a monomial is the sum of the exponents of all its variables.

Example: The degree of 4xy is 2, the degree of x^5z^6 is 11, the degree of a^3b^4 is 7, the degree of 5 is 0.

- Degree of a polynomial: The degree of a polynomial is the greatest degree of any term in the polynomial.

Example: $x^5 + 3x^2 + 5x + 1$, has degree 5, $x^3 + y^8 - 3x^2 = 3y$ has degree 8.

- The degree of a nonzero constant term is 0.
- Zero has not degree

2.4 Addition of polynomials
Addition of polynomials is just a matter of associating same terms, with some order of operations applications discharge. You can add polynomials in grouping like terms and then simplifying. Like terms is known as two terms with the same variables and same powers.

Examples: $3x^5$ and x^5 are like terms
          x and 4x are like terms
          $-6y^4z^3$ and y^4z^3 are like terms.

2.5 Two Different Rules of Adding Polynomials:
• Horizontal Addition of Polynomials.
• Vertical Addition of Polynomials.

2.5.1 Horizontal Addition of Polynomials
This technique is especially related to the adding of real number consequent the horizontal method. But it has its own rules of addition.

2.5.2 Vertical Addition of Polynomials
     Vertical technique of addition of polynomials is based the similar rule as the horizontal technique although then the arrangement of expressions is vertical different the other process.

Example 1) Simplify (2x+6y)+(3x+8y)
          Solution: first group like terms, and then simplify
                    (2x+6y) + (3x+8y)
                    = (2x+3x) + (8y+6y)

$$= (2+3)x + (8+6)y$$
$$= 5x + 14y$$

Example 2) Simplify $(3x^3+2x^2-5x+8) + (x^3-4x^2+2x+9)$

Solution: First group like terms and then simplify

$$(3x^3+2x^2-5x+8) + (x^3-4x^2+2x+9)$$
$$= 3x^3+x^3+2x^2-4x^2-5x+2x+8+9$$
$$= 4x^3-2x^2-3x+17$$

or vertically

$$\begin{array}{l} 3x^3+2x^2-5x+8 \\ + \ x^3-4x^2+2x+9 \\ \hline 4x^3-2x^2-3x+17 \end{array}$$

Either way, we get the same answer: $4x^3-2x^2-3x+17$

Note that each column in the vertical addition above contains only one degree of x: the first column was the x^3 column, the second column was the x^2 column, the third column was the x column, and the fourth column was the constants column. This is analogous to having a thousands column, a hundreds column, a tens column, and a ones column when doing strictly - numerical addition.

Example 3) Simplify $(6x^3+2x^2+x+1) + (x^3-x^2+2x+3) + (2x^3+3x^2+3x-1)$

Solution: It is possible to have three or more polynomials to add at once. We will go slowly and do each step thoroughly, and it should work out right.

Adding Horizontally:

$$(6x^3+2x^2+x+1) + (x^3-x^2+2x+3) + (2x^3+3x^2+3x-1)$$
$$= 6x^3+2x^2+x+1+x^3-x^2+2x+3+2x^3+3x^2+3x-1$$
$$= 6x^3+1x^3+2x^3+2x^2-1x^2+3x^2+x+2x+3x+1+3-1$$
$$=9x^3+4x^2+6x+3$$

Note that 1's in the third line. Any time you have a variable without a coefficient, there is an "understood" 1 as the coefficient. If you find it helpful to write that 1 in, then do so.

Adding Vertically:

$6x^3+2x^2+x+1$

x^3-x^2+2x+3

$2x^3+3x^2+3x-1$
_____

$9x^3+4x^2+6x+3$

Either way, we get the same result: $9x^3+4x^2+6x+3$

Example 4) Simplify $(6x^2-2x-4) + (x^2-5x+6) + (x^2+4x+9)$

Solution: Adding Horizontally

$(6x^2-2x-4) + (x^2-5x+6) + (4x^2+4x+9)$

$= 6x^2+x^2+4x^2-2x-5x+4x-4+6+9$

$= 11x^2-3x+11$

Adding Vertically:

$6x^2-2x-4$

x^2-5x+6

$4x^2+4x+9$
_____

$11x^2-3x+11$

Either way, we get the same answer: $11x^2-3x+11$

2.6 Subtraction of Polynomials

Subtracting polynomials is quite similar to adding polynomials, but you have that pesky minus sign to deal with.

2.7 Two Different Rules of Subtracting Polynomials:
• Horizontal Subtraction of Polynomials
• Vertical Subtraction of Polynomials.

Example 1) Simplify $(x^3+2x^2-5) - (4x^3+x^2+6)$

Solution: $(x^3+2x^2-5) - (4x^3+x^2+6)$

The first thing we have to do is to take that negative through the parentheses. Some times we find it helpful to put a "1" infront of the parentheses, to help them keep track of the minus sign;

2.7.1 Horizontal: Subtraction

$(x^3+2x^2-5) - (4x^3+x^2+6)$
$= (x^3+2x^2-5) -1(4x^3+x^2+6)$
$= x^3+2x^2-5) -1(4x^3) -1(x^2) - 1(6)$
$= x^3+2x^2-5-4x^3-x^2-6$
$= x^3-4x^3+2x^2-x^2-5-6$
$= -3x^3+x^2-11$

2.7.2 Vertical Subtraction

x^3+2x^2-5
$-(4x^3+x^2+6)$
_____

In the horizontal case, you may have noticed that running the negative through the the the parentheses changed the sign on each term inside the parentheses. The shortcut here is to not bother writing in the subtraction sign or the parentheses; instead, you just change all the signs in the second row.

We will change all the signs in the second row and add down:
x^3+2x^2-5
$-(4x^3+x^2+6)$
_____
x^3+2x^2-5
$-4x^3-x^2-6$
_____
$-3x^3+x^2-11$

Either way, we get the same answer: $-3x^3+x^2-11$

Example 2) Simplify $(5x^4-6x^3+3x^2-2) - (7x^4+3x^3+4)$
Solution: Horizontally
$(5x^4-6x^3+3x^2-2) - (7x^4+3x^3+4)$
$(5x^4-6x^3+3x^2-2) -1(7x^4+3x^3+4)$
$(5x^4-6x^3+3x^2-2) -1(7x^4)-1(3x^3)-1(4)$
$5x^4-6x^3+3x^2-2-7x^4-3x^3-4$
$5x^4-7x^4-6x^3-3x^3+3x^2-2-4$
$-2x^4-9x^3+3x^2-6$

Vertically

We will align like terms in a vertical columns and leaving gaps as necessary:
$5x^4-6x^3+3x^2-2$
$7x^4+3x^3 \quad +4$

Then we will change the signs in the second line, and add:
$5x^4-6x^3+3x^2-2$
$-(7x^4+3x^2 \quad +4)$

$5x^4-6x^3+3x^2-2$
$-7x^4-3x^3 \quad -4$

$-2x^4-9x^3+3x^2-6$

2.8   Multiplication of Polynomials.
- The general rule is that each term in the first factor has to multiply each term in the other factor.
- The number of products you get has to be the number of terms in the first factor times the number of terms in the second factor. For example, a binomial times a trinomial gives six products.
- Be very careful and methodical to avoid missing any terms.
- After the multiplication is complete you can try to collect like terms to simplify the result.

Example 1) product of a binomial and trinomial.
$$(x+3)(x^2-3x+2)$$
There are six possible products. We can start with x and multiply it by all three terms in the other factor, and then do the same with 3. It would look like this:
$(x+3)(x^2-3x+2)$
$= x(x^2)-x(3x)+x(2)+3(x^2)-3(3x)+3(2)$
$= x^3-3x^2+2x+3x^2-9x+6$
$= x^3-3x^2+3x^2+2x-9x+6$
$= x^3-7x+6$

We can multiply the above vertically as:-

$$x^2-3x+2$$
$$x+3$$
$$\overline{}$$
$$3x^2-9x+6$$
$$x^3-3x^2+2x$$
$$\overline{}$$
$$x^3-7x+6$$

Either way, we get the same answer: x^3-7x+6

Exercise 2.1

A)   For each of the following polynomial function
a) Find the degree
b) Find the leading coefficient
c) Find the constant term

1) $f(x) = 5x^3-3x^2+2x-1$ 6) $3+2x^3-7x^4-41x^8$

2) $f(x) = -4x+1$ 7) $2(x-7)^2$

3) $f(x) = 4x^5-6x^2-9$ 8) $4x^6-2x-\frac{1}{2}$

4) $f(x) = -5(x-1)^2$

5) $f(x) = 5$

B)   Determine each of the following, whether they are monomial, binomial, trinomial or multinomial.

1) $x+y+z+2$ 4) $6m+4n+x$

2) $5x$ 5) $2+x-y$

3) $6y+4$ 6) $x-6y+z+w$

C)   Determine the degree of each of the following:

(1) $f(x) = 0$ 3) $f(x) = 0x^5+6x+1$

(2) $f(x) = 5$ 4) $f(x) = 2x-1$

D)   Determine the degree of each of the following

(1) a^3b^4

(2) x^5+3x^2-1

(3) x^5z^6

(4) xyz

E)   Write the polynomial so that the exponents decrease from left to right. Identify the degree and the leading coefficient of the polynomial.

1) $8-2x-3x^2$
2) $7a+3a^5+4a^2-2a^3-8a^7$
3) $2b^2-5b+10-3b^4\cdot3b^4$
4) $-13c+7c^5+18c^2+6c^6$
5) $2y^7+4y^3+3y^5-4y+2$

F)   Find the product of:-
a) $(x+4)(x^3-2x+9)$
b) $(2x-1)(3x-7)$

G)   Find the sum of each of the following polynomials
   a) In horizontal format
   b) In vertical format

1) $(3x^2-2x+5)+(2x^2+4x+6)$
2) $(2x^4-3x^3+4x^2+2)+(6x^4+5x^2+6)$
3) $(2a^5+3a^2-6)+(4a^5+2a^4+4a^2+2a-1)$
4) $(d^2-16d+10)+(12d^2+8d+3)$
5) $(11y^2+6y-1)+(2y^2-7y+5)$

H)   Find the difference of each of the following polynomials.
   a) In horizontal format
   b) In vertical format

1) $(2x^2+7x+2)-(6x-2)$
2) $(5x^3+4x^2-6x+1)-(3x^2+3x+7)$
3) $(3y^2-9y+2)-(y^2-2)$
4) $(-4c^5-2c^4-3c^2+2)-(2c^4+c^2+4)$
5) $(11x^2+6x-1)-(3x-5)$

2.9 Quadratic Functions

- If a function can be described by a second-degree polynomial, then it is called quadratic. The following is a more precise definition.

> Definition: A quadratic function is a function that can be described as follows:
> $$f(x)=ax^2+bx+c, \text{ where } a\neq0$$

In this definition, we insist that $a\neq0$; otherwise the polynomial would not be of degree two. One or both of the constants b and c can be 0.

Example 1): $f(x) = 3x^2+2x-5$ or $y=3x^2+2x-5$ is a quadratic function where a=3, b=2 and c=-5

Example 2) $f(x)=-x^2+2x+1$ or $y=-x^2+2x+1$ is a quadratic function where a=-1, b=2 and c=1

What are some of the ordered pairs which belong to this function?

Let's construct a table to see some ordered pairs. This can be done by substituting some numbers for the independent variable x.

$y=-x^2+2x+1$	x	-3	-2	-1	0	1	2	3
	y	-14	-7	-2	1	2	1	-2

From this you can see some of the ordered pairs which belong to the function. Such as (-3, -14), (-2, -7), (-1, -2), (0, 1), (1, 2), (2, 1), (3, -2), ...

or when we evaluate the function at some values of x you can see: f(-3)=-14. (The function evaluated at x=-3 gives -14 or simply f of -3 is equal to -14), f(-2)=-7, f(-1)=-2, f(0)=1, f(1)=2, f(2)=1, f(3)=-2...

Note: In the above definition of quadratic function, we insist that $a\neq0$; otherwise the polynomial would not be of degree two. One or both of the constants b and c can be 0.

Consider $f(x)=x^2$. This is an even function, so the y-axis is a line of symmetry. The graph opens up, as shown below. If we multiply by a constant a to get $f(x)=ax^2$, we obtain a vertical stretching or shrinking, and a reflection if a is negative. Examples of these are shown in the figures. Bear in your mind that if a is negative, the graph opens down.

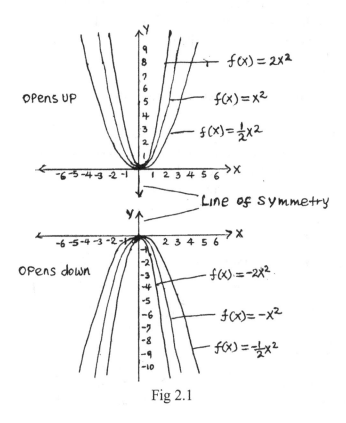

Fig 2.1

2.10 Graphs of Quadratic Functions.

2.10.1 Introduction

Graphs of quadratic function are called parabolas. The highest or lowest point at which the graph turns is called the vertex. In each parabola shown above, the point (0, 0) is the vertex. The line that runs down through the center of the parabola and divides the graph into two perfect halves is called the line of symmetry or axis of symmetry. In the above parabolas shown, the line x=0, or the y-axis, is the line of symmetry or axis of symmetry.

Note:- All graphs of quadratic functions have the same basic "U" shape.

2.10.2 Graph of $y=ax^2+k$, coefficient of x or b=0. Graph of $y=ax^2+k$ can be obtained by shifting the graph of $y=ax^2$ by k units up or down.
- If k>0, the graph of $y=ax^2$ will be shifted up by k units
- If k<0, the graph of $y=ax^2$ will be shifted down by k units.
- The vertex of the graph is (0, k)
- The graph opens upward for a>0.
- The graph opens downward for a<0.
- The minimum or maximum value is k.

Let us consider $f(x)=a(x-h)^2$. We have replaced x by x-h in $f(x)=ax^2$, and will therefore obtain a horizontal translation. The translation will be to the right if h is positive and to the left if h is negative. The examples in the figures below illustrate translations.

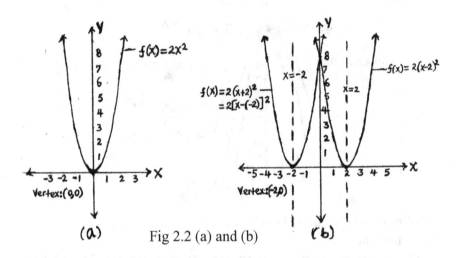

(a)   Fig 2.2 (a) and (b)   (b)

2.10.3 The graph of $f(x)=a(x-h)^2+k$:
a) opens up if a>0 and down if a<0
b) has x=h as a line of symmetry;
c) has (h, k) as a vertex;
d) has k as a minimum value of a>0 and has k as a maximum value of a<0.

Thus without graphing, we can determine a lot of information about a function described by $f(x)=a(x-h)^2+k$. The following table shown below is an example.

Function	Vertex	Line of symmetry	Maximum value	Minimum value
$f(x)=3\left(x-\dfrac{1}{5}\right)^2-4$ $=3\left(x-\dfrac{1}{5}\right)^2+(-4)$	$\left(\dfrac{1}{5},-4\right)$	$x=\dfrac{1}{5}$	No, graph extends up; 3>0	-4, graph extends up, 3>0
g(x)=-2(x+6)²+8 =-2[x-(-6)]²+8	(-6, 8)	x=-6	8, graph extends down; -2<0	No, graph extends down; -2<0

Remark: The vertex (h, k) is used to find the maximum or the minimum value. The maximum or the minimum value is the number k, not the ordered pair (h, k).

Now let us consider a quadratic function f(x)=ax²+bx+c. Note that it is not in the form f(x)=a(x-h)²+k. We can put it into that form by completing the square.

Example 1)   Use completing the square to put the function

$$f(x) = x^2 - 8x + 3$$
$$= x^2 - 8x + \left(\dfrac{8}{2}\right)^2 + 3 - \left(\dfrac{8}{2}\right)^2$$

...(Adding half of 8 squared and subtracting half of 8 squared)

=x²-8x+16-13....(The underlined part is a perfect square.)
f(x)=(x-4)²-13

Thus, the last form of writing f(x) will make the process of graphing easier.

The graph can be plotted by first shifting the graph of f(x)=x², 4 units to the right and then shifting it 13 units down.

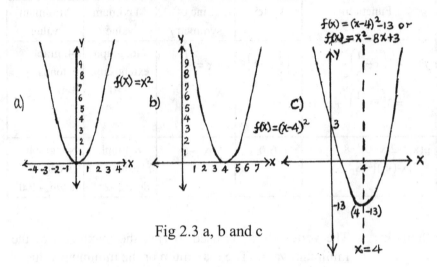

Fig 2.3 a, b and c

- The axis or line of symmetry is x=4
- The vertex is (4, -13)
- The range is {y:y≥-13}

2.11 VERTEX OF PARABOLAS

THEOREM 2-1

The vertex of the graph of the quadratic function $f(x)=ax^2+bx+c$

is $\left(\dfrac{-b}{2a}, \dfrac{4ac-b^2}{4a}\right)$.

Proof:- We begin by factoring out the x^2-coefficient, as follows:

$$f(x) = a\left(x^2 + \frac{b}{a}x\right) + c \cdot$$

Next, we proceed inside the parentheses. We take half the coefficient

of x and square it. That number is $\left(\dfrac{b}{2a}\right)^2$. We add $\left(\dfrac{b}{2a}\right)^2 - \left(\dfrac{b}{2a}\right)^2$, but do

it inside the parentheses:

$$f(x) = a\left(x^2 + \frac{b}{a}x + \left(\frac{b}{2a}\right)^2 - \left(\frac{b}{2a}\right)^2\right) + c$$

We now have an extra, unwanted term inside the parentheses, so we use the distributive law and multiplication to simplify, as follows:

$$f(x) = a\left(x^2 + \frac{b}{a}x + \left(\frac{b}{2a}\right)^2\right) - a\left(\frac{b}{2a}\right)^2 + c$$

$$= a\left(x^2 + \frac{b}{a}x + \left(\frac{b}{2a}\right)^2\right) - \frac{b^2}{4a} + c$$

$$= a\left(x^2 + \frac{b}{a}x + \left(\frac{b}{2a}\right)^2\right) - \frac{b^2 - 4ac}{4a}$$

$$= a\left(x + \frac{b}{2a}\right)^2 - \frac{b^2 - 4ac}{4a}$$

Thus, the vertex or the turning point is $\left(\dfrac{-b}{2a}, \dfrac{4ac - b^2}{4a}\right)$

2.11.1 More on graph of y=ax²+bx+c

- The y-intercept of y=ax²+bx+c is at c. (The graph crosses the y-axis at (0, c)

- The range is $y \geq \dfrac{4ac - b^2}{4a}$, if a>0 (upward parabola) and it has minimum value

- The range is $y \leq \dfrac{4ac - b^2}{4a}$ if a<0 (downward parabola) and it has maximum value.

- The graph is symmetrical about $x = -\dfrac{b}{2a}$

- The vertex or the turning point is $\left(\dfrac{-b}{2a}, \dfrac{4ac - b^2}{4a}\right)$

- The x-intercept is obtained by solving ax²+bx+c=0

Example 1) For the function f(x)=2x²+6x+4.
    (a) Find the vertex or turning point.
    (b) The the line of symmetry.
    (c) Determine whether there is a maximum or minimum value and find the value
    (d) Find the range

Solution: a) we note that a=2, b=6 and c=4. The first coordinate of vertex is

$$x = \frac{-b}{2a} = \frac{-6}{2x2} = \frac{-6}{4} = \frac{-3}{2}$$

We find the second coordinate of the vertex as follows:

$$\frac{4ac - b^2}{4a} = \frac{4(2)(4) - 6^2}{4(2)} = \frac{32 - 36}{8} = \frac{-4}{8} = -\frac{1}{2}$$

We can also substitute $\dfrac{-3}{2}$ in to the original function formula

to find the second coordinate, thus;

$$f\left(\frac{-3}{2}\right) = 2\left(\frac{-3}{2}\right)^2 + 6\left(\frac{-3}{2}\right) + 4$$

$$= 2\left(\frac{9}{4}\right) + 6\left(\frac{-3}{2}\right) + 4$$

$$f\left(\frac{-3}{2}\right) = \frac{-1}{2}$$

The vertex is $\left(\dfrac{-3}{2}, -\dfrac{1}{2}\right)$

b) The line of symmetry is $x = \dfrac{-b}{2a} = \dfrac{-6}{2(2)} = \dfrac{-6}{4} = \dfrac{-3}{2}$

(Thus the line of symmetry goes through the point, thus the line

is $x = \dfrac{-3}{2}$)

c) In the function f(x)=2x²+6x+4, the value of a is greater than 0. Its graph is upward parabola, this shows that the graph has minimum value.

The minimum value occurs at $\dfrac{4ac-b^2}{4a}$.

Minimum value $= \dfrac{4(2)(4)-6^2}{4(2)} = \dfrac{32-36}{8} = \dfrac{-4}{8} = -\dfrac{1}{2}$

d) The range is $y \geq \dfrac{4ac-b^2}{4a}$, which is $y \geq \dfrac{4(2)(4)-6^2}{4(2)}$

$y \geq \dfrac{32-36}{8}$

$y \geq -\dfrac{1}{2}$

Example 2) For the function f(x)=-2x²+10x-7.
   (a) Find the vertex
   (b) The line of symmetry
   (c) Determine whether there is a maximum or minimum value and find the value.
   (d) Find the range

Solution a) We note that a=-2, b=10 and c=-7. The first coordinate of the vertex is

$$x = \dfrac{-b}{2a} = \dfrac{-10}{2(-2)} = \dfrac{5}{2}$$

We find the second coordinate of the vertex as follows:

$$\dfrac{4ac-b^2}{4a} = \dfrac{4(-2)(-7)-10^2}{4(-2)} = \dfrac{11}{2}$$

We can also substitute $\dfrac{5}{2}$ into the original function formula to find the

second coordinate, thus;

$$f(x) = -2x^2 + 10x - 7$$

$$f\left(\frac{5}{2}\right) = -2\left(\frac{5}{2}\right) + 10\left(\frac{5}{2}\right) - 7 = \frac{11}{2}$$

Thus; the vertex is $\left(\dfrac{5}{2}, \dfrac{11}{2}\right)$

b) The line of symmetry is $x = \dfrac{-b}{2a} = \dfrac{-10}{2(-2)} = \frac{5}{2}$

c) Since the coefficient of x² is negative, the function f(x)=-2x²+10x-7 has the maximum value.

$$\text{Maximum value} = \frac{4ac - b^2}{4a} = \frac{4(-2)(-7) - 10^2}{4(-2)} = \frac{11}{2}$$

d) The coefficient of x² is negative, so that the range (value of y) is

$$y \le \frac{4ac - b^2}{4a}$$

$$y \le \frac{4(-2)(-7) - 10^2}{4(-2)}$$

$$y \le \frac{11}{2}$$

Exercise 2.2

1)   a) Graph f(x)=-2x²
     b) Does the graph open up or does it open down?
     c) What is the line of symmetry?
     d) What is the minimum value of the function?
     e) What is the vertex?
     f) What is the range?

2)   a) Graph f(x)=-x²+x+1
     b) Does the graph open up or does it open down?

c) What is the line of symmetry?
d) What is the maximum value of the function?
e) What is the vertex?
f) What is the range?

3) a) Graph $f(x)=x^2+1$
    b) Does the graph open up or does it open down?
    c) What is the line of symmetry?
    d) What is the minimum value of the function?
    e) What is the vertex?
    f) What is the range?

4) Use completing the square method and put the following function in to the form $f(x)=a(x-h)^2+k$.
    a) $f(x)=3x^2+24x+10$
    b) $f(x)=x^2-6x+4$
    c) $f(x)=3x^2-10x-8$
    d) $f(x)=-3x^2+9x+2$
    e) $f(x)=2x^2+6x-1$

5) For the function $f(x)=-\frac{1}{4}x^2-6x+8$, find the vertex and the maximum or minimum value.

6) Without graphing the function $f(x)=-3x^2-x+6$, answer each of the following:-
    a) What is the vertex?
    b) What is the line of symmetry?
    c) What is the maximum value?
    d) What is the range?

7) Consider the function $f(x)=\frac{1}{4}x^2+3x-1$, at what value of x is the minimum value occurs?

8) Use the completing the square method to put the function: $f(x)=2x^2+12x+5$, into the form $f(x)=a(x-h)^2+k$

2.12 Factorization methods of solving quadratic equation.

2.12.1 Zero product property

If the quadratic equation $ax^2+bx+c=0$ can be factorized as $m \cdot n=0$, where m and n are linear expressions then the equation can be solved by equating m=0 or n=0

Example) Find the solution set of each of the following equations

Remark: Use the fact that; if $m \cdot n=0$, then m=0 or n=0 or both m=0 and n=0

a) $(x-5)(x+8)=0$

x-5=0 or x+8=0

x=5 or x=-8

∴ s.s={5, -8}

(b) $(x+2)(3x-1)=0$

x+2=0 or 3x-1=0

x=-2 or x=⅓

∴ ss {-2, ⅓}

c) $(2x+3)(5x-2)=0$

2x+3=0 or 5x-2=0

$x = \dfrac{-3}{2}$ or $x = \dfrac{2}{5}$

$\therefore s.s = \left\{ \dfrac{-3}{2}, \dfrac{2}{5} \right\}$

d) $x(x-9)=0$

x=0 or x-9=0

x=0 or x=9

s.s={0, 9}

e) $(-4x+3)(3-x)=0$

-4x+3=0 or 3-x=0

-4x=-3 or -x=-3

$x = \dfrac{3}{4}$ x=3

$s.s = \left\{ \dfrac{3}{4}, 3 \right\}$

Exercise 2.3

Find the solution set using zero product property

1) $(x-\frac{1}{2})(2x+8)=0$

2) $(2x-1)(3x+2)=0$

3) $(x-7)(4-3x)=0$

4) $x(2x+9)=0$

5) $(-6x+3)(4+7x)$

2.13 Factorization by Removing Monomial Factor

In the quadratic equation $ax^2+bx+c=0$, if $c=0$, we can eliminate a monomial factor and solve the equation.

Find the solution set of each of the following questions

Example -1) $x^2+5x=0$, here the constant term $c=0$, we can eliminate x as a common factor.

$x^2+5x=0$

$x(x+5)=0$...(Here, x and $(x+5)$ are common factors of $x^2+5x=0$)

$x=0$ or $x+5=0$

$x=0$ or $x=-5$

s.s.=$\{0, -5\}$

Example -2) $3x^2 - \frac{1}{4}x = 0$ ... (Here, x and $\left(3x - \frac{1}{4}\right)$ are common

factors of $3x^2 - \frac{1}{4} = 0$)

$3x^2 - \frac{1}{4}x = 0$

$x\left(3x - \frac{1}{4}\right) = 0$

$x = 0$ or $3x - \frac{1}{4} = 0$

$x = 0$ or $3x = \frac{1}{4}$

$x = 0$ or $x = \frac{1}{12}$

$s.s = \left\{0, \frac{1}{12}\right\}$

Example 3) $5x^2+6x+5=4x+5$

$5x^2+6x-4x+5-5=0$

$5x^2+2x=0$....(Here, x and (5x+2) are common factors of $5x^2+2x=0$)

$x(5x+2)=0$

$x=0$ or $5x+2=0$

$5x=-2$

$x=0$ or $x = -\dfrac{2}{5}$

$s.s = \left\{0, -\dfrac{2}{5}\right\}$

Exercise 2.4

1)   Find the solution set of each of the following

a) $x^2-3x=0$

b) $\dfrac{1}{4}x^2 - 7x = 0$

c) $6x-2x^2=0$

d) $3x^2+2+5x-x^2=x^2+2$

e) $4x^2+2x=0$

2)   Factorize each of the following

a) $3x^2-5x$

b) $\dfrac{1}{4}x^2 - 9x$

c) x^2-25x

d) $2x^2 - \dfrac{1}{4}x$

e) $2x - \dfrac{3}{2}x^2$

2.14 Factorization by using difference of two squares. In the quadratic equation $ax^2+bx+c=0$, if $b=0$ and a and c have opposite sign. We can use the difference of two squares to factorize the quadratic part. Here use $a^2-b^2=(a+b)(a-b)$.

Consider the expression of $(a+b)(a-b)$

$(a+b)(a-b) = a(a-b)+b(a-b)$

$=a^2-ab+ba-b^2$

$=a^2-b^2$... (ab and ba are like terms)

Note that the expression of (a+b)(a-b) yields the difference of two squares in a^2 and b^2. Thus:

$a^2-b^2=(a+b)(a-b)$

An expression in the form of a^2-b^2 is called the difference of two squares.

Example 1) Factorize each of the following using $a^2-b^2=(a+b)(a-b)$

      a) x^2-25

      b) $4x^2-49$

      c) x^2-36

      d) $2x^2-\dfrac{1}{4}$

      $\left(\sqrt{2}x+\dfrac{1}{2}\right)\left(\sqrt{2}x-\dfrac{1}{2}\right)$

      e) $-9x^2+81$

Solution a) x^2-25

      $(x+5)(x-5)$

      b) $4x^2-49$

      $(2x+7)(2x-7)$

      c) x^2-36

      $(x+6)(x-6)$

      e) $-9x^2+81$

      $=81-9x^2$

      $=(9+3x)(9-3x)$

Example 2) Using the difference of two squares $a^2-b^2=(a+b)(a-b)$; find the solution set of each of the following expression.

a) $x^2-29=0$

b) $\dfrac{1}{9}x^2-81=0$

c) $1-25x^2=0$

d) $\dfrac{1}{4}x^2-8=0$

e) $x^2-49=0$

f) $4x^2-36=0$

Solution: a) $x^2 - 29 = 0$

$$= x^2 - \left(\sqrt{29}\right)^2 = 0$$

$$= \left(x + \sqrt{29}\right)\left(x - \sqrt{29}\right) = 0$$

$$x + \sqrt{29} = 0 \text{ or } x - \sqrt{29} = 0$$

$$x = -\sqrt{29} \text{ or } x = \sqrt{29}$$

$$s.s = \left\{\sqrt{29}, -\sqrt{29}\right\}$$

b) $\dfrac{1}{9}x^2 - 81 = 0$

$$\left(\frac{1}{3}x + 9\right)\left(\frac{1}{3}x - 9\right) = 0$$

$$\frac{1}{3}x + 9 = 0 \text{ or } \frac{1}{3}x - 9 = 0$$

$$\frac{1}{3}x = -9 \text{ or } \frac{1}{3}x = 9$$

$$x = -27 \text{ or } x = 27$$

$$s.s = \{27, -27\}$$

c) $1 - 25x^2 =$

$$(1 + 5x)(1 - 5x) = 0$$

$$1 + 5x = 0 \text{ or } -5x = -1$$

$$x = -\frac{1}{5} \text{ or } x = \frac{1}{5}$$

$$s.s = \left\{\frac{1}{5}, -\frac{1}{5}\right\}$$

d) $\frac{1}{4}x^2 - 8 = 0$

$\left(\frac{1}{2}x + \sqrt{8}\right)\left(\frac{1}{2}x - \sqrt{8}\right) = 0$

$\frac{1}{2}x + \sqrt{8} = 0$ or $\frac{1}{2}x - \sqrt{8} = 0$

$\frac{1}{2}x = -\sqrt{8}$ or $\frac{1}{2}x = \sqrt{8}$

$\frac{1}{2}x = -2\sqrt{2}$ or $\frac{1}{2}x = 2\sqrt{2}$

$x = -4\sqrt{2}$ or $x = 4\sqrt{2}$

$s.s = \left\{4\sqrt{2}, -4\sqrt{2}\right\}$

e) x²-49=0
(x+7)(x-7)=0
(x+7)=0 or (x-7)=0
x=-7 or x=7
s.s={7, -7}

f) 4x²-36=0
(2x+6)(2x-6)=0
2x+6=0 or 2x-6=0
2x=-6 or 2x=6
x=-3 or x=3
s.s={3, -3}

Exercise 2.5

1) Factorize each of the following using the difference of two squares: a²-b²=(a+b)(a-b)

a) $x^2 - \frac{1}{81}$
b) x²-2
c) 36- a²b²
d) 81-25x²
e) 0.25x²-0.64

f) $2x^2-144$

2)   Find the solution set of each of the following expression using the difference of two squares: $a^2-b^2=(a+b)(a-b)$

a) $x^2-3=0$

b) $81-4y^2=0$

c) $\frac{1}{4}a^2-6=0$

d) $z^2-16=0$

e) $0.36-x^2=0$

f) $m^2-25=0$

2.15 Factorization of trinomials with integral coefficients. In this lesson we are going to see how to factorize and solve some quadratic equations of the form $ax^2+bx+c=0$, where a, b and c are integers.

If we are able to find two integers r_1 and r_2 such that whose sum is b, or $(r_1+r_2=b)$ and whose product is ac or $(r_1+r_2=ac)$, then the quadratic trinomial can be factorized by writing bx, as r_1x+r_2x, i.e,

$ax^2+bx+c=0$

$=ax^2+r_1x+r_2x+c=0$

Example 1) Find the solution set of $1x^2+5x+6=0$

Solution:- Find two numbers whose sum is 5 and whose product $1x6=6$, these numbers are 2 and 3.

$1x^2+5x+6=0$

$1x^2+2x+3x+6=0$ ....$5x=2x+3x$

$(x^2+2x)+(3x+6)=0$ .... Grouping

$x(x+2)+3(x+2)=0$ .... Removing common factor x and 3

$(x+3)(x+2)=0$ ... Removing common factor (x+2)

$x+3=0$ or $x+2=0$ ..... Zero product property.

$x=-3$ or $x=-2$

s.s$=\{-2, -3\}$

Example 2) Find the solution set of $2x^2+7x+6=0$

Solution: First find two numbers whose sum is 7 and whose product is 12. (Note: $ac=2x6=12$)

These two numbers are 3 and 4, because $3+4=7$ and $3x4=12$

$2x^2+7x+6=0$

$2x^2+3x+4x+6=0$

$x(2x+3)+2(2x+3)=0$

$(x+2)(2x+3)=0$

$x+2=0$ or $2x+3=0$

$x=-2$ or $x=-3/2$

$s.s.=\left\{-2,-3/2\right\}$

Example 3) $x^2-6x+8=0$

Solution: The two numbers whose sum is -6 and whose product is 8 are -2 and -4

$x^2-6x+8=0$

$x^2-2x-4x+8=0$

$(x^2-2x)-(4x+8)=0$

$x(x-2)-4(x-2)=0$

$(x-4)(x-2)=0$

$x-4=0$ or $x-2=0$

$x=4$ or $x=2$

$s.s=\{2, 4\}$

Example 4) $2x^2-x-6=0$

Solution: Find two numbers whose sum is -1 and whose product is $(2x-6)=-12$, these two numbers are -4 and 3.

$2x^2-x-6=0$

$2x^2-4x+3x-6=0$

$(2x^2-4x)+(3x-6)=0$

$2x(x-2)+3(x-2)=0$

$(2x+3)(x-2)=0$

$2x+3=0$ or $x-2=0$

$x=-3/2$ or $x=2$

$s.s=\left\{-3/2,2\right\}$

Example 5) Find the solution set of $-4x^2+12x-5=0$

Solution:- Find two numbers whose sum is 12 and whose product is $(-4x-5)=20$, this numbers are 2 and 10.

$-4x^2+12x-5=0$

$-4x^2+2x+10x-5=0$

$(-4x^2+2x)+(10x-5)=0$

$-2x(2x-1)+5(2x-1)=0$

$(-2x+5)(2x-1)=0$

$-2x+5=0$ or $2x-1=0$

$x=\dfrac{5}{2}$ or $x=\dfrac{1}{2}$

s.s= $\left\{\dfrac{1}{2},\dfrac{5}{2}\right\}$

MORE ON SUM AND PRODUCT METHOD

In the quadratic equation $ax^2+bx+c=0$, if we find two integers r_1 and r_2 such that $r_1+r_2=b$ and product $r_1r_2=ac$, then the solution set of the quadratic equation is $\left\{\dfrac{-r_1}{a},\dfrac{-r_2}{a}\right\}$

This can be shown below

$ax^2+bx+c=0$

$a^2x^2+bax+ac=0$ ...... Multiplying both sides by a.

$a^2x^2+(r_1+r_2)ax+r_1r_2=0$ ... $b=r_1+r_2$ and $ac=r_1r_2$

$a^2x^2+r_1ax+r_2ax+r_1r_2=0$

$(a^2x^2+r_1ax)+(r_2ax+r_1r_2)=0$

$ax(ax+r_1)+r_2(ax+r_1)=0$

$(ax+r_1)+(ax+r_2)=0$

$ax+r_1=0$ or $ax+r_2=0$ .... Zero product property.

$ax=-r_1$ or $ax=-r_2$

$x=\dfrac{-r_1}{a}$ or $x=\dfrac{-r_2}{a}$

s.s= $\left\{\dfrac{-r_1}{a},\dfrac{-r_2}{a}\right\}$

Example 1) Find the solution set of $2x^2+8x+6=0$,

Solution:- The two numbers whose sum 8 and whose product $(2x6)=12$ are 2 and 6. To apply the above solution method let,

$r_1=2$ and $r_2=6$, then $\dfrac{-r_1}{a}=\dfrac{-2}{2}=-1$ and $\dfrac{-r_2}{a}=\dfrac{6}{2}=-3$

s.s $=\left\{\dfrac{-r_1}{a},\dfrac{-r_2}{a}\right\}=\{-1,-3\}$

Example 2) Find the solution set of $2x^2-3x-5=0$

Solution:- The two numbers whose sum -3 and whose product $(2x-5)=-10$ are -5 and 2. To apply the above solution method let,

$r_1=-5$ and $r_2=2$. Then $\dfrac{-r_1}{a}=\dfrac{-(-5)}{2}=\dfrac{5}{2}$ and $\dfrac{-r_2}{a}=\dfrac{-2}{2}=-1$

$s.s=\left\{\dfrac{-r_1}{a},\dfrac{-r_2}{a}\right\}=\left\{\dfrac{5}{2},-1\right\}$

Example 3) Find the solution set of $x^2-3x+2=0$

Solution:- The two numbers whose sum -3 and whose product $(1x2)=2$ are -1 and -2. To apply the above solution method, let,

$r_1=-1$ and $r_2=-2$, then $\dfrac{-r_1}{a}=\dfrac{-(-1)}{1}=1$ and $\dfrac{-r_2}{a}=\dfrac{-(-2)}{1}=2$

$s.s=\left\{\dfrac{-r_1}{a},\dfrac{-r_2}{a}\right\}=\{1,2\}$

Note:- We can use this method to factorize the quadratic equation $ax^2+bx+c=0$

If $r_1+r_2=b$ and $r_1r_2=ac$, then

$ax^2+bx+c=a\left(x+\dfrac{r_1}{a}\right)\left(x+\dfrac{r_2}{a}\right)$

Here, we can use (a) to multiply one of $\left(x+\dfrac{r_1}{a}\right)$ or $\left(x+\dfrac{r_2}{a}\right)$

In the quadratic ax^2+bx+c, if a, b and c are integers to make the factorization over integers you should choose the appropriate factor that should be multiplied by (a).

Example 1) Factorize $2x^2-3x-5$

Solution:- Find two numbers whose sum -3 and whose product $(2x-5)=-10$, these numbers are -5 and 2.

$r_1=-5$ and $r_2=2$,

from $ax^2+bx+c=a\left(x+\dfrac{r_1}{a}\right)\left(x+\dfrac{r_2}{a}\right)$

$$2x^2 - 3x - 5 = 2\left(x - \frac{5}{2}\right)\left(x + \frac{2}{2}\right)$$

$$= 2\left(x - \frac{5}{2}\right)(x + 1)$$

$$= \left(x - \frac{5}{2}\right)2(x + 1)$$

$$2x^2 - 3x - 5 = \left(x - \frac{5}{2}\right)(2x + 2)$$

Here note that we used 2 to multiply (x+1) to make the factorization over integers.

Example 2) Factorize x²-4x-5

Solution:- Find two numbers whose sum -4 and whose product (1x-5)=-5, these two numbers are -5 and 1

$r_1 = -5$ and $r_2 = 1$,

from $ax^2 + bx + c = a\left(x + \dfrac{r_1}{a}\right)\left(x + \dfrac{r_2}{a}\right)$

$$x^2 - 4x + 5 = 1\left(x - \frac{5}{1}\right)\left(x + \frac{1}{1}\right)$$

$$= (x - 5)1(x + 1)$$

$$x^2 - 4x + 5 = (x - 5)(x + 1)$$

Example 3) Find the solution set of 2x²+9x-5=0,

Solution:- Find two numbers whose sum 9 and whose product (2x-5)=-10, these numbers are 10 and -1

$r_1 = 10$ and $r_2 = -1$,

$$\frac{-r_1}{a} = \frac{-10}{2} = -5 \text{ and } \frac{-r_2}{a} = \frac{1}{2}$$

$$\text{s.s} = \left\{-5, \tfrac{1}{2}\right\}$$

The factorizing form of $2x^2 + 9x - 5 = 2\left(x + \dfrac{10}{2}\right)\left(x - \dfrac{1}{2}\right)$

$$= (x+5)2\left(x - \dfrac{1}{2}\right)$$

$$= (x+5)(2x-1)$$

Example) Find the solution set of 3x²-x-4=0

Solution: Find two numbers whose sum -1 and whose product (3x-4)=-12, these two numbers are -4 and 3.

$$r_1 = -4,\; r_2 = 3,\; \dfrac{-r_1}{a} = \dfrac{-(-4)}{3} = \dfrac{4}{3} \text{ and } \dfrac{-r_2}{a} = \dfrac{-3}{3} = -1$$

s.s $= \left\{\dfrac{4}{3}, -1\right\}$

Factorizing form of $3x^2 - x - 4 = 3\left(x - \dfrac{4}{3}\right)(x+1)$

$$\left(x - \dfrac{4}{3}\right)3(x+1)$$

$$\left(x - \dfrac{4}{3}\right)(3x+3)$$

Exercise 2.6

1) Find two integers whose
   a) sum 4 and product -21
   b) sum -1 and product -20
   c) sum 9 and product 18
   d) sum $\dfrac{3}{4}$ and product $\dfrac{1}{8}$

   e) sum -8 and product 15
   f) sum 6 and product -7
   g) sum 11 and product -42

2) Find the solution set of each of the following equations
   a) x²+7x+12=0
   b) x²-8x+15=0
   c) 3y²-2y-1=0
   d) y²+¾y+⅛=0
   e) 2x²-9x-5=0

3) Factorize the following quadratic equations
   a) x^2-25
   b) $x^2-12x+36$
   c) y^2-2y+2
   d) $x^2-8x+16$
   e) $x^2-9x+20$
   f) $x^2-7x+12$

4) Find the solution set of the following equations.
   a) $x^2-\frac{1}{4}=0$
   b) $2x^2+3x-2=0$
   c) $x^2-8x+15=0$
   d) $x^2-9x+20=0$
   e) $x^2-x-12=0$

5) Find the solution set of the following quadratic equation
   a) $6x^2-x-2=0$
   b) $-5x^2-8x+4=0$
   c) $7x^2-6x-1=0$
   d) $x^2+7x+12=0$
   e) $x^2-3x=28$

2.16 Factoring Special Products
2.16.1 Product of Binomials and Square of Binomials

The square of a binomial is always a trinomial. Let us see the following:
$$(a+b)^2=a^2+2ab+b^2$$
$$(a-b)^2=a^2-2ab+b^2$$

Note:- The square of any binomial produces the following three terms:
   1) The square of the first term of the binomial: a^2
   2) Twice the product of the two terms: $2ab$
   3) The square of the second term: b^2

The square of every binomial called perfect square trinomial has the form: $a^2+2ab+b^2$.

Note:- If the binomial has a minus sign, then the minus sign appears only in the middle term of the trinomial. Therefore, using the double sign ± (plus or minus"), we can state the rule as follows:

$$(a\pm b)^2 = a^2 \pm 2ab + b^2$$

This means: If the binomial is a+b, then the middle term will be +2ab; but if the binomial is a-b, then the middle term will be -2ab.

Example 1) $(x+2)^2 = (x+2)(x+2)$
$\qquad = x^2 + 2x + 2x + 4$
$\qquad = x^2 + 4x + 4$ ..... perfect square

Example 2) $(2x+3)^2 = (2x+3)(2x+3)$
$\qquad = 4x^2 + 6x + 6x + 9$
$\qquad = 4x^2 + 12x + 9$ .... perfect square

Example 3) $(x-2)^2 = (x-2)(x-2)$
$\qquad = x^2 - 3x - 3x + 9$
$\qquad = x^2 - 6x + 9$ ..... perfect square

Example 4) $(3x-4)^2 = (3x-4)(3x-4)$
$\qquad = 9x^2 - 12x - 12x + 16$
$\qquad = 9x^2 - 24x + 16$ ..... perfect square

Example 5) Is $x^2 + 10x + 25$, a perfect square trinomial?
$\qquad$ Solution: Yes, it is the square of $(x+5)$.
$\qquad$ x^2 is the square of x, 25 is the square of 5. And 10x is twice the product of x and 5.
$\qquad$ In other words, $x^2 + 10x + 25 = (x+5)^2$.

Note:- If the coefficient of x had been any number but 10, this would not have been a perfect square trinomial.

Example 6) Is $x^2 + 7x + 25$, a perfect square trinomial?
$\qquad$ Solution: No, it is not. Although x^2 is the square of x, and 25 is the square of 5, 7x is not twice the product of x and 5. (Twice their product is 10x).

Example 7) Use example 1-6 to factorize and solve the following equations.

    a) $x^2-14x+49=0$

    b) $x^2+6x+9=10$

    c) $16x^2+\dfrac{8}{3}x+\dfrac{1}{9}=0$

    d) $x^2-20x+100=0$

    e) $x^2+8x+16=0$

Solution a) $x^2-14x+49=0$

    $(x-7)^2=0$

    $x-7=0$

    $x=7$

    ss=\{7\}

    b) $x^2+6x+9=0$

    $(x+3)^2=0$

    $x+3=0$

    $x=-3$

    s.s=\{-3\}

    c) $16x^2+\dfrac{8}{3}x+\dfrac{1}{9}=0$

    $\left(4x+\dfrac{1}{3}\right)^2=0$

    $4x+\dfrac{1}{3}=0$

    $4x=-\dfrac{1}{3}$

    $x=\dfrac{-1}{12}$

    $s.s=\left\{-\dfrac{1}{12}\right\}$

d) $x^2-20x+100=0$
$(x-10)^2=0$
$x=10$
s.s=$\{10\}$

e) $x^2+8x+16=0$
$(x+4)^2=0$
$x+4=0$
$x=-4$
s.s=$\{-4\}$

2.17 Difference of Two Squares
Consider the expression of (a+b)(a-b).
(a+b)(a-b)
= a(a-b)+b(a-b) .... (Use distributive law)
= a^2-ab+ab-b^2 ... (Add like terms)
= a^2-b^2

Note that the expansion of (a+b)(a-b) yields the difference of two squares in a^2 and b^2. Thus:
a^2-b^2=(a+b)(a-b)
An expression of the form a^2-b^2 is called the difference of two squares.

Example 1) Expand each of the following
a) (x+y)(x-y)
b) (p+q)(p-q)
c) (g+h)(g-h)
d) (r+s)(r-s)
e) (d+e)(d-e)

Solution: a) (x+y)(x-y)
x(x-y)+y(x-y) ... (use distributive law)
= x^2-xy+yx-y^2 .... (Add like terms)
= x^2-y^2

b) (p+q)(p-q)
= p(p-q)+q(p-q)
= p^2-pq+qp-q2
= p^2-q^2

c) $(g+h)(g-h)$
$g(g-h)+h(g-h)$
$g^2-gh+hg-h^2$
g^2-h^2

d) $(r+s)(r-s)$
$r(r-s)+s(r-s)$
$r^2-rs+sr-s^2$
r^2-s^2

e) $(d+e)(d-e)$
$d(d-e)+e(d-e)$
$d^2-de+ed-e^2$
d^2-e^2

Example 2) Find the product of the following
  a) $(x+3)(x-3)$
  b) $(3y+5)(3y-5)$
  c) $(7a+2)(7a-2)$

Solution a) $(x+3)(x-3)$
  $x^2+3x-3x-9$
  x^2-9

b) $(3y+5)(3y-5)$
$9y^2-15y+15y-25$
$9y^2-25$

c) $(7a+2)(7a-2)$
$7a(7a-2)+2(7a-2)$
$4a^2-14a+14a-4$
$4a^2-4$

Exercise 2.7
1) Find the product of each of the following
  a) $(2y+3)(2y-3)$
  b) $(4x+1)(4x-1)$

c) $(a+\frac{1}{4})(a-\frac{1}{4})$

d) $(7x+3)(7x-3)$

2) Expand each of the following
    a) $(2x+2y)(2x-2y)$
    b) $(2-6x)(2+6x)$
    c) $(5+4x)(5-4x)$

3) Factorize each of the following
    a) x^2-6x+9

    b) $x^2 - \frac{3}{4}x + \frac{1}{8}$

    c) y^2-2y+1
    d) $x^2+8x+16$

4) Determine whether each of the following is a perfect square or not.
    a) $x^2-20x+100$
    b) $x^2+8x+16$

    c) $x^2 - \frac{3}{4}x - \frac{1}{8}$

    d) $x^2+7x+14$
    e) $y^2+25x+100$
    f) y^2-2y+4

2.18 COMPLETING THE SQUARE METHOD

Completing the square refers to the process of creating a perfect trinomial square.

    If we try to solve this quadratic equation by factoring
        $x^2+8x+4=0$
    We cannot. Therefore, we use a technique called completing the square. This means to make the quadratic into a <u>perfect square trinomial</u>, i.e. the form $a^2+2ab+b^2=(a+b)^2$.
    The technique is valid only when the coefficient of x^2 is 1.

1) <u>Transpose</u> the constant term to the right:

$$x^2+8x=-4$$

2) Add the square number to both sides. Add the square of half the coefficient of x. In this case, add the square of 4:

$$x^2+8x+16=-4+16$$

The left-hand side is now the perfect square of (x+4)

$$(x+4)^2=12$$

4 is half of the coefficient 8.

This equation has the form

$$a^2=b$$

which implies

$$a = \pm\sqrt{b}$$

Therefore, $x+4 = \pm\sqrt{12}$

$$x = -4 \pm \sqrt{12}$$

That is, the solution set to

$$x^2+8x+4=0 \text{ are}$$

Conjugate pair, $-4+\sqrt{12}$, $-4-\sqrt{12}$

We can check this. The sum of those roots is -8, which is the negative of the coefficient of x. And the product of the roots

is $(-4)^2 -\left(\sqrt{12}\right)^2 = 4$, which is the constant term. Thus both

conditions on the roots are satisfied. These are the two roots of the quadratic.

Example 1) Use completing the square method to solve the equation
$x^2+6x+7=0$
Solution; $x^2+6x+7=0$ ... (The coefficient of x^2 is 1, so we can start solving)

$x^2+6x=-7$ .... (Transpose the constant term to the right)

$x^2+6x+9=-7+9$ ...(Add the square of half of the coefficient of x, in this case, add the square of 3)

$x^2+6x+9=2$ .... (x^2+6x+9) is a perfect square it is the square of the binomial (x+3))

$(x+3)^2=2$ ... x^2+6x+9 is a perfect square

$x+3=\pm\sqrt{2}$ ... Take the square root.

$x=-3\pm\sqrt{2}.....s.s=\left\{-3+\sqrt{2},-3-\sqrt{2}\right\}$

Example 2) Use completing the square method to solve $3x^2+12x+6=0$

Solution; Here the coefficient of x^2 is 3. To make the coefficient of x^2, 1 divided both sides of the equation by 3.

$$\frac{3x^2}{3}+\frac{12x}{3}+\frac{6}{3}=\frac{0}{3}$$

$x^2+4x+2=0$

$x^2+4x=-2$ .... Transpose the constant term to the right.

$x^2+4x+4=-2+4$.... Add 4 to make the left side a perfect square

$(x+2)^2=2$ ..... x^2+4x+4 is a perfect square

$x+2=\pm\sqrt{2}$

$x=-2\pm\sqrt{2}......s.s=\left\{-2+\sqrt{2},-2-\sqrt{2}\right\}$

Example 3) Use completing the square method to solve $\frac{3}{2}x^2+\frac{11}{3}x+\frac{1}{2}=0$

Solution: Here the coefficient of x^2 is $\frac{3}{2}$. to make the coefficient of x^2, 1 multiply both sides of the equation by $\frac{2}{3}$.

$$\frac{2}{3}\left(\frac{3}{2}x^2\right)+\frac{2}{3}\left(\frac{11}{3}x\right)+\frac{2}{3}\left(\frac{1}{2}\right)=\frac{2}{3}(0)$$ .... Multiplying by $\left(\frac{2}{3}\right)$

$x^2+\frac{22}{9}x+\frac{1}{3}=0$

$x^2+\frac{22}{9}x=-\frac{1}{3}$ .... Transpose the constant term to the right.

$x^2+\frac{22}{9}x+\frac{121}{81}=-\frac{1}{3}+\frac{121}{8}$ ..... To make the left side a perfect

square, add half of $\frac{22}{9}$ squared or $\left[\frac{1}{2}\left(\frac{22}{9}\right)\right]^2=\frac{121}{81}$ on both sides.

$$\left(x+\frac{11}{9}\right)^2=\frac{355}{24} \dots \left(x+\frac{11}{9}\right)^2=x^2+\frac{22}{9}x+\frac{121}{81} \text{ is a perfect square.}$$

$$x+\frac{11}{9}=\pm\sqrt{\frac{355}{24}}$$

$$x=\frac{-11}{9}\pm\sqrt{\frac{355}{24}} \dots ss=\left\{\frac{-11}{9}+\sqrt{\frac{355}{24}},\frac{-11}{9}-\sqrt{\frac{355}{24}}\right\}$$

Example 4) Use completing the square method to solve $x^2+4x+7=0$

Solution: $x^2+4x+7=0$

$\qquad x^2+4x=-7$ ..... (Transpose the constant term to the right)

$\qquad x^2+4x+4=-7+4$... (To make the left side a perfect square add
$\qquad\qquad$ half of the square of the coefficient of x. In this case,
$\qquad\qquad$ add the square of 2)

$\qquad x^2+4x+4=-3$ .... x^2+4x+4 is a perfect square.

$\qquad (x+2)^2=-3$ ... s.s.={ }

Here the square of a real number can not be negative, therefore, the solution is the EMPTY SET.

Example 5) Solve $(x+3)^2=-5$

Solution:- The square of a real number cannot be negative so that the
$\qquad$ solution set is:- Empty set. i.e.

$\qquad (x+3)^2=-5$ .... s.s.={ }

Example 6) $(x+4)^2=0$

Solution: $x+4=0 \Rightarrow x=-4$

$\qquad$ s.s.={-4}

Exercise 2.8

1) Solve each of the following equations

$\qquad$ a) $x^2=49$

$\qquad$ b) $y^2=12$

$\qquad$ c) $x^2=\frac{5}{4}$

$\qquad$ d) $y^2=64$

$\qquad$ e) $x^2=\frac{1}{121}$

2) Solve each of the following equations
a) $(x-2)^2=17$
b) $(x+3)^2=-4$
c) $(4x-1)^2=16$
d) $(x-2)^2=-25$
e) $(3y+4)^2=1$

3) Using completing the square method, solve each of the following equations.
a) $x^2+6x+3=0$
b) $x^2+4x+6=0$
c) $x^2-11x+3=0$
d) $x^2+8x+5=0$
e) $x^2 - \dfrac{3}{2}x - \dfrac{1}{2} = 0$
f) $x^2+3x+2=0$

4) Using completing the square method, solve each of the following equations.
a) $2x^2+8x-4=0$
b) $3x^2+6x+5=0$

c) $-3x^2 - \dfrac{2}{3}x + 1 = 0$

d) $2x^2+4x+12=0$
e) $\frac{2}{3}x^2+9x+3=0$

2.19 Quadratic Formula

2.19.1 Driving the quadratic formula:

The quadratic formula is obtained by solving the general quadratic equation, $ax^2+bx+c=0$, by applying the completing the square method. This is one way to drive the quadratic formula:

$$ax^2+bx+c=0$$

Divide each side of the equation by a.

$$x^2 + \frac{b}{a}x + \frac{c}{a} = 0$$

Transpose $\dfrac{c}{a}$ to the right side of the equation.

$$x^2 + \dfrac{b}{a}x = -\dfrac{c}{a}$$

Add $\left(\dfrac{b}{2a}\right)^2$ to each side of the equation (to complete the square).

$$x^2 + \dfrac{b}{a}x + \left(\dfrac{b}{2a}\right)^2 = -\dfrac{c}{a} + \left(\dfrac{b}{2a}\right)^2$$

$$\left(x + \dfrac{b}{2a}\right)^2 = -\dfrac{c}{a} + \dfrac{b^2}{4a^2}$$

Find the common denominator for the right side of the equation.

$$\left(x + \dfrac{b}{2a}\right)^2 = -\dfrac{4ac}{4a^2} + \dfrac{b^2}{4a^2}$$

$$\left(x + \dfrac{b}{2a}\right)^2 = \dfrac{-4ac + b^2}{4a^2}$$

Take the square root of each side of the equation,

$$x + \dfrac{b}{2a} = \pm\sqrt{\dfrac{b^2 - 4ac}{4a^2}} \dots\dots b^2 - 4ac \geq 0$$

Transpose $\dfrac{b}{2a}$ to the right side of the equation.

$$x = \dfrac{-b}{2a} \pm \sqrt{\dfrac{b^2 - 4ac}{4a^2}}$$

$$x = \dfrac{-b \pm \sqrt{b^2 - 4ac}}{2a}$$

The plus or minus sign shows that there are two possible solutions.

$$x = \dfrac{-b \pm \sqrt{b^2 - 4ac}}{2a} \quad \text{.... is called quadratic formula.}$$

In this quadratic formula the expression in the radical sign b^2-4ac is called the discriminant of the quadratic equation ax^2+bx+c=0.

2.19.2 The possible cases of the discriminant (b^2-4ac)

1) If b^2-4ac>0, the quadratic equation will have two real and distinct roots. i.e. $\dfrac{-b+\sqrt{b^2-4ac}}{2a}$ or $\dfrac{-b-\sqrt{b^2-4ac}}{2a}$

2) If b^2-4ac=0, the quadratic equation will have exactly one root. It is $\dfrac{-b}{2a}$
     Note: If b^2-4ac=0, then ax^2+bx+c is a perfect square

3) If b^2-4ac<0 (negative), then the quadratic equation has no real solution or its solution set is { }.
     Note: If b^2-4ac<0 or if b^2-4ac is negative, ax^2+bx+c is not factorized.

Note: It is advisable first to find the discriminant (b^2-4ac) while using the quadratic formula.

Example-1) use the quadratic formula to find the truth set of the following equations.
a) $3x^2$+5x+4=0
b) $2x^2$-3x+1=0
c) $-4x^2$-2x+6=0
d) $5x^2$-3x+8=0
e) x^2-2x+1=0

Solution a) $3x^2$+5x+4=0
Here a=3, b=5, c=4
b^2-4ac=25-4(3)(4)=25-48=-23
The discriminant b^2-4ac<0, i.e. -23<0, therefore the solution set is { }

b) $2x^2$-3x+1=0
Solution: Here a=2, b=-3, c=1
     b^2-4ac=9-4(2)(1)=9-8=1

The discriminant b²-4ac>0, therefore we expect two distinct real roots.

$$x = \frac{-b + \sqrt{b^2 - 4ac}}{2a} \text{ or } x = \frac{-b - \sqrt{b^2 - 4ac}}{2a}$$

$$x = \frac{-(-3) + \sqrt{(-3)^2 - 4(2)(1)}}{2(2)} \text{ or } x = \frac{-(-3) - \sqrt{(-3)^2 - 4(2)(1)}}{2(2)}$$

$$x = \frac{3 + \sqrt{9 - 8}}{4} \text{ or } x = \frac{3 - \sqrt{9 - 8}}{4}$$

$$x = \frac{3 + 1}{4} \text{ or } x = \frac{3 - 1}{4} = \frac{1}{2}$$

$$x = 2 \text{ or } \frac{1}{2}$$

$$\text{s.s} = \left\{ 2, \frac{1}{2} \right\}$$

c) -4x²-2x+6=0

Solution: Here a=-4, b=-2, c=6

$$b^2-4ac=(-2)^2-4(-4)(6)=4+96=100$$

The discriminant b²-4ac>0, i.e. 100>0, therefore we expect two distinct real roots.

$$x = \frac{-b + \sqrt{b^2 - 4ac}}{2a} \text{ or } x = \frac{-b - \sqrt{b^2 - 4ac}}{2a}$$

$$= \frac{-(-2) + \sqrt{(-2)^2 - 4(-4)(6)}}{2(-4)} \text{ or } x = \frac{-(-2) - \sqrt{(-2)^2 - 4(-4)(6)}}{2(-4)}$$

$$x = \frac{2 + \sqrt{4+96}}{-8} \text{ or } x = \frac{2 - \sqrt{4+96}}{-8}$$

$$= \frac{2 + \sqrt{100}}{-8} \text{ or } x = \frac{2 - \sqrt{100}}{-8}$$

$$x = \frac{2 + 10}{-8} \text{ or } x = \frac{2 - 10}{-8}$$

$$x = \frac{-3}{2} \text{ or } x = 1$$

$$\text{s.s} = \left\{ 1, \frac{-3}{2} \right\}$$

d) $5x^2 - 3x + 8 = 0$
Solution: Here a=5, b=-3, c=8
$b^2 - 4ac = (-3)^2 - 4(5)(8) = 9 - 160 = -151$
The discriminant $b^2 - 4ac < 0$, i.e. -151<0, therefore the solution set is { }.

e) $x^2 - 2x + 1 = 0$
Solution: Here a=1, b=-2, c=1
$b^2 - 4ac = (-2)^2 - 4(1)(1) = 4 - 4 = 0$
The discriminant $b^2 - 4ac = 0$, showing $x^2 - 2x + 1$ is a perfect square therefore we expect one distinct real root.

$$x = \frac{-b + \sqrt{b^2 - 4ac}}{2a} \text{ or } x = \frac{-b - \sqrt{b^2 - 4ac}}{2a}$$

$$x = \frac{-(-2) + \sqrt{(-2)^2 - 4(1)(1)}}{2(1)} \text{ or } x = \frac{-(-2) - \sqrt{(-2)^2 - 4(1)(1)}}{2(1)}$$

$$x = \frac{2 + \sqrt{4-4}}{2} \text{ or } x = \frac{2 - \sqrt{4-4}}{4}$$

$$x = \frac{2 + 0}{2} \text{ or } x = \frac{2 - 0}{2}$$

x=1
s.s.={1}

Example 2) A man jumping rope leaves the ground at an initial vertical velocity of 8 feet per second. After how many seconds does the man land on the ground?

Solution: Let $h=-16t^2+vt+s$ be the model for the height above the ground.

$h=-16t^2+vt+s$ ... vertical motion equation

$h=-16t^2+8t+0$ ... substitute 8 for v and 0 for s.

$h=-16t^2+8t$

Substitute 0 for h. When the man lands. The man's height above the ground is 0 feet. Solve for t.

$0=-16t^2+8t$ ... vertical motion model

$0=8t(-2t+1)$ ... factor the right side

∴ The man lands on the ground ½ second after the man jumps.

Exercise 2.9

1 Use the quadratic formula to find the solution set of the following equation

a) $x^2-6x+9=0$

b) $2x^2-4x+8=0$

c) $-3x^2+5x-1=0$

d) $\frac{1}{2}x^2+3x-1=0$

e) $4x^2+6x+2=0$

2 Determine whether the following equations have two real roots, one real root or no root. (Hint: use discriminant)

a) $x^2-3x+4=0$

b) $-3x^2-8x+1=0$

c) $x^2-8x+16=0$

d) $\frac{1}{4}x^2-2x+9=0$

e) $x^2+3x+2=0$

3) What is the value of k, such that $4x^2+24x+k$ to be a perfect square?

4) What should be the value of k, if the quadratic expression is a perfect square?

a) $x^2+2kx+4$

b) kx^2+3x+5

c) $3x^2+9x+k$

d) $-2x^2+kx-1$

5) Find the value of k such that the quadratic equation:-
    a) $2x^2+8kx+8=0$, has two distinct roots
    b) $x^2-2kx+6=0$, has exactly one real root.
    c) $3kx^2-2x+5=0$, has no real root.
    d) $-4x^2-3x-k=0$, has two roots.

2.20 Graphical Method of Solving Quadratic Equations

We define a zero of a function to be any input value that yields zero as the output value. More formally, a zero of a function f is any number c such that f(c)=0. It follows that if c is a zero of f, then (c, 0) is an x-intercept of the graph of f. Thus, we can find the zeros of a function f graphically by identifying the x-intercept of the graph of $y=ax^2+bx+c$.

Example: Solve $x^2+x-6=0$ by graphing.
    Solution:
        Step 1:
Given the equation: $x^2+x-6=0$
Here a=1, b=1, and c=-6
Graph the related function $f(x)=x^2+x-6$.
The equation of the axis of symmetry is $x = \dfrac{-b}{2a}$

$$x = \frac{-b}{2a} = \frac{-1}{2(1)} = \frac{-1}{2} = -0.5$$

$\Rightarrow x = -\dfrac{1}{2}$ is the axis of symmetry of the equation.

        Step 2:

When $x = -\dfrac{1}{2} = -0.5$

$$f\left(\frac{-1}{2}\right) = \left(\frac{-1}{2}\right)^2 + \left(\frac{-1}{2}\right) - 6$$

$$f\left(\frac{-1}{2}\right) = -6.25$$

So, the coordinate of the vertex is (-0.5, -6.25).

Step 3:

Now, make the table with some points.

$f(x)=x^2+x-6$	x	-4	-3	-2	-1	-0.5	0	1	2	3
	f(x)	6	0	-4	-6	-6.25	-6	-4	0	6

Step 4:

Now, plot the ordered pairs in a coordinate plane and join the points.

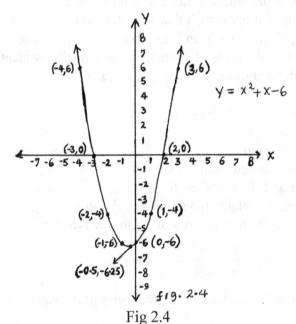

Fig 2.4

The point where the related function f(x)=0 is the solution to this quadratic equation. This occurs at the x-intercepts. The x-intercepts are the points where the graph touches the x-axis. So, the graph has -3 and 2 as x-intercepts.

So, s.s.= {2, -3}

Exercise 2.10

Use graphical method to solve the following equations.

a) $x^2-2x+1=0$

b) $x^2-3x+6=0$

2.21 Additional methods on graph of y=ax²+bx+c and the equation ax²+bx+c=0

To draw the graphs from the equation ax²+bx+c=0

Find the x-intercepts by solving ax²+bx+c=0 the graph crosses the x-axis at

$$x_1 = \frac{-b + \sqrt{b^2 - 4ac}}{2a} \text{ and } x_2 = \frac{-b - \sqrt{b^2 - 4ac}}{2a}$$

- If b²-4ac<0, the graph will not cross the x-axis
- If b²-4ac=0, the graph will not cross the x-axis but touch it (will be tangent) at the point $\frac{-b}{2a}$.

- The y-intercept of y=ax²+bx+c is at c. (The graph crosses the y-axis at (0, c))

- The vertex or turning point can be obtained by using completing the square method. It is $\left(\frac{-b}{2a}, \frac{4ac - b^2}{4a} \right)$

- The range or y-value is $y \geq \frac{4ac - b^2}{4a}$, if a>0 (The parabola is open upward)

- The range is $y \leq \frac{4ac - b^2}{4a}$, if a<0 (The parabola is open downward)

- The graph is symmetrical about $x = \frac{-b}{2a}$

- Then by taking some additional points to the right and left of $x = \frac{-b}{2a}$, you can draw the graph.

Example 1) Observe the following rough sketches on perfect square quadratics.

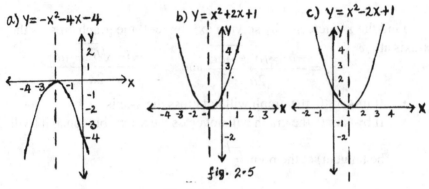

Fig 2.5 a, b and c

- If ax^2+bx+c, for a, b, c$\neq$0 is not a perfect square, this time you can use

  Vertex $\left(\dfrac{-b}{2a}, \dfrac{4ac-b^2}{4a}\right)$,

- x-intercepts by solving $ax^2+bx+c=0$
- y-intercept which is at (0, c)
- Additional tabular values.

Example: 2) plot the graph of $f(x)=x^2+x-6$
Solution: Here, a=1, b=1 and c=-6

- a>0, parabola opens upward

- vertex $\left(\dfrac{-1}{2}, \dfrac{-25}{4}\right) = (-0.5, -6.25)$

- x-intercepts are 2 and -3
- f(-4)=6, f(3)=6 (Additional points)
- The rough sketch is given in fig (2.6) below.

Range: $\left\{y : y \geq -25\big/4\right\}$

Minimum value is $\dfrac{-25}{4}$

The Graph is symmetrical with respect to $x = \dfrac{-1}{2}$

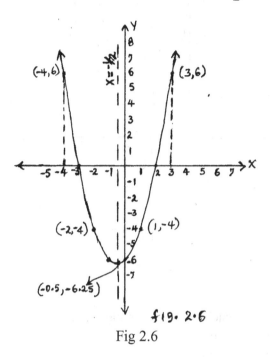

Fig 2.6

Example 3) Plot the graph of $y=x^2-2x+1$

Solution: Vertex $= \left(\dfrac{-b}{2a}, \dfrac{4ac-b^2}{4a} \right) = (1,0)$

- The y-intercept is 1
- Parabola opens upwards.
- Range: {y: y≥0}
- Minimum value is 0, it is obtained at x=1
- Axis of symmetry: x=1
- Additional points

	x	-3	-2	-1	0	1	2	3	4
y=x²-2x+1	y	16	9	4	1	0	1	4	9

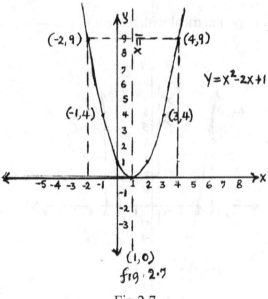

Fig 2.7

Exercise 2.11

1) Plot the graph of each of the following and find
    a) The x-intercept
    b) The y-intercept
    c) The axis of symmetry
    d) The vertex of the parabola
    e) The range
    f) The maximum value
        1a) $y=x^2+6x+8$
        1b) $y=x^2-5x+6$

2.22 Sum and product of the roots of a quadratic equation:- If the quadratic equation $ax^2+bx+c=0$ has roots r_1 and r_2, it is possible to find the sum and product of the roots before actually calculating the roots.

    Let's investigate:

Consider the general quadratic equation:
    $ax^2+bx+c=0$, where $a\neq0$

Multiply both sides by $\dfrac{1}{a}$ to create a leading coefficient of 1.

$$\frac{1}{a}\left(ax^2 + bx + c\right) = \frac{1}{a}(0)$$

$$x^2 + \frac{b}{a}x + \frac{c}{a} = 0$$

Let us represent the roots of the equation as r_1 and r_2:

$$(x\text{-}r_1)(x\text{-}r_2)=0$$
$$x^2\text{-}(r_1\text{+}r_2)x\text{+}r_1r_2=0$$

Comparing the equations, it can be seen that:

$$-\left(r_1 + r_2\right) = \frac{b}{a} \text{ or } r_1 + r_2 = \frac{-b}{a}$$

$$\text{and } r_1 \bullet r_2 = \frac{c}{a}$$

Hence, $(r_1\text{+}r_2)$ is the sum of the roots. and $r_1 \bullet r_2$ is the product of the roots.

$$\boxed{\begin{array}{l} \text{Sum of the roots: } \left(r_1 + r_2\right) = \dfrac{-b}{a} \\[2em] \text{Product of the roots: } r_1 \bullet r_2 = \dfrac{c}{a} \end{array}}$$

Example 1) Write a quadratic equation whose roots are -4 and 3.
Solution: This question could be answered by simply multiplying the factors formed by the roots.

$$(x\text{-}r_1)(x\text{-}r_2)=0$$
$$(x\text{-}(\text{-}4))(x\text{-}3)=0$$
$$(x\text{+}4)(x\text{-}3)=0$$
$$x^2\text{-}3x\text{+}4x\text{-}12=0$$
$$x^2\text{+}x\text{-}12=0$$

But, we can also arrive at the answer by utilizing the relationship between the roots and coefficients and constants. The sum of the roots is -4+3=-1

So the coefficient of the second term will be 1 (the negation of the sum)
The product of the roots is $(-4)(3)=-12$
So the constant term will be -12.
Therefore, the equation is $x^2+x-12=0$

Example 2) Find the sum of roots and the product of roots of the following equations without calculating the actual roots.

a) $3x^2-5x+6=0$
b) $-2x^2-4x+3=0$
c) $x^2+2x-1=0$
d) $2x^2+7x+4=0$

Solution: a) $3x^2-5x+6=0$

Here a=3, b=-5 and c=6

$$\text{Sum of roots} = \frac{-b}{a} = \frac{-(-5)}{3} = \frac{5}{3}$$

$$\text{Product of roots} = \frac{c}{a} = \frac{6}{3} = 2$$

b) $-2x^2-4x+3=0$
Solution:- Here a=-2, b=-4 and c=3

$$\text{Sum of roots} = \frac{-b}{a} = \frac{-(-4)}{-2} = \frac{4}{-2} = -2$$

$$\text{Product of roots} = \frac{c}{a} = \frac{3}{-2} = -\frac{3}{2}$$

c) $x^2+2x-1=0$
Solution: Here, a=1, b=2 and c=-1

$$\text{Sum of roots} = \frac{-b}{a} = \frac{-2}{1} = -2$$

$$\text{Product of roots} = \frac{c}{a} = \frac{-1}{1} = -1$$

d) $2x^2+7x+4=0$
Solution: Here, a=2, b=7 and c=4

$$\text{Sum of roots} = \frac{-b}{a} = \frac{-7}{2}$$

$$\text{Product of roots} = \frac{c}{a} = \frac{4}{2} = 2$$

Example 3) What is the solution set of the equation $2x^2-6x+k=0$, if the product of the roots is 2?

Solution:- First determine the value of k. We have seen that the product of the roots is $\dfrac{c}{a}=\dfrac{k}{2}=2$, k=4

So the equation is $2x^2-6x+4=0$, a=2, b=-6 and c=4

$b^2-4ac=36-4(2)(4)=36-32=4>0$

We expect two possible solution set.

$$x=\frac{-(-6)+\sqrt{36-4(2)(4)}}{2(2)} \quad \text{or } x=\frac{-(-6)-\sqrt{36-4(2)(4)}}{2(2)}$$

$$x=\frac{6+\sqrt{4}}{4} \quad \text{or } x=\frac{6-\sqrt{4}}{2}$$

x=2 or x=1

s.s=$\{1, 2\}$

Example 4) In the quadratic equation $kx^2-2kx+3=0$. Find the value of k, such that it has two solution set.

Solution: If the quadratic equation has two solution set, the discriminant b^2-4ac must be greater than zero.

$b^2-4ac>0$

$(-2k)^2-4(k)(3)>0$

$4k^2-12k>0$

$4k(k-3)>0$

$k(k-3)>0$

To find the value of k, let us use sign chart method.

	0	3	
k	-	+	+
(k-3)	-	-	+
k(k-3)	+	-	+

So the possible values of k is k<0 or k>3

Example 5) Find the possible values of k in the quadratic equation $x^2-2kx+k=0$, such that it has only one root.

Solution: If the quadratic equation has only one root the discriminant b^2-4ac must be equal to zero.

i.e $b^2-4ac=0$

    $(-2k)^2-4(1)(k)=0$

    $4k^2-4k=0$

    $4k(k-1)=0$

    $4k=0$ or $k=1$

    $k=0$ or $k=1$

$\therefore$ The possible values of k are 0 and 1.

Example 6) For what values of k is the quadratic equation $3kx^2-kx+4=0$ doesn't have solution set?

Solution: If the quadratic equation has not solution set, the discriminant b^2-4ac must be less than zero.

i.e, $b^2-4ac<0$

    $(-k)^2-4(3k)(4)<0$

    $k^2-48k<0$

    $k(k-48)<0$

To find the values of k, let us use sign chart method

	0		48	
k	-	+		+
(k-48)	-	-		+
k(k-48)	+	-		+

$\therefore$ The possible values of k are all real numbers between 0 and 48.

Example 7) If f(2) is one of the roots of the equation $2x^2-x-k=0$, what is the value of the second root?

Solution: If f(2) is the root of the equation $2x^2-x-k=0$, then $2(2)^2-2-k=0$

$\Rightarrow k=6$

    i.e, $2x^2-x-6=0$

    $2x^2-4x+3x-6=0$

    $2x(x-2)+3(x-2)=0$

    $(2x+3)(x-2)=0$

2x+3=0 or x-2=0

2x=-3 or x=2

$x = -\frac{3}{2}$ or x=2

∴ The value of the second root is $-\frac{3}{2}$.

Alternate Method

$$r_1 + r_2 = \frac{-b}{a} \text{ (sum of roots)}$$

$$2 + r_2 = \frac{-(-1)}{2} = \frac{1}{2}$$

$$2 + r_2 = \frac{1}{2}$$

$$r_2 = \frac{1}{2} - 2 = \frac{-3}{2}$$

Example 8) If the sum of the roots of the equation $2x^2-kx+9=0$ is $1\frac{1}{2}$, what is the value of k?

Solution: The sum of the roots of the quadratic equation is given by $\frac{-b}{a}$.

In the equation $2x^2-kx+9=0$, a=2, b=-k and c=9, then

$$\frac{-b}{a} = \frac{-(-k)}{2} = 1\frac{1}{2}$$

$$\Rightarrow \frac{k}{2} = \frac{11}{2}$$

k=11

Example 9) What is the product of the roots of the quadratic equation

$$\frac{1}{4}x^2 - \frac{1}{3}x + 6 = 0$$

Solution: product of roots $= \frac{c}{a} = \frac{6}{\frac{1}{4}} = 24$

Example 10) What is the solution set of the equation $2x^2-7x+k=0$, if the product of its roots is -2?

Solution: First determine the value of k. We have seen that product of

roots is $\frac{c}{a} = \frac{k}{2} = -2$

$\Rightarrow$ k=-4

$\Rightarrow$ 2x²-7x-4=0

2x²-8x+x-4=0

2x(x-4)+1(x-4)=0

(2x+1)(x-4)=0

2x+1=0 or x-4=0

$x = -\frac{1}{2}$ or x=4

$\therefore$ s.s $= \left\{ -\frac{1}{2}, \ 4 \right\}$

Example 11) The roots of 2x²+3x-2=0 are r_1 and r_2, then find:-

a) $r_1^2 + r_2^2$

b) $\dfrac{1}{r_1 r_2^2} + \dfrac{1}{r_2 r_1^2}$

Solution: a) If r_1 and r_2 are the roots of 2x²+3x-2=0 then, the sum of roots,

$$r_1 + r_2 = \frac{-b}{a} = \frac{-3}{2}$$

$$r_1 + r_2 = -\frac{3}{2}$$

and the product of the roots, $r_1 r_2 = \dfrac{c}{a} = \dfrac{-2}{2} = -1$

$$\left(r_1 + r_2 \right)^2 = r_1^2 + 2r_1 r_2 + r_2^2$$

$$r_1^2 + r_2^2 = \left(r_1 + r_2 \right)^2 - 2r_1 r_2$$

$$= \left(\frac{-3}{2} \right)^2 - 2(-1)$$

$$r_1^2 + r_2^2 = \frac{9}{4} + 2 = \frac{17}{4}$$

$$\therefore r_1^2 + r_2^2 = \frac{17}{4}$$

b) $\dfrac{1}{r_1 r_2^2} + \dfrac{1}{r_2 r_1^2} = \dfrac{r_1 + r_2}{r_1^2 r_2^2} = \dfrac{-3/2}{(-1)^2} = -3/2$

$\therefore \dfrac{r_1 + r_2}{r_1^2 r_2^2} = -3/2$

Exercise 2.12

1) Find the sum of the roots of the following equations without solving.

a) $x^2+4x+4=0$

b) $x^2-4x+4=0$

c) $2x^2-9x-11=0$

d) $-x^2-7x-12=0$

e) $\dfrac{1}{2}x^2 + \dfrac{1}{3}x + \dfrac{1}{20} = 0$

2) Find the product of the roots of the following equations without solving.

a) $x^2-5x+6=0$

b) $-3x^2+7x+10=0$

c) $\dfrac{1}{2}x^2 + \dfrac{1}{3}x + \dfrac{1}{20} = 0$

d) $x^2-4=0$

e) $x^2+7x-18=0$

3) What is the value of k in the equation $x^2-2kx+9=0$ if the sum of the roots is 3?

4) Find two numbers such that --

a) Whose sum is 6 and product -16

b) Whose sum is -1 and product -12

c) Whose sum is -18 and product 77

d) Whose sum is $\dfrac{11}{18}$ and product $\dfrac{1}{18}$

e) Whose sum -14 and product 45.

5) What are the roots of $3x^2-kx-8=0$, if the sum of the roots is $\dfrac{5}{3}$?

6) If (-3) is one of the roots of the equation $2x^2+kx-15=0$ find the value of k and the second root.

7) If the sum of the roots of the equation $kx^2-18x+27=0$ is 6, what is the value of k?

8) The roots of $x^2-5x+6=0$ are m and n, then find

     a) m^2+n^2

     b) $\dfrac{1}{mn^2}+\dfrac{1}{nm^2}$

     c) $\dfrac{1}{m}+\dfrac{1}{n}$

9) If the equation $-3x^2+(k+1)x+2=0$ has the sum of roots equal to $\frac{1}{3}$. What is the value of k?

2.23 THE BINOMIAL THEOREM

2.23.1 Binomial Expansion Using Pascal's Triangle.

The numbers are in Pascal's triangle can be used to find coefficients in binomial expansions $(a+b)^n$, where $(a+b)$ is any binomial and n is a positive integer.

Binomial Expansion   Pascal's Triangle

Binomial Expansion	Pascal's Triangle						
$(a+b)^0=1$				1			n=0 (0th row)
$(a+b)^1=1a+1b$			1		1		n=1 (1st row)
$(a+b)^2=1a^2+2ab+1b^2$		1		2		1	n=2 (2nd row)
$(a+b)^3=1a^3+3a^2b+3ab^2+1b^3$	1		3		3	1	n=3 (3rd row)
$(a+b)^4=1a^4+4a^3b+6a^2b^2+4ab^3+1b^4$	1	4	6	4	1		n=4 (4th row)

. . .   .

Each expansion is a polynomial. There are some patterns to be noted in the expansions.

1. In each term, the sum of the exponents is n.
2. The exponents of a start with n and decrease to 0. The last term, b^n, has no factor of a. The first term has no factor of b. The exponents

of b start in the second term with 1 and increase to n, or we can think of them starting in the first term with 0 and increasing to n.

3. There is one more term than the power n. That is, there are n+1 terms in the expansion of $(a+b)^n$.

4. Now we consider the coefficients. The first and last coefficients are 1, and the coefficients have a symmetry to them. they start at 1 and increase through certain values about "half"-way and then decrease through these same values back to 1.

Example 1) Use the fifth row of Pascal's triangle and find the numbers in the sixth row of Pascal's triangle.

Solution: Write the sixth row of Pascal's triangle by adding numbers from the fifth row.

n=5 (5th row)   1   5   10   10   5   1
n=6 (6th row)   1   6   15   20   15   6   1

The first and the last numbers are 1, so the numbers in the sixth row of Pascal's triangle are:

1, 6, 15, 20, 15, 6 and 1.

Example 2) Use the Binomial Theorem and Pascal's triangle to write the binomial expansion of $(x+2)^4$.

Solution: The binomial coefficients from the fourth row of Pascal's triangle are 1, 4, 6, 4 and 1. So, the expansion is as follows:

$$(x+2)^4=(1)(x^4)+(4)(x^3)(2^1)+(6)(x^2)(2^2)+(4)(x^1)(2^3)+(1)(2^4)$$
$$= x^4+8x^3+24x^2+32x+16$$

Example 3) Use the Binomial Theorem and Pascal's triangle to write the binomial expansion of $(x+3)^5$.

Solution:- The Binomial coefficients from the fifth row of Pascal's triangle are 1, 5, 10, 10, 5 and 1. So the expansion is as follows:-

$$(x+3)^5=(1)(x^5)+(5)(x^4)(3^1)+(10)(x^3)(3^2)+(10)(x^2)(3^3)+(5)(x^1)(3^4)+(1)(3^5)$$
$$= x^5+15x^4+90x^3+270x^2+405x+243$$

Example 4) Use the Binomial theorem and Pascal's triangle to write the binomial expansion of $(x-2)^4$.

Solution:- The binomial from the fourth row of Pascal's triangle are 1, 4, 6, 4 and 1, so the expansion is as follows:-

$(x-2)^4=[x+(-2)]^4$

$=(1)(x^4)+(4)(x^3)(-2)^1+(6)(x^2)(-2)^2+(4)(x)(-2)^3+(1)(-2)^4$

$= x^4-8x^3+24x^2-32x+16$

Example 5) Use the Binomial Theorem and Pascal's triangle to write the binomial expansion of $(2x+3)^5$

Solution:- The Binomial coefficients from the fifth row of Pascal's triangle are 1, 5, 10, 10, 5 and 1. So the expansion is as follows:-

$(2x+3)^5=(1)(2x)^5+(5)(2x)^4(3)+(10)(2x)^3(3^2)+(10)(2x)^2(3^3)+(5)(2x)(3^4)+(1)(3^5)$

$= 32x^5+240x^4+720x^3+1080x^2+810x+243$

Example 6) Find the coefficient of x in the expansion of $(3x+2)^5$.

Solution:- The binomial coefficients from the fifth row of Pascal's triangle are 1, 5, 10, 10, 5 and 1. From the Binomial Theorem, you know the expansion has the following form:

$(3x+2)^5=(1)(3x)^5+(5)(3x)^4(2)+(10)(3x)^3(2^2)+(10)(3x)^2(2^3)+(5)(3x)(2^4)+(1)(2^5)$

The coefficient of the x-term is $(5)(3)(2^4)$

$= 15+16$

$=240$

Example 7) Find the coefficient of x^2 in the expansion of $(5-x)^6$.

Solution:- The Binomial coefficients from the sixth row of Pascal's triangle are 1, 6, 15, 20, 15, 6 and 1. From the Binomial Theorem, you know the expansion has the following form.

$(5-x)^6=[(-x)+5]^6$

$=(1)(-x)^6+(6)(-x)^5(5)+(15)(-x)^4(5^2)+(20)(-x)^3(5^3)+(15)(-x)^2(5^4)+(6)(-x)(5^5)+(1)(5^6)$

The coefficient of the x^2 term is $(15)(5^4) = 9375$

Example 8) What is the sum of the numbers in row six of Pascal's triangle?

Solution:- The Binomial coefficients from the sixth row of Pascal's triangle are 1, 6, 15, 20, 15, 6 and 1. Therefore the sum of the numbers in this row is 64.

Example 9) Find the coefficient of x^3 in the expansion of $(2x+3)^4$

Solution:- The binomial coefficients from the fourth row of Pascal's triangle are 1, 4, 6, 4, and 1. So, the expansion is as follows.

$(2x+3)^4=(1)(2x)^4+(4)(2x)^3(3)+(6)(2x)^2(3^2)+(4)(2x)(3^3)+(1)(3^4)$

$= 16x^4+96x^3+216x^2+216x+81$

$\therefore$ The coefficient of x^3 is 96.

Example 10) Use the Binomial Theorem and Pascal's triangle to write the binomial expansion of $(a+3b)^4$

Solution:- $(a+3b)^4=[a+(3b)]^4$

$=(1)(a^4)+(4)(a^3)(3b)+(6)(a^2)(3b)^2+(4)(a)(3b)^3+(1)(3b)^4$

$=a^4+12a^3b+54a^2b^2+108ab^3+81b^4$

Exercise 2.13

1) Find the numbers in the sixth row of Pascal's triangle
2) Find the numbers in the ninth row of Pascal's triangle.
3) What is the sum of the numbers in row seven of Pascal's triangle?
4) Find the coefficient of x^2 in the expansion of $(2x-3)^5$?
5) Use the Binomial Theorem and Pascal's triangle to write the binomial expansion of each of the following.
   a) $(x-4)^4$
   b) $(4-y)^5$
   c) $(3x-2)^3$
   d) $(2x+6)^4$
   e) $(4x+2)^4$
6) What is the coefficient of x^3 in the expansion of $(3x-2)^6$?
7) Find the coefficient of x in the expansion of $(3-x)^5$

CHAPTER-THREE

3. CUBIC, SQUARE ROOT AND RATIONAL FUNCTIONS

3.1 CUBIC FUNCTIONS

Any function of the form $f(x)=x^3$, is referred to as a cubic function. We shall also refer to this function as the "Parent" and the following graph is a sketch of the parent graph.

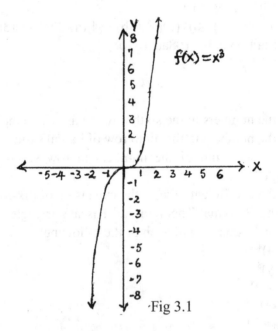

·Fig 3.1

We also want to consider factors that may alter the graph. Let's begin by considering the functions.

$f(x)=x^3$

$g(x)=x^3+2$

$h(x)=x^3-2$

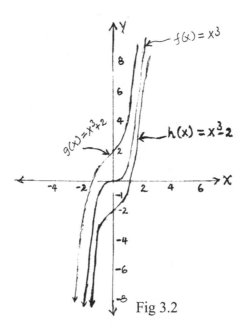

Fig 3.2

As you have seen from the above graphs, the graph of g(x)=x³+2 is a vertical shift of 2 units up along the y-axis of the parent graph (f(x)=x³) and the graph of h(x)=x³-2 is a vertical shift of 2 units down along the y-axis of the parent graph (f(x)=x³).

It is now easy to generalize:
- If y=f(x)+c and c>0, the graph undergoes a vertical shift c units up along the y-axis.
- If y=f(x)+c and c<0, the graph undergoes a vertical shift c units down along the y-axis.

It is also necessary to evaluate the functions at specific values and examine their graphs. Let's investigate the changes to the graph for the following values (x+1), and (x-1) given these values, our new functions would be:-

f(x)=x³
g(x)=(x+1)³
h(x)=(x-1)³

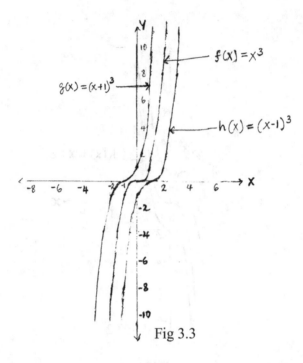

Fig 3.3

As before, our parent graph is $f(x)=x^3$, $f(x+1)$ is shown by $g(x)=(x+1)^3$, and $f(x-1)$ is shown by $h(x)=(x-1)^3$. Let's make our observations:

- If $y=f(x+c)$ and $c>0$, the graph undergoes a horizontal shift c units to the left.
- If $y=f(x+c)$ and $c<0$, the graph undergoes a horizontal shift c units to the right.

Consider the function $f(x)=(x+c)^3+d$

1) If $d>0$, the graph shifts d units up, if $d<0$, the graph shifts d units down.
2) If $c>0$, the graph shifts c units to the left, if $c<0$, the graph shifts c units to the right.

3.2 Function Notation: Even and Odd Functions

Definition: Let $k \subseteq R$, then we say that k is a symmetric set if and only if $-x \in k$ when ever $x \in k$.

Example 1) a) The set of integer is symmetric.
          b) The set of rational number is symmetric.
          c) The set of irrational numbers is symmetric.
          d) The set of real numbers is symmetric.

Example 2) The set of whole numbers is <u>not</u> symmetric. Because $2 \in W$, but $-2 \notin W$.

Definition:- A function f is said to be even if and only if its domain is a symmetric set and $f(-x)=f(x)$, for all x in the domain.

Example 3) $f(x)=x^2$ is an even function, you can see for any x, $f(x)=f(-x)$. You can consider the following to convince yourself.
   $f(1)=1$, $f(-1)=1$ and $f(1)=f(-1)$
   $f(2)=4$, $f(-2)=4$ and $f(2)=f(-2)$
   $f(3)=9$, $f(-3)=9$ and $f(3)=f(-3)$

Example 4) Define $f{:}R{\to}R$ by $f(x)=x^4$. Then f is even. This can be shown as:
i) The domain of f is R and R is symmetric.
ii) $f(-x)=f(-x)^4$
     $=x^4$
     $=f(x)$
For instance, $f(-2)=(-2)^4=f(2)$
        $f(-3)=(-3)^4=81=f(3)$, etc

Example 5) $f(x)=x^2-8$ is an even function.

Example 6) $f(x)=x^2+3x+2$: is not an even function, because $f(1){\neq}f(-1)$

Example 7) Show that $f(x)=2x^2-3$ is an even function.
     Solution: The domain of f=R, a symmetric set.
       $f(-x) = 2(-x)^2-3$
            $= 2x^2-3$
            $= f(x)$
   $\therefore$ $f(x)=2x^2-3$ is an even function.

Example 8)   Show that $h(x) = \dfrac{|x|}{x^2 - 4}$ is an even.

Solution: The domain of h=R| {2, -2}, symmetric.

$$h(-x) = \frac{|-x|}{(-x)^2 - 4} = \frac{|x|}{x^2 - 4} = h(x)$$

$\therefore h(x) = \dfrac{|x|}{x^2 - 4}$ is an even function.

Example 9)   Show that g(x)=x²+2x+3 is not an even function.
Solution:- The domain of g=R, symmetric.

But g(-x)=(-x)²+2(-x)+3

g(-x)=x²-2x+3

g(-x)≠g(x)

For instance g(-1)=(-1)²+(2)(-1)+3

=1-2+3

=2

g(1)=1²+2(1)+3

=6

g(1)=6

$\therefore$ g(-1)≠g(1)

Hence g(x)=x²+2x+3, is not an even function.

Note:- The graphs of even functions are symmetric about the y-axis.

Definition:-   A function f is said to be odd if and only if the domain of f is a symmetric set and f(-x)=-f(x), for all values of x in its domain.

Example 10) Show that f(x)=x³ is an odd function.
          Solution: The domain of f is R, which is symmetric,
          f(-x)=-x=-f(x)

Example 11) Show that g(x)=-3x³ is an odd function.
Solution:- Domain of g=R, which is symmetric.

g(-x)=-3(-x)³

=3x³

=-(-3x³)

=-g(x)

∴ g(x)=-3x³ is an odd function.

Example 12) f(x)=x is an odd function because,
- Its domain is the set of real numbers, which is symmetric.
  f(-x)=-x=-f(x).

Example 13) Show that $f(x) = \dfrac{x}{x^2 - 4}$ is an odd function.

Solution: The domain of f={x: x²-4≠0}

={x: x≠±2}, symmetric

$$f(-x) = \dfrac{-x}{(-x)^2 - 4}$$

$$= \dfrac{-x}{x^2 - 4}$$

$$= -\left(\dfrac{x}{x^2 - 4}\right)$$

$$= -f(x)$$

Note: The graphs of odd functions are symmetric about the origin.
- The end behavior of a function's graph is the behavior of the graph as x approaches positive infinity (+∞) or negative infinity (-∞).
- If the degree is odd and the leading coefficient is positive: f(x)→-∞ as x→-∞ and f(x)→+∞ as x→+∞.
- If the degree is odd and the leading coefficient is negative: f(x)→+∞ as x→-∞ and f(x)→-∞ as x→+∞.

3-3 Graph y=ax³

Example 1) Graph g(x)=-3x³ and compare with the graph f(x)=x³
Solution: Make a table of values for g(x)=-3x³

g(x)=-3x³	x	-2	-1	0	1	2
	y	24	3	0	-3	-24

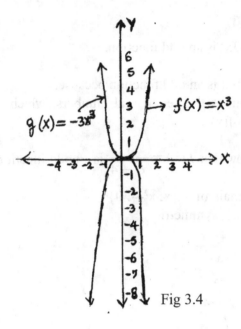

Fig 3.4

plot points from the table and connect them with a smooth curve. The degrees of $f(x)=x^3$ and $g(x)=-3x^3$ are odd but the leading coefficients do not have the same sign, so the graphs have different end behavior. The graph of $g(x)=-3x^3$ is narrower than the graph of $f(x)=x^3$. This is because the graph of $g(x)=-3x^3$ is a vertical stretch (by a factor of 3) with a reflection in the x-axis of the graph of $f(x)=x^3$. The graphs could also be viewed as being reflected in the y-axis.

Example 2) Determine whether $f(x)=x^2-4x^3$ is even, odd, or neither even nor odd.

Solution:- a) Find $f(-x)$ and $-f(x)$ and simplify.

$$f(x)=x^2-4x^3$$
$$f(-x)=(-x)^2-4(-x)^3$$
$$=x^2+4x^3$$
$$-f(x)=-x^2+4x^3$$

b) Compare $f(x)$ and $f(-x)$ to determine whether f is even. Since $f(x)$ and $f(-x)$ are not the same for all x in the domain. f is not even.

c) Compare $f(-x)$ and $-f(x)$ to determine whether f is odd. Since $f(-x)$ and $-f(x)$ are not the same for all x in the domain, f is not odd.

Thus f is neither even nor odd.

Exercise 3.1

1) Determine each of the following functions whether they are even, odd or neither even nor odd.

a) $f(x) = \frac{1}{3}x^3 + x$

b) $g(x) = 3x^3 - 2x^2 + 1$

c) $h(x) = -4x^3 - 3$

d) $f(x) = x^2 + 3$

e) $g(x) = x^4 + x^2 + 5$

f) $h(x) = 2x + 3$

g) $f(x) = \dfrac{x}{x^2 + 2}$

h) $g(x) = \dfrac{|2x|}{x^3 + 3x}$

2) Draw the graph of $f(x) = -x^3$ and write its behavior.

3) Determine whether each of the following functions are symmetrical with respect to the y-axis or the origin.

a) $f(x) = 3x^4 + 2x^2 + 3$

b) $g(x) = x^3 + x$

c) $h(x) = \dfrac{x^3}{x^2 + 2}$

d) $f(x) = x^2 + 1$

e) $f(x) = \frac{1}{3}x^3$

3.4 Increasing and Decreasing Functions

3.4.1 Definition: A function f is said to be increasing on an interval I if whenever x_1, $x_2 \in I$ and $x_1 < x_2$, then $f(x_1) < f(x_2)$.

3.4.2 A function f is said to be decreasing on an interval J if when ever x_1, $x_2 \in J$ and $x_1 < x_2$, then $f(x_2) < f(x_1)$.

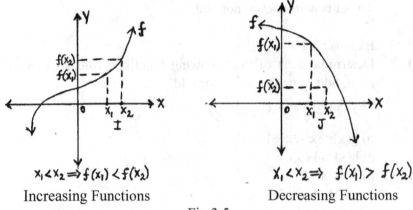

$$x_1 < x_2 \Rightarrow f(x_1) < f(x_2)$$

Increasing Functions

$$x_1 < x_2 \Rightarrow f(x_1) > f(x_2)$$

Decreasing Functions

Fig 3.5

Example 1) The function f(x)=3x-1 is an increasing function. In this function as x is increasing y is also increasing.
Suppose x_1=2 and x_2=5
$f(x_1)$=f(2)=3x2-1=5 and $f(x_2)$=f(5)=3x5-1=14
Therefore 14>5 and $f(x_2)$>f(x_1), which is f(5)>f(2)

Example 2) The function f(x)=-3x+2 is decreasing function. In this function as x is increasing y will be decreasing suppose x_1=4 and x_2=6
$f(x_1)$=f(4)=-3x4+2=-10 and $f(x_2)$=f(6)=-3x6+2=-16
Therefore 6>4, but f(6)<f(4).

Example 3) a) f(x)=x^5 is increasing.
b) f(x)=-x+1 is decreasing.
c) f(x)=2x+3 is increasing.

Example 4) Let f(x)=-x^2+4. Find the largest interval on which
a) f is increasing
b) f is decreasing
Solution:- The graph of f is a downward parabola.

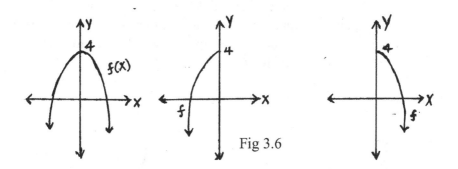

Fig 3.6

Note: Let f be an even function which is increasing on [0, ∞⁺), then it is decreasing on (-∞, 0]

- If f is even and decreasing on [0, ∞⁺), then it will be increasing on (-∞, 0].
- If f is odd and decreasing on [0, ∞⁺), then it is also decreasing on (-∞, 0].

Example 5)   Find the largest interval on which
        a) $f(x)=x^2$ is decreasing.
        b) $f(x)=x^2$ is increasing.

Solution:- From the graph of $f(x)=x^2$, we see that f is decreasing on (-∞, 0] and increasing on [0, ∞⁺).

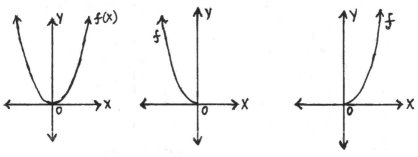

decreasing on (-∞, 0]   increasing on [0, ∞⁺)

Fig 3.7

Note: $f(x)=x^2$ is neither increasing nor decreasing on the interval (-∞, ∞⁺).

More on increasing and decreasing functions

If the graph of a function rises from left to right, it is said to be an increasing function. If the graph of a function drops from left to right, it is said to be a decreasing function.

Example 6)

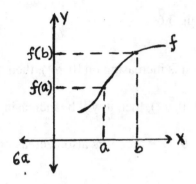

An increasing function
If a<b, then f(a)<f(b).

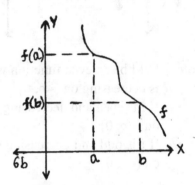

A decreasing function
If a<b, then f(a)>f(b).

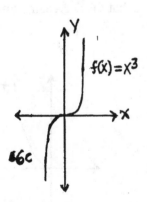

f is increasing

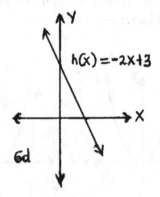

h is decreasing

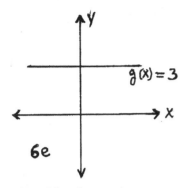

6e

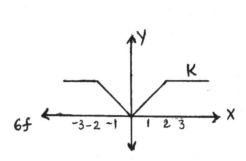

6f

g is neither increasing nor decreasing - It is a constant function

k is neither increasing nor decreasing

fig. 3.8

In example 6f, the function k is neither increasing nor decreasing on the entire real line. But it is increasing on the interval [0, 2] decreasing on the interval [-2, 0], and constant on the interval (-∞, -2] and [2, ∞⁺).

Exercise 3.2
Determine the following function whether they are increasing, decreasing or neither increasing nor decreasing.
a) $f(x)=3x^2$.
b) $f(x)=-2x^3+2$
c) $f(x)=2x+5$
d) $f(x)=-x+4$
e) $f(x)=x^2$, $D=\{x: x\geq0\}$
f) $f(x)=x^2$, $D=\{x: x\leq0\}$
g) $f(x)=x^2$

3.5 Special Products and Factoring of Cubics
Some products occur often enough in mathematical work that one should be familiar with their result. These products are called special products. Alternatively, given the result of a special product, one should be able to recognize how to rewrite the expression or equation as a product or factors. The list below shows some special products and factoring patterns of cubics. In each example, a, b, c, and d are real numbers.

3.5.1 Cubes of Binomials (Perfect Cube Binomials)

$(x+a)^3=(x+a)(x+a)(x+a)=x^3+3ax^2+3a^2x+a^3$
$(x-a)^3=(x-a)(x-a)(x-a)=x^3-3ax^2+3a^2x-a^3$

3.5.2 Difference of two cubes

$x^3-a^3=(x-a)(x^2+ax+a^2)$

3.5.3 Sum of Two Cubes

$x^3+a^3=(x+a)(x^2-ax+a^2)$

Example 1) Factorize each of the following expression using special product patterns.
  a) $x^3-9x^2+27x-27$
  b) $x^3+6x^2+12x+8$
  c) x^3-3x^2+3x-1
  d) $x^3-6x^2+12x-8$
  e) $x^3+12x^2+48x+64$

Solution; a) $x^3-9x^2+27x-27$
$\qquad = x^3-3x^2(3)+3x(3^2)-3^3$ 　　　Rewrite
$\qquad = (x-3)^3$ 　　　use pattern.
  b) $x^3+6x^2+12x+8$
$\qquad = x^3+3x^2(2)+3x(2^2)+2^3$ 　　　Rewrite.
$\qquad = (x+2)^3$ 　　　use pattern
  c) $x^3-3x^2+3x-1 = x^3-3x^2(1)+3x(1^2)-1^3$ 　　Rewrite
$\qquad\qquad = (x-1)^3$ 　　　use pattern
  d) $x^3-6x^2+12x-8 = x^3-3x^2(2^1)+3x(2^2)-2^3$ 　Rewrite
$\qquad\qquad = (x-2)^3$ 　　　use pattern
  e) $x^3+12x^2+48x+64 = x^3+3x^2(4)+3x(4^2)+4^3$ 　　Rewrite
$\qquad\qquad = (x+4)^3$ 　　　use pattern

Example 2)　Factor each of the following expressions
  a) $2x^3-6x^2+6x-2$
  b) $-3x^318x^2-36x+24$
  c) $x^4+15x^3+75x^2+25x$
  d) $x^4-6x^3+12x^2-8x$

Solution: a) $2x^3-6x^2+6x-2$

$= 2(x^3-3x^2+3x-1)$ Factor common monomial

$= 2[(x^3-3x^2(1)+3x(1^2)-1^3)]$ rewrite

$= 2(x-1)^3$ use pattern

b) $-3x^3+18x^2-36x+24$

$= -3[x^3-6x^2+12x-8]$ Factor common monomial

$= -3[x^3-3x^2(2)+3x(2^2)-2^3]$ Rewrite

$= -3(x-2)^3$ use pattern

c) $x^4+15x^3+75x^2+125x$

$= x[x^3+15x^2+75x+125]$ Factor common monomial

$= x[(x^3+3x^2(5)+3x(5^2)+5^3)]$ Rewrite

$= x(x+5)^3$ use pattern

d) $x^4-6x^3+12x^2-8x$

$= x[x^3-6x^2+12x-8]$ Factor common monomial

$= x[(x^3-3x^2(2)+3x(2^2)-(2^3))]$ Rewrite

$= x(x-2)^3$ use pattern

Example 3)   Factor each of the following expressions

a) $x^3+y^3+12x^2y^2+48xy+64$

b) $343a^3-147a^2b+21ab^2-b^3$

c) $p^3q^3+15p^2q^2+75pq+125$

d) $16x^3+24x^2y+12xy^2+2y^3$

Solution a) $x^3+y^3+12x^2y^2+48xy+64$

$= (xy)^3+3(xy)^2(4)+3(xy)(4^2)+4^3$ Rewrite

$= (xy+4)^3$ use pattern.

b) $343a^3-147a^2b+21ab^2-b^3$

$= (7a)^3-3(7a)^2(b)+3(7a)(b^2)-b^3$ Rewrite

$= (7a-b)^3$ use pattern

c) $p^3q^3+15p^2q^2+75pq+125$

$= (pq)^3+3(pq)^2(5)+3(pq)(5^2)+5^3$ Rewrite

$= (pq+5)^3$ use pattern

d) $16x^3+24x^2y+12xy^2+2y^3$
$= 2[8x^3+12x^2y+6xy^2+y^3]$ Rewrite
$= 2(2x+y)^3$ use pattern

Example 4)   The amount x (in dollars) that Hana's savings account after t years can be modeled by the equation $x=A(1+r)^t$, where A is the initial amount and r is the annual interest rate expressed as a decimal. The polynomial $1000+3000r+3000r^2+1000r^3$ represents the amount of money in her account after 3 years. What was the initial amount of her investment?

Solution:- Factor the polynomial so that it is of the form $A(1+r)^t$.

$$1000+3000r+3000r^2+1000r^3=1000(1+3r+3r^2+r^3)$$
$$=1000(1+r)^3$$

The polynomial factors to $1000(1+r)^3$, so the initial amount of her investment was $1000.

Exercise 3.3

(1)   Factor each of the following expression.
        (a) $x^3+9x^2+27x+27$
        (b) $x^4+21x^3+147x^2+243x$
        (c) $-4x^3-36x^2-108x-108$
        (d) $3x^4-18x^3+36x^2-24x$
        (e) $x^3+18x^2+108x+216$
        (f) $-2x^3+18x^2-54x+54$

(2)   Factor each of the following expression.
        (a) $-4a^3b^3-48a^2b^2-192ab-256$
        (b) $343x^3y^3+147x^2y^2z+21xyz^2+z^3$
        (c) $r^3s^3t^3+15r^2s^2t^2+75rst+125$
        (d) $3p^3q^3-54p^2q^2+324pq-648$

(3)   Determine the value of k for which the expression can be factored using a special product pattern.
        a) x^3+6x^2+kx+8
        b) $64x^3-kx^2+108x-27$

4)   The amount y (in dollars) that is in Maria's savings account after t years can be modeled by the equation $y=A(1+r)^t$, where A is the initial amount and r is the annual interest rate expressed as a decimal. The polynomial:-
$900+2700r+2700r^2+900r^3$ represents the amount of money in the account after 3 years. What was the initial amount of her investment?

3.6 Graphs of Square Root Functions

- A radical expression is an expression that contains a radical, such as a square root, cube root, or other root.
- A radical function contains a radical expression with the independent variable in the radicand.
- If the radical is a square root, then the function is called a square root function.
- The most basic square root function in the family of all square root functions, called the parent square root function, is $y=\sqrt{x}$

i.e, The function defined by the equation $f(x)=\sqrt{x}$, where x≥0 is called square root function.

We begin the section by drawing the graph of the function, then we address the domain (the value of x) and the range (the value of y). After that we'll see a number of different transformation of the function.

Example 1) Graph the function $f(x)=\sqrt{x}$
Solution:- First let's create a table points that satisfy the equation of the function, then plot the points from the table on a Cartesian coordinate system on graph paper.

We know we cannot take the square root of a negative number. Therefore, we don't want to put any negative x-values in our table. To further simplify our computations, let's use numbers whose square root is easily calculated. These numbers are perfect square roots such as 0, 1, 4, 9, 16, 25, and so on. We have placed these numbers as x-values in the table below fig 3.9 (b). Then calculated the square root of each. In fig 3.9 (a), you see each of the points from the table plotted as a solid dot. If we continue to add points to the table, plot them, the graph will eventually fill in and take the shape of the solid curve shown in fig 3.9 (c).

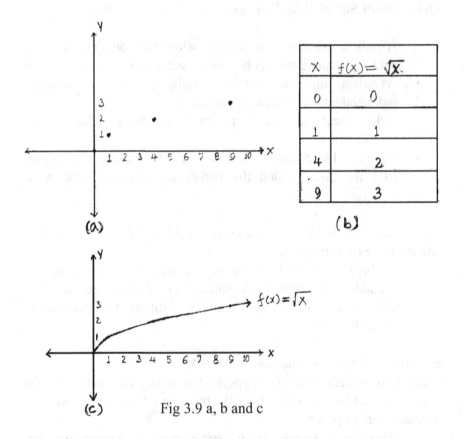

X	$f(x) = \sqrt{x}$
0	0
1	1
4	2
9	3

(a)

(b)

(c)

Fig 3.9 a, b and c

Note:- We can determine the domain and range of the square root function by projecting all points on the graph onto the x- and y-axes, as shown in the figure 3.10(a) and (b) respectively.

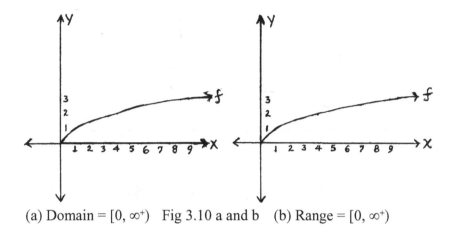

(a) Domain = [0, ∞⁺)   Fig 3.10 a and b   (b) Range = [0, ∞⁺)

i.e, The domain of the function $f(x) = \sqrt{x}$, is the set of all values of x which is greater than or equal to zero. And the range is the set of all values of y which is greater than or equal to zero.

3.7 TRANSLATIONS

Note:- If we shift the graph of $f(x) = \sqrt{x}$, right and left, or up and down, the domain and/or range are affected

Example)   Sketch the graph of $f(x) = \sqrt{x-3}$, use your graph to determine the domain and range.

Solution:- We know that the basic equation $f(x) = \sqrt{x}$ has the graph shown in figure 3.9(c). If we replace x with x-3, the basic equation $f(x) = \sqrt{x}$ becomes $f(x) = \sqrt{x-3}$. From our previous work with geometric transformations. We know that this will shift the graph three units to the right, as shown in figures fig 3.11(a) and (b).

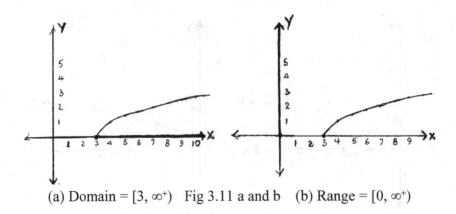

(a) Domain = [3, ∞⁺)   Fig 3.11 a and b   (b) Range = [0, ∞⁺)

i.e. The domain (the value of x) of the function $f(x) = \sqrt{x-3}$, is the set of all values of x which is greater than or equal to 3. And the range (the value of y) is the set of all values of y which is greater than or equal to zero.

Note:- To draw the graph of $f(x) = \sqrt{x-3}$, shift the graph of $f(x) = \sqrt{x}$, three units to the right.

To find the domain, we project each point on the graph of f onto the x-axis, as shown in the figure fig 3.11(a) above. Note that all points to the right of or including 3 are shaded on the x-axis. Consequently, the domain of f is:

Domain = [3, ∞⁺) = {x: x≥3}

As there has been no shift in the vertical direction (y-axis), the range remains the same. To find the range, we project each point on the graph onto the y-axis, as shown in figure 3.11(b) above. Note that all points at and above zero are shaded on the y-axis. Thus, the range is:

Range = [0, ∞⁺) = {y: y≥0}.

We can find the domain of this function algebraically by examining its defining equation $f(x) = \sqrt{x-3}$. We understand that we cannot take the square root of a negative number. Therefore, the expression under the radical must be non negative (positive or zero) number. That is,

x-3≥0

Solving this inequality for x,

x≥3.

Thus, the domain of f is Domain = [3, ∞⁺), which matches the graphical solution above.

Let's look at another example,

Example)   Sketch the graph of $f(x) = \sqrt{x+4} + 3$. Use your graph to determine the domain and range of f.

Again, we know that the basic equation $f(x) = \sqrt{x}$ has the graph shown in figure 3.9(c). If we replace x with x+4, the basic equation $f(x) = \sqrt{x}$ becomes $f(x) = \sqrt{x+4}$. From our previous experience with geometric transformations, we know that this will shift the graph of $f(x) = \sqrt{x}$ four units to the left, as shown in Figure 3.12.

If we now add 3 to the equation $f(x) = \sqrt{x+4}$ to produce the equation $f(x) = \sqrt{x+4} + 3$, this will shift the graph of $f(x) = \sqrt{x+4}$ three units upward, as shown in figure 3.13

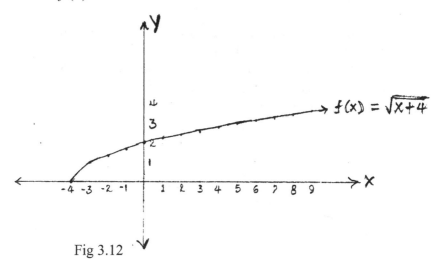

Fig 3.12

To draw the graph of $f(x) = \sqrt{x+4}$, shift the graph of $f(x) = \sqrt{x}$ four units to the left.

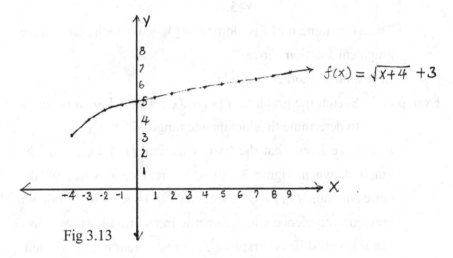

Fig 3.13

To draw the graph of $f(x) = \sqrt{x+4} + 3$, shift the graph of $f(x) = \sqrt{x+4}$ three units upward.

- Translating the original equation $f(x) = \sqrt{x}$ to get the graph of $f(x) = \sqrt{x+4} + 3$

To identify the domain of $f(x) = \sqrt{x+4} + 3$, we project all points on the graph of f onto the x-axis, as shown in figure 3.14(a). Note that all points to the right of or including -4 are shaded on the x-axis. Thus, the domain of $f(x) = \sqrt{x+4} + 3$ is:

Domain = [-4, ∞) = {x: x≥-4}

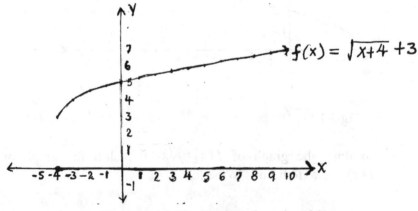

Fig 3.14 (a) Shading the domain of f

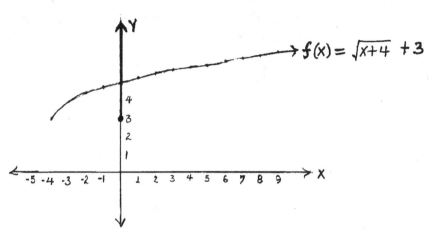

Fig 3.14 (b) Shading the range of f.

- Project points of f onto the axes to determine the domain and range.

  Similarly, to find the range of f, project all points on the graph of f on to the y-axis, as shown in figure 3.14(b). Note that all points on the y-axis greater than or including 3 are shaded. Consequently, the range of f is:

  Range = $[3, \infty^+)$ = {y: y≥3}.

We can also find the domain of f algebraically by examining the equation $f(x) = \sqrt{x+4} + 3$. We can not take the square root of a negative number, so the expression under the radical must be nonnegative (zero or positive). Consequently,

  x+4≥0

Solving this inequality for x,

  x≥-4

Thus, the domain of f is Domain = $[-4, \infty^+)$, which matches the graphical solution presented above.

Example 1) Graph the function $f(x) = 3\sqrt{x}$ and identify its domain and range. Compare the graph with the graph of $f(x) = \sqrt{x}$

Solution:-   First make a table, because the square root of a negative
             number is undefined, the value of x must be non-negative.
             So the domain is x≥0.

$f(x) = 3\sqrt{x}$	x	0	1	2	3	4
	y	0	3	4.2	5.2	6

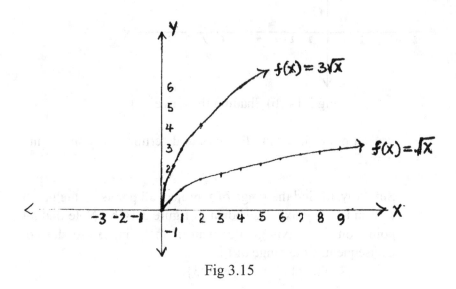

Fig 3.15

Plot the point and draw a smooth curve through the points. From either
the table or the graph, you can see the range of the function is y≥0. When
we compare the graph with the graph of $f(x) = \sqrt{x}$. The graph of $3\sqrt{x}$
is a vertical stretch (by a factor of 3) of the graph of $f(x) = \sqrt{x}$.

Example 2)   Graph the function $y = \frac{-1}{2}\sqrt{x}$ and identify its domain and
             range. Compare the graph with the graph of $y = \sqrt{x}$.

Solution:- To graph the function, first make a table, plot the points, and
draw a smooth curve through the points. The domain is x≥0.

$y = \frac{-1}{2}\sqrt{x}$	x	0	1	2	3	4	5
	y	0	-0.5	-0.7	-0.9	-1	-1.1

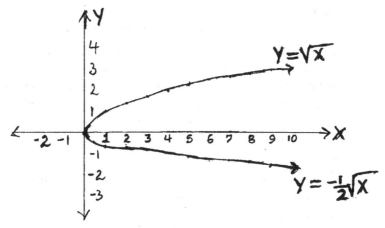

Fig. 3.16

The range is y≤0. The graph of $y = \dfrac{-1}{2}\sqrt{x}$ is a vertical shrink (by a factor of $\tfrac{1}{2}$) with a reflection in the x-axis of the graph of $y = \sqrt{x}$.

EXERCISE 3.4
Graph each of the following functions and identify the domain and range.

a) $y = \sqrt{x-2}$

b) $y = -2\sqrt{x}$

3.8 Radical Notation
A number m is said to be a square root of d if $m^2 = d$, thus -5 is a square root of 25 because $(-5)^2 = 25$, similarily, 5 is also a square root of 25 because $5^2 = 25$. A number m is said to be an nth root of d if $m^n = d$. For example, 3 is a third root (a cube root) of 27 because $3^3 = 27$. The number 27 has no other real number cube root. Any real number has one real-number cube root.

The symbol $\sqrt{a}$ denotes the non negative square root of the number a. The symbol $\sqrt[3]{a}$ denotes the real-number cube root of a, and $\sqrt[n]{a}$ denotes the nth root of a, that is, a number whose n^{th} power is a. The symbol $\sqrt[n]{}$ is called a radical, and the symbol under the radical is called the radicand. The number n (which is omitted when it is 2) is called the index. Examples of nth roots are

$$\sqrt[3]{729} = 9 \quad \text{and} \quad -\sqrt[4]{81} = -3$$

3.8.1 Odd and Even Roots

Any positive real number has two square roots, one positive and one negative. The same is true for fourth roots, or roots of any even index. The positive root is called the principal root. When a radical such as $\sqrt{9}$ or $\sqrt[4]{12}$ is used, it is understood to represent the principal (nonnegative) root. To denote a nonpositive root, we use $-\sqrt{16}$, $-\sqrt[4]{81}$, and so on.

3.8.2 Principal Root

Definition:- A radical expression $\sqrt[n]{a}$, where n is even, represents the principal (nonnegative) n^{th} root of a. The nonpositive root is denoted $-\sqrt[n]{a}$.

Note that, again, keep in mind that when the index is an even number E, then $\sqrt[E]{x}$ never represents a negative number. For example, $\sqrt{25}$ represents 5, and not -5, which is represented by $-\sqrt{25}$!

3.8.3 Simplifying Radical Expressions

Consider the expression $\sqrt{(-5)^2}$. This is equivalent to $\sqrt{25}$, which simplifies to 5. Similarly, $\sqrt{5^2} = 5$. This illustrates an important general principle for simplifying radical expressions of even index.

Theorem 3-1

For any radicand R, $\sqrt{R^2} = |R|$. Similarly, for any even index n, $\sqrt[n]{R^n} = |R|$.

Examples

1) $\sqrt{x^2} = |x|$

2) $\sqrt{x^2 - 2ax + a^2} = \sqrt{(x-a)^2} = |x-a|$

3) $\sqrt{x^2 y^6} = \sqrt{(xy^3)^2} = |xy^3| = |y^2 xy| = y^2 |xy|$

If an index is odd, no absolute-value signs are necessary, because there is only one real root and it has the same sign as the radicand.

3.9 Prime factorization to find Radicals

To simplify some expressions under radical sign we can use prime factorization.

Example-1) Simplify $\sqrt{49}$

Solution:- $\sqrt{49} = \sqrt{(7)(7)} = \sqrt{7^2} = 7$

Example 2) Simplify $\sqrt{80}$

Solution:- $\sqrt{80} = \sqrt{(2)(2)(2)(2)(5)} = \sqrt{(2^4)(5)} = 4\sqrt{5}$

Example 3) Simplify $\sqrt{625}$

Solution:- $\sqrt{625} = \sqrt{(5)(5)(5)(5)} = \sqrt{5^4} = 25$

Example 4) Simplify $\sqrt{15,000}$

Solution:- $\sqrt{15,000}$

Solution:

$$\sqrt{15,000} = \sqrt{(2)(2)(2)(3)(5)(5)(5)(5)} = \sqrt{(2^3)(3)(5^4)} = 50\sqrt{6}$$

Definition:- An integer which is a square of another integer is called a perfect square.

Example 5) a) 49 is the square of 7, this shows 49 is a perfect square
  b) 27 is not a perfect square because it is not the square of an integer.
  c) 81 is the square of 9, this shows 81 is a perfect square.
  d) some of the perfect square integers are: 1, 4, 9, 16, 25, 36, 49, 64, 81, 100, 121, 125, ...

Exercise 3.5

Simplify each of the following

a) $\sqrt{1024}$

b) $\sqrt{2916}$

c) $\sqrt{9,000}$

d) $\sqrt{1,944}$

e) $\sqrt{98}$

f) $\sqrt{126}$

g) $\sqrt{576}$

Theorem 3-2

For any radicand R and any odd index n, $\sqrt[n]{R^n} = R$.

Example 4) Simplify $\sqrt[3]{(15xy)^3}$

Solution: $\sqrt[3]{(15xy)^3} = 15xy$

Example 5) Simplify: $\sqrt[5]{(-2ab^2)^5}$

Solution: $\sqrt[5]{(-2ab^2)^5} = -2ab^2$

Example 6) Simplify $\sqrt[7]{(4xy)^7}$

Solution: $\sqrt[7]{(4xy)^7} = 4xy$

Theorem 3-3

For any nonnegative real numbers a and b and any index n,

$$\sqrt[n]{a} \cdot \sqrt[n]{b} = \sqrt[n]{a \cdot b}$$

Examples:- Multiply.

8) $\sqrt{3} \cdot \sqrt{5} = \sqrt{3 \cdot 5} = \sqrt{15}$

9) $\sqrt{x+2} \cdot \sqrt{x-2} = \sqrt{(x+2)(x-2)} = \sqrt{x^2 - 4}$

10) $\sqrt[3]{4} \cdot \sqrt[3]{5} = \sqrt[3]{4 \cdot 5} = \sqrt[3]{20}$

Examples:- Simplify.

11) $\sqrt{50} = \sqrt{25 \cdot 2} = \sqrt{25} \cdot \sqrt{2} = 5\sqrt{2}$

12) $\sqrt{5x^2} = \sqrt{x^2 \cdot 5} = \sqrt{x^2} \cdot \sqrt{5} = |x|\sqrt{5}$

13) $\sqrt[3]{32} = \sqrt[3]{8 \cdot 4} = \sqrt[3]{8} \cdot \sqrt[3]{4} = 2\sqrt[3]{4}$

14) $\sqrt[5]{32x^5 y^4} = \sqrt[5]{32x^5} \cdot \sqrt[5]{y^4} = 2x\sqrt[5]{y^4}$

15) $\sqrt{144x^5 y^3} = \sqrt{144 \cdot x^4 \cdot x \cdot y^2 \cdot y} = |12x^2 y|\sqrt{xy} = 12x^2 |y|\sqrt{xy}$

16) $\sqrt{2x^2 - 4x + 2} = \sqrt{2(x^2 - 2x + 1)} = \sqrt{2(x-1)^2} = |x-1|\sqrt{2}$

17) $\sqrt[3]{(a-b)^4} = \sqrt[3]{(a-b)^3 \cdot (a-b)} = \sqrt[3]{(a-b)^3} \cdot \sqrt[3]{(a-b)} = (a-b) \cdot \sqrt[3]{(a-b}$

Theorem 3-4

For any nonnegative number a and any positive number b, and any index n, $\sqrt[n]{\dfrac{a}{b}} = \dfrac{\sqrt[n]{a}}{\sqrt[n]{b}}$

This property can be used to divide and to simplify radical expressions.

Examples:- simplify.

18) $\sqrt{25x^3 y^{-4}} = \sqrt{\dfrac{25x^3}{y^4}} = \dfrac{\sqrt{25x^3}}{\sqrt{y^4}} = \dfrac{\sqrt{25x^2 \cdot x}}{\sqrt{y^4}} = \dfrac{5|x|\sqrt{x}}{y^2}$

19) $\sqrt[3]{\dfrac{8a^5}{343x^3 y^3}} = \dfrac{\sqrt[3]{8a^5}}{\sqrt[3]{343x^3 y^3}} = \dfrac{\sqrt[3]{8a^3 \cdot a^2}}{\sqrt[3]{343x^3 y^3}} = \dfrac{2a\sqrt[3]{a^2}}{7xy}$

20) $(16)^{3/2} = \left(\sqrt{16}\right)^3 = 4^3 = 64$

21) $\sqrt[3]{27} = (27)^{1/3} = \left(3^3\right)^{1/3} = 3$

22) $\dfrac{36\sqrt{81}}{12\sqrt{6}} = 3\sqrt{\dfrac{81}{6}} = 3\sqrt{\dfrac{27}{2}} = 3\sqrt{\dfrac{9 \cdot 3}{2}} = 3 \cdot 3\sqrt{\dfrac{3}{2}} = 9\sqrt{3/2}$

23) $\dfrac{\sqrt[3]{32}}{\sqrt[3]{4}} = \sqrt[3]{\dfrac{32}{4}} = \sqrt[3]{8} = 2$

Theorem 3-5

For any nonnegative number a and any index n and any natural number m,

$$\sqrt[n]{a^m} = \left(\sqrt[n]{a}\right)^m$$

Examples:- Simplify

24) $\sqrt[3]{8^5} = \left(\sqrt[3]{8}\right)^5 = 2^5 = 32$

25) $\left(\sqrt{3}\right)^4 = \sqrt{3^4} = \sqrt{\left(3^2\right)^2} = 3^2 = 9$

26) $3\sqrt{8} - 5\sqrt{2} = 3\sqrt{4 \cdot 2} - 5\sqrt{2}$

$= 3 \cdot 2\sqrt{2} - 5\sqrt{2}$

$= 6\sqrt{2} - 5\sqrt{2}$

$= (6-5)\sqrt{2}$ Here we use a distributive law.

$= \sqrt{2}$

27) $\left(4\sqrt{3} + \sqrt{2}\right)\left(\sqrt{3} - 5\sqrt{2}\right)$

$= 4\sqrt{3} \cdot \sqrt{3} - 4\sqrt{3} \cdot 5\sqrt{2} + \sqrt{2} \cdot \sqrt{3} - \sqrt{2} \cdot 5\sqrt{2}$

$= 4\sqrt{3 \cdot 3} - 20\sqrt{6} + \sqrt{6} - 5\sqrt{2 \cdot 2}$

$= 4\sqrt{9} - 19\sqrt{6} - 5\sqrt{4}$

$= 4 \cdot 3 - 19\sqrt{6} - 5 \cdot 2$

$= 12 - 19\sqrt{6} - 10$

$= 2 - 19\sqrt{6}$

Exercise 3.6

Simplify

1) $\sqrt{(x+3)^2}$

2) $\sqrt{y^2(y-2)^2}$

3) $\sqrt[4]{(x+1)^4}$

4) $\sqrt{x^2+10x+25}$

5) $\sqrt[3]{(-2xy)^3}$

6) $\sqrt[5]{(243x^5y^5)}$

7) $\sqrt{121(x^2-10x+25)}$

8) $\sqrt{x^2y^2z^3w^3}$

9) $(\sqrt{96})(\sqrt{72})$

10) $\sqrt[3]{16}$

11) $\sqrt[3]{(a+b)^4}$

12) $\sqrt[3]{27^{10}}$

Simplify

13) $\sqrt{\dfrac{12}{y^3}}$

14) $\dfrac{\sqrt{75}}{\sqrt{3}}$

15) $(\sqrt{5})^4$

16) $\dfrac{\sqrt{2x^5}}{\sqrt{50x^2}}$

17) $\dfrac{\sqrt[3]{24x^3y}}{\sqrt[3]{3y^4}}$

18) $7\sqrt{5}+3\sqrt{5}-16\sqrt{40}$

19) $6\sqrt[3]{16y^4x}+7\sqrt[3]{2xy}$

20) $(\sqrt{2}-5\sqrt{3})(2\sqrt{2}+\sqrt{3})$

21) $\sqrt[3]{3/25}$

22) $\dfrac{\sqrt{3x}}{\sqrt{7y}}$

23) $\dfrac{\sqrt[3]{48y^3}}{\sqrt[3]{9x^3}}$

3.10 Rationalizing Denominators or Numerators

To convert the irrational denominator or numerator to rational we will multiply by a convenient number which is called rationalizing factor.

3.10.1 Rationalizing Denominators

The rationalizing factor we use to rationalize the denominator of

$$\frac{1}{\sqrt{b}+a} \quad \text{is} \quad \begin{cases} \dfrac{\sqrt{b}-a}{\sqrt{b}-a} & \text{if } b > a^2 \\[2ex] \dfrac{a-\sqrt{b}}{a-\sqrt{b}} & \text{if } a^2 > b \end{cases}$$

Example 1)   Let us practice the following simplifications which will help us in the process of rationalization.

(a) $\left(\sqrt{7}\right)^2 = 7, \quad \left(\sqrt{11}\right)^2 = 11, \quad \left(\sqrt{13}\right)^2 = 13$

(b) $\sqrt{2} \times \sqrt{11} = \sqrt{22}, \quad \sqrt{3} \times \sqrt{5} = \sqrt{15}, \quad \sqrt{5} \times \sqrt{7} = \sqrt{35}$

(c) $\left(\sqrt{7} - 3\right)\left(\sqrt{7} + 3\right) = \sqrt{7}\left(\sqrt{7} + 3\right) - 3\left(\sqrt{7} + 3\right)$

$$= \sqrt{7} \times \sqrt{7} + 3\sqrt{7} - 3\sqrt{7} - 9$$

$$= 7 - 9$$

$$= -2$$

(d) $\left(\sqrt{5} - 1\right)\left(\sqrt{5} + 1\right) = \sqrt{5}\left(\sqrt{5} + 1\right) - 1\left(\sqrt{5} + 1\right)$

$$5 + \sqrt{5} - \sqrt{5} + 1$$

$$= 5 + 1$$

$$= 6$$

(e) $\left(\sqrt{7} - 3\right)\left(\sqrt{7} + 3\right) = \left(\sqrt{7}\right)^2 - 3^2 = 7 - 9 = -2$

(f) $\left(\sqrt{2} - 1\right)\left(\sqrt{2} + 1\right) = \left(\sqrt{2}\right)^2 - 1^2 = 2 - 1 = 1$

(g) $\left(\sqrt{15} - 3\right)\left(\sqrt{15} + 3\right) = \left(\sqrt{15}\right)^2 - 3^2 = 15 - 9 = 6$

(h) $\left(\sqrt{11} - \sqrt{7}\right)\left(\sqrt{11} + \sqrt{7}\right) = \left(\sqrt{11}\right)^2 - \left(\sqrt{7}\right)^2 = 11 - 7 = 4$

(i) $\left(\sqrt{13} - \sqrt{6}\right)\left(\sqrt{13} + \sqrt{6}\right) = \left(\sqrt{13}\right)^2 - \left(\sqrt{6}\right)^2 = 13 - 6 = 7$

Pairs of expressions like $c - \sqrt{b}$, $c + \sqrt{b}$ and $\sqrt{a} - \sqrt{b}$, $\sqrt{a} + \sqrt{b}$ are called conjugates. The product of a pair of conjugates has no radicals in it. Thus when we wish to rationalize a denominator that has two terms and one or more of them involves a square-root radical, we multiply by 1 using the conjugate of the denominator to write a symbol for 1.

Example 2) Rationalize the denominator of $\dfrac{\sqrt{3}}{\sqrt{5} - \sqrt{2}}$.

Solution: $\dfrac{\sqrt{3}}{\sqrt{5} - \sqrt{2}} = \dfrac{\sqrt{3}}{\sqrt{5} - \sqrt{2}} \times \dfrac{\sqrt{5} + \sqrt{2}}{\sqrt{5} + \sqrt{2}}$

$$= \frac{\sqrt{3}\left(\sqrt{5} + \sqrt{2}\right)}{\left(\sqrt{5}\right)^2 - \left(\sqrt{2}\right)^2}$$

$$= \frac{\sqrt{3} \cdot \sqrt{5} - \sqrt{3} \cdot \sqrt{2}}{5 - 2}$$

$$= \frac{\sqrt{15} - \sqrt{6}}{3}$$

Example 3) Rationalize the denominator of $\dfrac{\sqrt{2a}}{\sqrt{5b}}$

Solution: $\dfrac{\sqrt{2a}}{\sqrt{5b}} = \dfrac{\sqrt{2a}}{\sqrt{5b}} \times \dfrac{\sqrt{5b}}{\sqrt{5b}}$

$$= \frac{\sqrt{10ab}}{\sqrt{\left(5b\right)^2}}$$

$$= \frac{\sqrt{10ab}}{|5b|}$$

$$= \frac{\sqrt{10ab}}{5b}$$

The absolute-value sign in the denominator is not necessary since $\sqrt{5b}$ would not exist at the outset unless b > 0.

Example 4) Rationalize the denominator of $\dfrac{6}{\sqrt{5x}}$

Solution: $\dfrac{6}{\sqrt{5x}} = \dfrac{6}{\sqrt{5x}} \times \dfrac{\sqrt{5x}}{\sqrt{5x}}$

$$= \dfrac{6\sqrt{5x}}{\sqrt{25x^2}}$$

$$= \dfrac{6\sqrt{5x}}{5|x|}$$

$$= \dfrac{6\sqrt{5x}}{5x}$$

Example 5) Rationalize the numerator of $\dfrac{\sqrt{6}}{\sqrt{5}-\sqrt{3}}$.

Solution: $\dfrac{\sqrt{6}}{\sqrt{5}-\sqrt{3}} = \dfrac{\sqrt{6}}{\sqrt{5}-\sqrt{3}} \times \dfrac{\sqrt{6}}{\sqrt{6}}$

$$= \dfrac{\sqrt{6 \cdot 6}}{\sqrt{6}\left(\sqrt{5}-\sqrt{3}\right)}$$

$$= \dfrac{6}{\sqrt{30}-\sqrt{18}}$$

$$= \dfrac{6}{\sqrt{30}-3\sqrt{2}} = \dfrac{6}{\sqrt{2}\left(\sqrt{15}-3\right)}$$

Example 6) Rationalize the denominator of $\dfrac{6}{3-\sqrt{5}}$

Solution: $\dfrac{6}{3-\sqrt{5}} = \dfrac{6}{3-\sqrt{5}} \times \dfrac{3+\sqrt{5}}{3+\sqrt{5}}$

$$= \dfrac{6\left(3+\sqrt{5}\right)}{3^2 - \left(\sqrt{5}\right)^2}$$

$$= \dfrac{18+6\sqrt{5}}{9-5}$$

$$= \dfrac{18+6\sqrt{5}}{4}$$

$$\dfrac{2\left(9+3\sqrt{5}\right)}{4}$$

$$= \dfrac{9+3\sqrt{5}}{2}$$

Example 7) Rationalize the numerator of $\dfrac{\left(\sqrt{6}+7\right)}{\sqrt{2}-3}$

Solution: $\dfrac{\sqrt{6}+7}{\sqrt{2}-3} = \dfrac{\sqrt{6}+7}{\sqrt{2}-3} \times \dfrac{\sqrt{6}-7}{\sqrt{6}-7}$

$$= \dfrac{\left(\sqrt{6}\right)^2 - 7^2}{\sqrt{2}\left(\sqrt{6}-7\right) - 3\left(\sqrt{6}-7\right)}$$

$$= \dfrac{6-49}{\sqrt{12} - 7\sqrt{2} - 3\sqrt{6} + 21}$$

$$= \dfrac{-43}{\sqrt{4} \times \sqrt{3} - 7\sqrt{2} - 3\sqrt{6} + 21}$$

$$= \dfrac{-43}{2\sqrt{3} - 7\sqrt{2} - 3\sqrt{6} + 21}$$

Example 8) Rationalize the denominator of $\dfrac{\sqrt{3}}{\sqrt{2}+\sqrt{3}-\sqrt{5}}$

Solution: $\dfrac{\sqrt{3}}{\sqrt{2}+\sqrt{3}-\sqrt{5}} = \dfrac{\sqrt{3}}{\sqrt{2}+\sqrt{3}-\sqrt{5}} \times \dfrac{\left(\sqrt{2}+\sqrt{3}\right)+\sqrt{5}}{\left(\sqrt{2}+\sqrt{3}\right)+\sqrt{5}}$

$$= \dfrac{\sqrt{3}\left[\left(\sqrt{2}+\sqrt{3}\right)+\sqrt{5}\right]}{\left(\sqrt{2}+\sqrt{3}\right)^2 - \left(\sqrt{5}\right)^2}$$

$$= \dfrac{\sqrt{6}+3+\sqrt{15}}{5+2\sqrt{6}-5}$$

$$= \dfrac{\sqrt{6}+\sqrt{15}+3}{2\sqrt{6}}$$

$$= \left[\dfrac{\sqrt{6}+\sqrt{15}+3}{2\sqrt{6}}\right] \times \dfrac{\sqrt{6}}{\sqrt{6}}$$

$$= \dfrac{\sqrt{36}+\sqrt{6\times15}+3\sqrt{6}}{2\sqrt{36}}$$

$$= \dfrac{6+3\sqrt{10}+3\sqrt{6}}{12}$$

$$= \dfrac{3\left(2+\sqrt{10}+\sqrt{6}\right)}{12}$$

$$= \dfrac{2+\sqrt{10}+\sqrt{6}}{4}$$

Note:- $(a-b)(a+b)=a^2-b^2$

Example 9) Rationalize the denominator of $\dfrac{8}{7-\sqrt{3}}$

Solution: $\dfrac{8}{7-\sqrt{3}}$, here, the rationalizing factor is $\dfrac{7+\sqrt{3}}{7+\sqrt{3}}$

$$\frac{8}{7-\sqrt{3}} = \frac{8}{7-\sqrt{3}} \times \frac{7+\sqrt{3}}{7+\sqrt{3}} = \frac{8\left(7+\sqrt{3}\right)}{7^2-\left(\sqrt{3}\right)^2}$$

$$= \frac{56+8\sqrt{3}}{7-3}$$

$$= \frac{56+8\sqrt{3}}{4}$$

$$= \frac{4\left(14+2\sqrt{3}\right)}{4}$$

$$= \left(14+2\sqrt{3}\right)$$

Example 10) Rationalize the numerator of $\dfrac{\sqrt{6}-\sqrt{5}}{2-\sqrt{8}}$

Solution: $\dfrac{\sqrt{6}-\sqrt{5}}{2-\sqrt{8}}$, here, the rationalizing factor is $\sqrt{6}+\sqrt{5}$

$$\frac{\sqrt{6}-\sqrt{5}}{2-\sqrt{8}} = \frac{\sqrt{6}-\sqrt{5}}{2-\sqrt{8}} \times \frac{\sqrt{6}+\sqrt{5}}{\sqrt{6}+\sqrt{5}}$$

$$= \frac{\left(\sqrt{6}\right)^2 - \left(\sqrt{5}\right)^2}{2\left(\sqrt{6}+\sqrt{5}\right) - \sqrt{8}\left(\sqrt{6}+\sqrt{5}\right)}$$

$$= \frac{6-5}{2\sqrt{6}+2\sqrt{5}-\sqrt{48}-\sqrt{40}}$$

$$= \frac{1}{2\sqrt{6}+2\sqrt{5}-4\sqrt{3}-2\sqrt{10}}$$

$$= \frac{1}{2\left(\sqrt{6}+\sqrt{5}-2\sqrt{3}-\sqrt{10}\right)}$$

Examples. Rationalize the denominator. Assume that all letters represent positive numbers.

11) $\dfrac{1}{\sqrt{2}+\sqrt{5}} = \dfrac{1}{\sqrt{2}+\sqrt{5}} \cdot \dfrac{\sqrt{2}-\sqrt{5}}{\sqrt{2}-\sqrt{5}}$, The number $\sqrt{2}-\sqrt{5}$ is the conjugate of $\sqrt{2}+\sqrt{5}$. It is found by changing the middle sign. We use the conjugate to form the symbol for 1.

$$= \dfrac{\sqrt{2}-\sqrt{5}}{\left(\sqrt{2}+\sqrt{5}\right)\left(\sqrt{2}-\sqrt{5}\right)}$$

$$= \dfrac{\sqrt{2}-\sqrt{5}}{\left(\sqrt{2}\right)^2 - \left(\sqrt{5}\right)^2}$$

$$= \dfrac{\sqrt{2}-\sqrt{5}}{2-5}$$

$$= \dfrac{\sqrt{2}-\sqrt{5}}{-3}$$

$$= \dfrac{-\left(-\sqrt{2}+\sqrt{5}\right)}{-3}$$

$$= \dfrac{\left(\sqrt{5}-\sqrt{2}\right)}{3}$$

12) $\dfrac{\sqrt{x}+\sqrt{y}}{\sqrt{x}-\sqrt{y}} = \dfrac{\sqrt{x}+\sqrt{y}}{\sqrt{x}-\sqrt{y}} \times \dfrac{\sqrt{x}+\sqrt{y}}{\sqrt{x}+\sqrt{y}}$, The conjugate of $\sqrt{x}-\sqrt{y}$ is

$$\sqrt{x}+\sqrt{y}$$

$$= \dfrac{\left(\sqrt{x}+\sqrt{y}\right)^2}{\left(\sqrt{x}\right)^2 - \left(\sqrt{y}\right)^2}$$

$$= \dfrac{x+2\sqrt{xy}+y}{x-y}$$

Examples:- Rationalize the numerator. Assume that all letters represent positive numbers and that all radicals are positive.

13) $\dfrac{2-\sqrt{3}}{5}$

$\dfrac{2-\sqrt{3}}{5} = \dfrac{2-\sqrt{3}}{5} \cdot \dfrac{2+\sqrt{3}}{2+\sqrt{3}}$ The conjugate of $2-\sqrt{3}$ is $2+\sqrt{3}$

$= \dfrac{\left(2-\sqrt{3}\right)\left(2+\sqrt{3}\right)}{5\left(2+\sqrt{3}\right)}$

$= \dfrac{2^2 - \left(\sqrt{3}\right)^2}{10+5\sqrt{3}}$

$= \dfrac{4-3}{10+5\sqrt{3}}$

$= \dfrac{1}{5\left(2+\sqrt{3}\right)}$

14) $\dfrac{\sqrt{x+y}-\sqrt{x}}{y} = \dfrac{\sqrt{x+y}-\sqrt{x}}{y} \cdot \dfrac{\sqrt{x+y}+\sqrt{x}}{\sqrt{x+y}+\sqrt{x}}$ The conjugate of

$\sqrt{x+y} - \sqrt{x}$ is $\sqrt{x+y}+\sqrt{x}$

$= \dfrac{\left(\sqrt{x+y}\right)^2 - \left(\sqrt{x}\right)^2}{y\left(\sqrt{x+y}+\sqrt{x}\right)}$

$= \dfrac{x+y-x}{y\left(\sqrt{x+y}+\sqrt{x}\right)}$

$= \dfrac{y}{y\left(\sqrt{x+y}+\sqrt{x}\right)}$

$= \dfrac{1}{\sqrt{x+y}+\sqrt{x}}$

15) $\dfrac{\sqrt{a+b}-\sqrt{a-b}}{ab} = \dfrac{\sqrt{a+b}-\sqrt{a-b}}{ab} \cdot \dfrac{\sqrt{a+b}+\sqrt{a-b}}{\sqrt{a+b}+\sqrt{a-b}}$

The conjugate of $\sqrt{a+b}-\sqrt{a-b}$ is $\sqrt{a+b}+\sqrt{a-b}$

$$= \dfrac{\left(\sqrt{a+b}\right)^2 - \left(\sqrt{a-b}\right)^2}{ab\left(\sqrt{a+b}+\sqrt{a-b}\right)}$$

$$= \dfrac{a+b-(a-b)}{ab\left(\sqrt{a+b}+\sqrt{a-b}\right)}$$

$$= \dfrac{2b}{ab\left(\sqrt{a+b}+\sqrt{a-b}\right)}$$

$$= \dfrac{2}{a\left(\sqrt{a+b}+\sqrt{a-b}\right)}$$

Examples:-

Rationalize the denominators of each of the following assume that all letters represent positive numbers and all radicands are positive.

16) $\dfrac{3}{\sqrt[3]{5}}$

$$\dfrac{3}{\sqrt[3]{5}} = \dfrac{3}{\sqrt[3]{5}} \times \dfrac{\sqrt[3]{5}\times\sqrt[3]{5}}{\sqrt[3]{5}\times\sqrt[3]{5}} = \dfrac{3\times\sqrt[3]{5\times5}}{\sqrt[3]{5x5x5}} = \dfrac{3\sqrt[3]{25}}{5}$$

17) $\dfrac{2}{\sqrt[5]{6}}$

$$\dfrac{2}{\sqrt[5]{6}} = \dfrac{2}{\sqrt[5]{6}} \times \dfrac{\sqrt[5]{6}\times\sqrt[5]{6}\times\sqrt[5]{6}\times\sqrt[5]{6}}{\sqrt[5]{6}\times\sqrt[5]{6}\times\sqrt[5]{6}\times\sqrt[5]{6}} = \dfrac{2\times\sqrt[5]{6\times6\times6\times6}}{\sqrt[5]{6\times6\times6\times6\times6}} = \dfrac{2\sqrt[5]{1296}}{6}$$

$$= \dfrac{\sqrt[5]{1296}}{3}$$

18) $\dfrac{1}{\sqrt{3}+\sqrt{2}}$

$\dfrac{1}{\sqrt{3}+\sqrt{2}} = \dfrac{1}{\sqrt{3}+\sqrt{2}} \times \dfrac{\sqrt{3}-\sqrt{2}}{\sqrt{3}-\sqrt{2}} = \dfrac{\sqrt{3}-\sqrt{2}}{\left(\sqrt{3}\right)^2 - \left(\sqrt{2}\right)^2} = \dfrac{\sqrt{3}-\sqrt{2}}{3-2} = \sqrt{3}-\sqrt{2}$

, $\sqrt{3}-\sqrt{2}$ is the conjugate of $\sqrt{3}+\sqrt{2}$

19) $\dfrac{1}{x+\sqrt{x}}$

$\dfrac{1}{x+\sqrt{x}} = \dfrac{1}{x+\sqrt{x}} \cdot \dfrac{\left(x-\sqrt{x}\right)}{\left(x-\sqrt{x}\right)} = \dfrac{x-\sqrt{x}}{x^2-\left(\sqrt{x}\right)^2} = \dfrac{x-\sqrt{x}}{x^2-x}$, $x-\sqrt{x}$ is the

conjugate of $x+\sqrt{x}$

20) $\dfrac{2+\sqrt{5}}{4-\sqrt{5}}$

$\dfrac{2+\sqrt{5}}{4-\sqrt{5}} = \dfrac{2+\sqrt{5}}{4-\sqrt{5}} \cdot \dfrac{4+\sqrt{5}}{4+\sqrt{5}} = \dfrac{8+2\sqrt{5}+4\sqrt{5}+\sqrt{25}}{16+4\sqrt{5}-4\sqrt{5}-\sqrt{25}} = \dfrac{13+6\sqrt{5}}{11}$

21) $\dfrac{a}{\sqrt{b}-\sqrt{c}}$

$\dfrac{a}{\sqrt{b}-\sqrt{c}} = \dfrac{a}{\sqrt{b}-\sqrt{c}} \cdot \dfrac{\sqrt{b}+\sqrt{c}}{\sqrt{b}+\sqrt{c}} = \dfrac{a\sqrt{b}+a\sqrt{c}}{b-c}$, $\sqrt{b}+\sqrt{c}$ is the

conjugate of $\sqrt{b}-\sqrt{c}$

22) $\sqrt[3]{\dfrac{7x}{2}}$

$\sqrt[3]{\dfrac{7x}{2}} = \sqrt[3]{\dfrac{7x}{2}} \cdot \dfrac{\sqrt[3]{4}}{\sqrt[3]{4}} = \dfrac{\sqrt[3]{28x}}{\sqrt[3]{8}} = \dfrac{\sqrt[3]{28x}}{2}$, $\sqrt[3]{49x^2}$ is the conjugate

of $\sqrt[3]{7x}$

SUMMARY

3.10.2 Properties of Radicals

We can use the following properties while simplifying radicals.

1) $\left(\sqrt[n]{x}\right)^m = \sqrt[n]{x^m}$

2) $\left(\sqrt[n]{x}\right)^n = \left(x^n\right)^{1/n} = x$

3) $\sqrt[m]{\sqrt[n]{x}} = \sqrt[mn]{x}$

4) $\sqrt[n]{x} \cdot \sqrt[m]{x} = \sqrt[mn]{x^{m+n}}$

5) $\sqrt[pn]{x^{pm}} = \sqrt[n]{x^m}$

6) $\dfrac{\sqrt[n]{x}}{\sqrt[n]{y}} = \sqrt[n]{\dfrac{x}{y}}$

7) $\sqrt[n]{x} \cdot \sqrt[n]{y} = \sqrt[n]{xy}$

Example 1) a) $\left(\sqrt[5]{32}\right)^2 = \sqrt[5]{(32)^2} = 4$

  b) $\left(\sqrt[5]{32}\right)^5 = \left(32^5\right)^{1/5} = 32$

  c) $\sqrt[3]{\sqrt[4]{4096}} = \sqrt[12]{4096} = 2$

  d) $\sqrt[3]{4096} \times \sqrt[4]{4096} = \sqrt[12]{(4096)^7} = 128 \text{ or } 2^7$

Note):- $2^{12} = 4096$

  e) $\sqrt[3\times4]{16^{3\times5}} = \sqrt[4]{16^5} = \sqrt[4]{\left(2^4\right)^5} = \left(2^{4\times5}\right)^{1/4} = 2^5 = 32$

  f) $\dfrac{\sqrt[5]{32}}{\sqrt[5]{243}} = \sqrt[5]{\dfrac{32}{243}} = \dfrac{2}{3}$

  g) $\sqrt[5]{32} \times \sqrt[5]{243} = 2 \times 3 = 6$

  h) $\sqrt[3]{1000} \times \sqrt[3]{-125} = 10 \times (-5) = -50$

Exercise 3.7

1)   Simplify each of the following

a) $\sqrt{5^2 + 3^2 + \left(\dfrac{3}{2}\right)^2 - \left(\dfrac{1}{2}\right)^2}$

b) $\sqrt{1^2 + 2^2 + 3^2 + 5^2}$

c) $\sqrt[3]{x^6 y^6 z^5 w^3}$

d) $\sqrt{0.00001 x^4 y^6 z^2}$

e) $\sqrt[3]{x^{27} y^{36} x^{-12}}$

f) $\dfrac{1}{2}\sqrt{28} - 3\sqrt{56}$

g) $\sqrt{x^2 - 6x + 9}$

h) $\sqrt{(2c - 3)^2}$

i) $\sqrt{xyz^2}$

j) $-3\sqrt[4]{3^8 \cdot 4^8 \cdot 2^{16}}$

k) $\sqrt{p^2 q^2 z^2}$

l) $\sqrt{\dfrac{72}{144}x^6y^7z^{-4}}$

m) $\sqrt{(p+6)^2}$

n) $\sqrt{\dfrac{(x-3)^8}{(x-3)^6}}, \quad x \neq 3$

o) $\sqrt[3]{\dfrac{81}{8}} - \sqrt{1024x^2y^2} + \sqrt{16xy}$

p) $\dfrac{2+\sqrt{5}}{2-\sqrt{5}} \times \dfrac{2+\sqrt{5}}{2+\sqrt{5}}$

q) $\sqrt[3]{81x^5y} - x\sqrt[3]{24x^2y}$

r) $4\sqrt{12r^2s^2k^2} + \sqrt{27r^2s^2} - \sqrt{rst}$

s) $\sqrt{(-5)(-250)m^2n^2r^2}$

2)   Rationalize the numerator. Assume that all letters represent positive numbers and that all radicands are positive

a) $\dfrac{\sqrt{x+1}+1}{\sqrt{x+1}+1}$

b) $\dfrac{\sqrt{3x}+5}{3}$

c) $\dfrac{\sqrt[3]{4x^2y}}{\sqrt{2x}}$

d) $\dfrac{\sqrt{y}-\sqrt{3}}{\sqrt{y}+\sqrt{3}}$

e) $\dfrac{2-\sqrt{6+y}}{y+3}$

3)   Rationalize the denominator. Assume that all letters represent positive numbers and that all radicands are positive.

a) $\sqrt[3]{\dfrac{5y}{2}}$

b) $\dfrac{1}{\sqrt[3]{ab}}$

c) $\dfrac{7}{9+\sqrt{10}}$

d) $\dfrac{-3\sqrt{2}}{\sqrt{3}-\sqrt{5}}$

e) $\dfrac{8-3\sqrt{5}}{\sqrt{2}-2\sqrt{7}}$

f) $\dfrac{\sqrt{a}-\sqrt{b}}{\sqrt{a}+\sqrt{b}}$

g) $\dfrac{3}{\sqrt{x-1}+\sqrt{x+2}}$

h) $\dfrac{x}{2+\sqrt{x}}$

i) $\dfrac{2}{\sqrt[3]{5}}$

j) $\dfrac{x-1}{\sqrt{x+3}-2}$

k) $\dfrac{5}{4\sqrt{3}}-\dfrac{3}{2\sqrt{2}}$

l) $\dfrac{1}{\sqrt{3}-2}-\dfrac{2}{\sqrt{3}+2}$

m) $\dfrac{x-1}{\sqrt{x+3}+2}$

n) $\dfrac{\sqrt{x+h}+\sqrt{x}}{\sqrt{x+h}-x}$

3.11 SOLVING RADICAL EQUATIONS

An equation that contains a radical expression with a variable in the radicand is a radical equation.

As you know, radical equations involving square roots are typically solved by isolating a radical and squaring both sides of the resulting equation. This process of squaring may lead to an answer that is actually not one of the roots of the original equation. This "extra" answer is called an extraneous root.

Example 1) Solve the following equation algebraically and check.

$x-3=\sqrt{30-2x}$

Solution: $(x-3)=\sqrt{30-2x}$

$$(x-3)^2=\left(\sqrt{30-2x}\right)^2$$

$= x^2-6x+9=30-2x$

$= x^2-6x+9-30+2x=0$

$= x^2-4x-21=0$

$= (x-7)(x+3)=0$

$= x=7; x=-3$

Check:

$$7-3=\sqrt{30-14}$$

$$4=\sqrt{16}$$

$$4=4$$

Check

$$-3-3=\sqrt{30-(-6)}$$

$$-6\neq\sqrt{36}$$

-3 is extraneous root

Note:- For any positive number n, if an equation a=b is true, then $a^n=b^n$ is true, this is called the principle of power.

Example 2) Solve $7+\sqrt{a-5}=1$

Solution: Isolate the radical expression

$$7+\sqrt{a-5}=1$$

$$\sqrt{a-5}=-6$$

There is no solution, since $\sqrt{a-5}$ cannot have a negative value.

Example 3) Solve $\sqrt{3x-5}+\sqrt{x-1}=2$

Isolate one of the radical expression.

$$\sqrt{3x-5}=2-\sqrt{x-1}$$

Raise both sides to the index of the radical; in this case, square both sides.

$$\left(\sqrt{3x-5}\right)^2=\left(2-\sqrt{x-1}\right)^2$$

$$3x-5=\left(2-\sqrt{x-1}\right)\left(2-\sqrt{x-1}\right)$$

$$=4-2\sqrt{x-1}-2\sqrt{x-1}+x-1$$

$$=3-4\sqrt{x-1}+x$$

This is still a radical equation. Isolate the radical equation.

$$3x - 5 = 3 - 4\sqrt{x-1} + x$$
$$4\sqrt{x-1} = 8 - 2x$$

Raise both sides to the index of the radical; in this case, square both sides.

$$\left(4\sqrt{x-1}\right)^2 = (8-2x)^2$$
$$16(x-1)=64-32x+4x^2$$
$$16x-16=64-32x+4x^2$$
$$4x^2-48x+80=0$$

This can be solved either by factoring or by applying the quadratic formula.

Applying the quadratic formula,

$$x = \frac{-b \pm \sqrt{b^2 - 4ac}}{2a}$$

$4x^2-48x+80=0$

a=4, b=-48, c=80

$$x = \frac{48 \pm \sqrt{(-48)^2 - 4(4)(80)}}{2(4)}$$

$$x = \frac{48 \pm \sqrt{2304 - 1280}}{8}$$

$$= \frac{48 \pm \sqrt{1024}}{8}$$

$$= \frac{48 \pm 32}{8}$$

x=10 or x=2

Check the solution

If x=10

$$\sqrt{3x-5} + \sqrt{x-1} = 2$$

$$\sqrt{3(10)-5} + \sqrt{10-1} = 2$$

If x=2

$$\sqrt{3x-5} + \sqrt{x-1} = 2$$

$$\sqrt{3(2)-5} + \sqrt{2-1} = 2$$

$\sqrt{25}+\sqrt{9}=2$ $\qquad$ $\sqrt{6-5}+\sqrt{1}=2$

5+3=2 $\qquad\qquad\qquad\qquad\qquad$ 1 + 1 = 2

8≠2 $\qquad\qquad\qquad$ 2 = 2

10 is an extraneous solution

$\qquad$ The solution set is {2}

Example 4) Solve $\sqrt[3]{4x^2+1}=5$

$\qquad$ Solution: $\left(\sqrt[3]{4x^2+1}\right)^3=5^3$, $\qquad$ cubing both sides

$\qquad\qquad$ 4x²+1=125

$\qquad\qquad$ 4x²=124

$\qquad\qquad$ x²=31

$\qquad\qquad$ $x=\pm\sqrt{31}$

Both $\sqrt{31}$ and $-\sqrt{31}$ check. The solution set is $\left\{\pm\sqrt{31}\right\}$

Example 5) Solve $x-\sqrt{x+7}=5$

Solution: We must first get the radical alone on one side. That is, we isolate the radical,

$\qquad$ $x-5=\sqrt{x+7}$ $\qquad$ Adding -5 and $\sqrt{x+7}$ on both

$\qquad$ sides to isolate the radical.

$\qquad$ $(x-5)^2=\left(\sqrt{x+7}\right)^2$ $\qquad$ squaring both sides

$\qquad$ x²-10x+25=x+7

$\qquad$ x²-11x+18=0

$\qquad$ (x-9)(x-2)=0

$\qquad$ x=9 or x=2

Check the solution

$\qquad$ $x-\sqrt{x+7}=5$ $\qquad\qquad\qquad$ $x-\sqrt{x+7}=5$

$\qquad$ If x=9 $\qquad\qquad\qquad\qquad\qquad$ If x=2

$\qquad$ $9-\sqrt{9+7}=5$ $\qquad\qquad\qquad$ $2-\sqrt{2+7}=5$

$\qquad$ $9-\sqrt{16}=5$ $\qquad\qquad\qquad$ $2-\sqrt{9}=5$

$$9 - 4 = 5 \qquad\qquad 2 - 3 = 5$$

5=5✓   -1≠5

2 is an extraneous solution

The only solution is 9. Thus the solution set is {9}.

Example 6) Solve $\sqrt{x-3} + \sqrt{x+5} = 4$

Solution:- $\sqrt{x-3} = 4 - \sqrt{x+5}$ Adding $-\sqrt{x+5}$, which isolates one of the radical terms.

$$\left(\sqrt{x-3}\right)^2 = \left(4 - \sqrt{x+5}\right)^2$$ Principle of power, squaring both sides.

> Here we are squaring the binomial. We square 4, then we subtract twice the product of 4 and $\sqrt{x+5}$, and then we odd the square of $\sqrt{x+5}$. Recall that $(A-B)^2 = A^2 - 2AB + B^2$

$$x - 3 = 16 - 8\sqrt{x+5} + x + 5$$

$-3 = 21 - 8\sqrt{x+5}$ Adding -x and collecting like terms.

$-24 = -8\sqrt{x+5}$ Isolating the remaining radical terms.

$3 = \sqrt{x+5}$

$3^2 = \left(\sqrt{x+5}\right)^2$ Squaring both sides.

9=x+5

x=4

Check the solution

$$\sqrt{x-3} + \sqrt{x+5} = 4$$

x=4

$$\sqrt{4-3} + \sqrt{4+5} = 4$$

$$\sqrt{1} + \sqrt{9} = 4$$

1+3=4

4=4

The solution set is {4}

Example 7) $x = \sqrt{20 - x}$
   Solution: $x = \sqrt{20 - x}$

$x^2 = \left(\sqrt{20-x}\right)^2$ squaring both sides.
x²=20-x
x²+x-20=0
x²-4x+5x-20=0
x(x-4)+5(x-4)=0
(x+5)(x-4)=0
x+5=0 or x-4=0   zero product property
x=-5 or x=4
Check the solution

$x = \sqrt{20 - x}$
If x=-5   If x=4
$-5 = \sqrt{20 - (-5)} = \sqrt{25}$ $4 = \sqrt{20 - 4}$
-5≠5 $4 = \sqrt{16}$
                                       4=4
-5 is an extraneous solution
The only solution is 4. Thus the solution set is {4}

Example 8) Solve: $\sqrt{2x - 5} = 1 + \sqrt{x - 3}$

Solution:- $\left(\sqrt{2x-5}\right)^2 = \left(1 + \sqrt{x-3}\right)^2$ one radical is already isolated,
we square both sides

$$2x - 5 = 1 + 2\sqrt{x-3} + (x - 3)$$
$$x - 3 = 2\sqrt{x - 3}$$ Isolating the remaining radicals.
$$(x-3)^2 = \left(2\sqrt{x-3}\right)^2$$ squaring both sides.

x²-6x+9=4(x-3)
x²-6x+9=4x-12
x²-10x+21=0
x²-7x-3x+21=0
x(x-7)-3(x-7)=0
(x-3)(x-7)=0
x-3=0 or x-7=0
x=3 or x=7   Zero product property.

After we checked the numbers 3 and 7 satisfy the equation, thus the solution set is $\{3, 7\}$.

Example 9) Solve $s = \sqrt{1 + \dfrac{x^2}{y^2}}$

for x, assume the variable represent positive numbers.

Solutions: $s^2 = 1 + \dfrac{x^2}{y^2}$ squaring both sides

$$s^2 = \frac{y^2 + x^2}{y^2}$$

$y^2s^2 = y^2 + x^2$ multiplying both sides by y^2

$y^2s^2 - y^2 = x^2$

$\sqrt{y^2s^2 - y^2} = x$

$\sqrt{y^2(s^2 - 1)} = x$

$y\sqrt{s^2 - 1} = x$

Example 10) Solve $\sqrt{5x - 12} - \sqrt{2x + 9} = 0$

Solution:- $\sqrt{5x - 12} = \sqrt{2x + 9}$,   Adding $\sqrt{2x + 9}$, which isolates one of the radical terms.

$\left(\sqrt{5x - 12}\right)^2 = \left(\sqrt{2x + 9}\right)^2$ Principle of power, squaring both sides.

5x-12=2x+9

5x-2x=9+12

3x=21

x=7

Check the solution

$\sqrt{5(7) - 12} = \sqrt{2(7) + 9}$

$\sqrt{35 - 12} = \sqrt{14 + 9}$

$\sqrt{23} = \sqrt{23}$

The solution set is $\{7\}$

Example 11) Solve and check the equation $\sqrt{x+6}+x=14$

Solution: $\sqrt{x+6}+x=14$

$\sqrt{x+6}=14-x$ Isolate the radicals

$\left(\sqrt{x+6}\right)^2=(14-x)^2$

square both sides

x+6=196-28x+x²

x²-29x+190=0   Simplify

(x-19)(x-10)=0

x-19=0 or x-10=0   Zero product property.

x=19 or x=10

Check the solution

$\sqrt{x+6}+x=14$

If x=19   If x=10

$\sqrt{19+6}+19=14$ $\sqrt{10+6}+10=14$

5+19=14   4+10=14

24≠14

Thus the solution set is {10}

Example 12) Solve and check the equation $\sqrt[3]{3y+2}-2=0$

Solution: $\sqrt[3]{3y+2}-2=0$

$\sqrt[3]{3y+2}=2$ Isolate the radicals.

$\left(\sqrt[3]{3y+2}\right)^3=2^3$ Cube both sides

3y+2=8

3y=6

y=2

Check: $\sqrt[3]{3y+2}-2=0$

$\sqrt[3]{3(2)+2}-2=0$

$\sqrt[3]{8}-2=0$

2-2=0

0=0

Thus, the solution set is {2}

Example 13) Solve $\sqrt{x-8} - \sqrt{x} = -2$

Solution: $\sqrt{x-8} - \sqrt{x} = -2$

$\sqrt{x-8} = \sqrt{x} - 2$ Isolate one radical

$(\sqrt{x-8})^2 = (\sqrt{x} - 2)^2$

$x - 8 = x - 4\sqrt{x} + 4$

$-12 = -4\sqrt{x}$

$3 = \sqrt{x}$ Isolate the radical

$9 = x$ square both sides

Check the solution

$$\sqrt{x-8} - \sqrt{x} = -2$$

$$\sqrt{9-8} - \sqrt{9} = -2$$

$$\sqrt{1} - \sqrt{9} = -2$$

$$1 - 3 = -2$$

$$-2 = -2$$

Thus, the solution set is {9}

Example 14) Solve $\sqrt{2x-1} = -2$

Solution:- Here the square root of any number can not be negative. So, the solution set is empty. (It doesn't have solution set).

EXERCISE 3.8

Solve each of the following equation and check for extraneous solution.

1) $x + 1 = \sqrt{3x-1}$

2) $\sqrt{4x+3} + \sqrt{4x} = 3$

3) $\sqrt{3x-2} = -1$

4) $\sqrt{x} + 3 = \sqrt{x+12}$

5) $\sqrt{\sqrt{x+2}} = 6$

6) $\sqrt[3]{4x^2-1} = -3$

7) $\sqrt{3x+1} = 1 + \sqrt{x+4}$

8) $\sqrt{\sqrt{x+25} - \sqrt{x}} = 5$

9) $\sqrt{x+2} = \sqrt{3x+1}$

10) $\dfrac{1}{\sqrt[3]{2x}} = \dfrac{-1}{2}$

11) A students solves the equation $\sqrt{x+12} = x$ and finds that x=-3 or x=4, without checking by substituting into the equation, which is the extraneous solution, -3 or 4?

12) Solve for x

$$x + \sqrt{1-3x} = -5$$

13) Solve for x

$$\sqrt{20-x} = \sqrt{8+x}$$

3.12 GRAPHS OF RATIONAL FUNCTIONS

A rational function is a function definable as a quotient of two polynomials. Here are some examples:

$$f(x) = \frac{x^2+3x-5}{x+4}, \quad f(x) = \frac{6}{x^2+2}, \quad f(x) = \frac{5x^2-4x+1}{3}$$

3.12.1 Definition Rational Function

A rational function is a function f that can be described by:-

$$f(x) = \frac{P(x)}{Q(x)}$$

Where P(x) and Q(x) are polynomials with no common factor other than 1 and -1, and where Q(x) is not the zero polynomial. The domain of f consists of all values of x for which Q(x)≠0.

The domain of a rational function is the set of all real numbers except the x-values that make the denominator zero. For example, the domain of the rational function

$$f(x) = \frac{x^2 + 7x + 8}{x(x-3)(x+2)}$$

is the set of all real numbers except 0, 3, and -2.

Example 1) Finding the domain of a rational function:- Find the domain of each rational function:

a) $f(x) = \dfrac{x^2 - 9}{x - 3}$

b) $g(x) = \dfrac{x}{x^2 - 4}$

c) $h(x) = \dfrac{x + 5}{x^2 + 25}$

Solution:- Rational functions contain division. Because division by 0 is undefined, we must exclude from the domain of each function values of x that cause the polynomial function in the denominator to be 0.

a)   The denominator of $f(x) = \dfrac{x^2 - 9}{x - 3}$ is 0 if x=3. Thus, x cannot

be equal 3. The domain of f consists of all real numbers except 3. We can express the domain in set-builder or interval notation:
Domain of f={x: x≠3}
Domain of f=(-∞, 3) ∪ (3, ∞⁺).

b)   The denominator of $g(x) = \dfrac{x}{x^2 - 4}$ is 0 if x=2 or x=-2. Thus,

the domain of g consists of all real numbers except -2 and 2. We can express the domain in set-builder or interval notation:
Domain of g={x: x≠-2, x≠2}
Domain of g=(-∞, -2) ∪ (-2, 2) ∪ (2, ∞⁺).

c)   No real numbers cause the denominator of $h(x) = \dfrac{x + 5}{x^2 + 25}$ to

equal 0. The domain of h consists of all real numbers.
Domain of h=(-∞, ∞⁺).

The most basic rational function is the reciprocal function, defined by $f(x) = \dfrac{1}{x}$. The denominator of the reciprocal function is zero when x=0, so the domain of f is the set of all real numbers except 0.

Let's look at the behavior of f near the excluded value 0. We start by evaluating f(x) to the left of 0.

x approaches 0 from the left

$f(x) = \dfrac{1}{x}$	x	-1	-0.5	-0.1	-0.01	-0.001
	f(x)	-1	-2	-10	-100	-1000

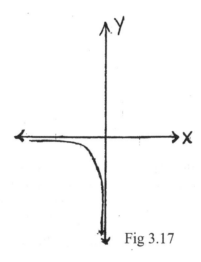

Fig 3.17

Mathematically, we say that "x approaches 0 from the left." From the table and the accompanying graph, it appears that as x approaches 0 from the left, the function values, f(x), decrease without bound. We say that "f(x) approaches negative infinity." We use a special arrow notation to describe this situation symbolically:

As $x \to 0^-$, $f(x) \to -\infty$ {As x approaches 0 from the left, f(x) approaches negative infinity (that is, the graph falls).

Observe that the minus (-) superscript on the 0 (x→0⁻) is read "from the left."

Next we evaluate f(x) to the right of 0.

x approaches 0 from the right.

$f(x)=\dfrac{1}{x}$	x	0.001	0.01	0.1	0.5	1
	f(x)	1000	100	10	2	1

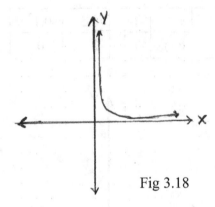

Fig 3.18

Mathematically, we say that "x approaches 0 from the right." From the table and the accompanying graph, it appears that as x approaches 0 from the right, the function values, f(x), increase without bound. We say that "f(x) approaches positive infinity." We again use a special arrow notation to describe this situation symbolically:

As x→0⁺, f(x)→∞⁺. {As x approaches 0 from the right, f(x) approaches positive infinity (that is, the graph rises}.

Observe that the plus (+) superscript on the 0 (x→0⁺) is read "from the right."

Now let's see what happens to the function values, f(x), as x gets farther away from the origin. The following tables suggest what happens to f(x) as x-increases or decreases without bound.

x increases without bound:

$f(x)=\dfrac{1}{x}$	x	1	10	100	1000
	f(x)	1	0.1	0.01	0.001

x-decreases without bound:

$f(x)=\dfrac{1}{x}$	x	-1	-10	-100	-1000
	f(x)	-1	-0.1	-0.01	-0.001

It appears that as x-increases or decreases without bound, the function values, f(x), are getting progressively closer to 0.

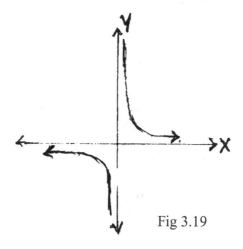

Fig 3.19

f(x) approaches 0 as x increases or decreases without bound.

The above figure 3.19 illustrates the end behavior of $f(x)=\dfrac{1}{x}$ as x increases or decreases without bound. The graph shows that the function values, f(x), are approaching 0. This means that as x increases or decreases without bound, the graph of f is approaching the horizontal line y=0 (that is, the x-axis). We use arrow notation to describe this situation.

As x→∞⁺, f(x)→0   and   as x→-∞, f(x)→0.

As x approaches positive infinity (that is, increases without bound), f(x) approaches 0.

As x approaches negative infinity (that is, decreases without bound), f(x) approaches 0.

Thus, as x approaches positive infinity (x→∞⁺) or as x approaches negative infinity (x→-∞), the function values are approaching zero: f(x)→0.

The graph of the reciprocal function $f(x) = \dfrac{1}{x}$ is shown in figure 3.20. Unlike the graph of a polynomial function, the graph of the reciprocal function has a break and is composed of two distinct branches.

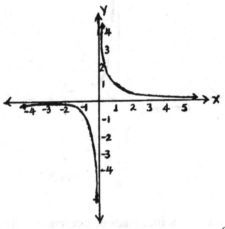

Fig 3.20 The graph of the reciprocal function $f(x) = \dfrac{1}{x}$

If x is far from 0, then $\dfrac{1}{x}$ is close to 0. By contrast, if x is close to 0, then $\dfrac{1}{x}$ is far from 0.

The arrow notation used throughout our discussion of the reciprocal function is summarized in the following box:

Arrow Notation

Symbol	Meaning
$x \to a^+$	x approaches a from the right.
$x \to a^-$	x approaches a from the left.
$x \to \infty^+$	x approaches positive infinity; that is, x increases without bound.
$x \to -\infty$	x approaches negative infinity; that is, x decreases without bound.

Another basic rational function is $f(x) = \dfrac{1}{x^2}$. The graph of this even function, with y-axis symmetry and positive function values, is shown in the fig below. Like the reciprocal function, the graph has a break and is composed of two distinct branches.

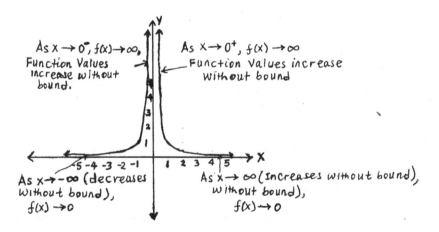

Fig 3.21 The graph of $f(x) = \dfrac{1}{x^2}$

3.13 Vertical Asymptotes of Rational Functions.

Look again at the graph of $\dfrac{1}{x^2}$ in the above fig. The curve approaches, but doesn't touch, the y-axis. The y-axis, or x=0, is said to be a vertical asymptote of the graph. A rational function may have no vertical asymptotes. The graph of a rational function never intersects a vertical asymptote. We will use dashed lines to show asymptotes.

3.13.1 Definition of a Vertical Asymptote

The line x=a is a vertical asymptote of the graph of a function f if f(x) increases or decreases without bound as x approaches a.

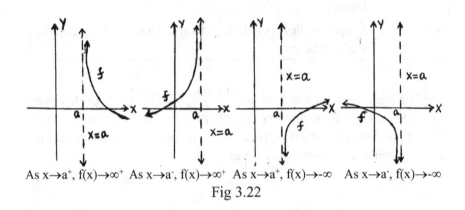

As x→a⁺, f(x)→∞⁺ As x→a⁻, f(x)→∞⁺ As x→a⁺, f(x)→-∞ As x→a⁻, f(x)→-∞

Fig 3.22

Thus, as x approaches a from either the left or the right, f(x)→∞⁺ or f(x)→-∞

If the graph of a rational function has vertical asymptotes, they can be located using the following theorem:

3.13.2 Locating Vertical Asymptotes

If $f(x) = \dfrac{P(x)}{Q(x)}$ is a rational function in which P(x) and q(x) have no common factors and a is a zero of q(x), the denominator, then x=a is a vertical asymptote of the graph of f.

Example 2)   Find the vertical asymptotes of a rational function.

a) $f(x) = \dfrac{x}{x^2 - 9}$

b) $g(x) = \dfrac{x+3}{x^2 - 9}$

c) $h(x) = \dfrac{x+3}{x^2 + 9}$

Solution:- Factoring is usually helpful in identifying zeros of denominators.

(a) $f(x) = \dfrac{x}{x^2 - 9} = \dfrac{x}{(x+3)(x-3)}$

There are no common factors in the numerator and denominator. The zeros of the denominator are -3 and 3. Thus, the line x=3 and x=3 are the vertical asymptotes for the graph of f. [see the figure 3.23 (a)]

(b)   We will use factoring to see if there are common factors.

$g(x) = \dfrac{x+3}{x^2 - 9} = \dfrac{(x+3)}{(x+3)(x-3)} = \dfrac{1}{x-3}$

There is a common factor, for (x+3) and (x²-9), which is x+3, so we can simplify it. After we simplified it, the only zero of the denominator of g(x) in simplified form is 3. Thus, the line x=3 is the only vertical asymptote of the graph of g. [See figure 3.23(b).]

(c)   We can not factor the denominator of h(x) over the real numbers.

$h(x) = \dfrac{x+3}{x^2 + 9}$

The denominator x²+9, has no real zeros. Thus, the graph of h has no vertical asymptotes. [see fig 3.23(c).]

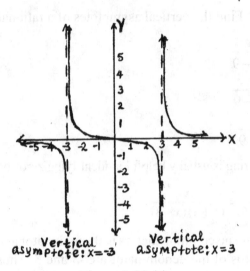

Figure 3.23 (a)

The graph of $f(x) = \dfrac{x}{x^2 - 9}$ has two vertical asymptotes.

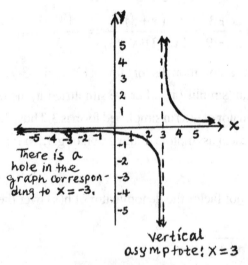

Figure 3.23 (b)

The graph of $g(x) = \dfrac{x+3}{x^2 - 9}$ has one vertical asymptote.

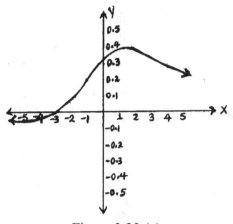

Figure 3.23 (c)

The graph of $h(x)=\dfrac{x+3}{x^2+9}$ has no vertical asymptotes.

Exercise 3.9

Find the vertical asymptotes, if any, of the graph of each rational function:

a) $f(x)=\dfrac{x}{x-2}$

b) $g(x)=\dfrac{x-2}{x^2-4}$

c) $h(x)=\dfrac{x-2}{x^2+4}$

3.14 Horizontal Asymptotes of Rational Functions

The graph of the reciprocal function $f(x)=\dfrac{1}{x}$, shown below in the fig (3.24). As x→∞⁺ and as x→-∞, the function values are approaching 0: f(x)→0. The line y=0 (that is, the x-axis) is a horizontal asymptote of the graph. Many, but not all, rational functions have horizontal asymptotes.

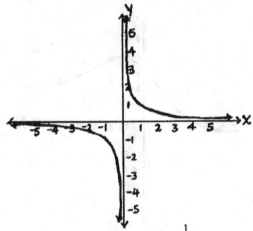

Fig 3.24. The graph of $f(x) = \dfrac{1}{x}$, (repeated)

3.14.1 Definition of a Horizontal Asymptote

The line y=b is a Horizontal Asymptote of the graph of a function f if f(x) approaches b as x increases or decreases without bound.

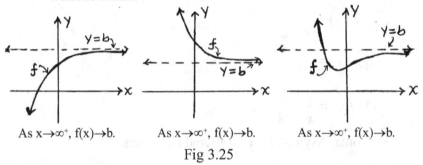

| As x→∞⁺, f(x)→b. | As x→∞⁺, f(x)→b. | As x→∞⁺, f(x)→b. |

Fig 3.25

Recall that a rational function may have several vertical asymptotes. By contrast, it can have at most one horizontal asymptote. Although a graph can never intersect a vertical asymptote, it may cross its horizontal asymptote.

If the graph of a rational function has a horizontal asymptote, it can be located using the following theorem:

3.14.2 Locating Horizontal Asymptotes

Let f be the rational function given by

$$f(x) = \frac{a_n x^n + a_{n-1} x^{n-1} + \ldots + a_1 x + a_0}{b_m x^m + b_{m-1} x^{m-1} + \ldots + b_1 x + b_0}, \ a_n \neq 0, b_m \neq 0.$$

The degree of the numerator is n. The degree of the denominator is m.

1)   If n<m, the x-axis, or y=0, is the horizontal asymptote of the graph of f.

2)   If n=m, the line $y = \dfrac{a_n}{b_m}$ is the horizontal asymptote of the graph of f.

3)   If n>m, the graph of f has no horizontal asymptote.

Example 3) Finding the horizontal asymptote of a rational function.

a) $f(x) = \dfrac{8x^2 - 3x + 7}{2x^2 + 7x}$

b) $f(x) = \dfrac{8x - 3}{x^2 - 9}$

c) $f(x) = \dfrac{10x^2 - 8x + 1}{5x + 1}$

d) $f(x) = \dfrac{4x}{2x^2 + 1}$

e) $g(x) = \dfrac{4x^2}{2x^2 + 1}$

f) $h(x) = \dfrac{4x^3}{2x^2 + 1}$

Solution:

a) $\dfrac{8x^2 - 3x + 7}{2x^2 + 7x}$

The degree of the numerator is the same as the degree of the denominator. Therefore, the horizontal asymptote is given by y=quotient of leading coefficients.

$$y = \frac{8}{2} = 4$$

Therefore, y=4 is the horizontal asymptote of f(x).

(b)   In this case, the degree of the numerator is less than the degree of the denominator, so the horizontal asymptote of f(x) is y=0.

(c)   There is no horizontal asymptote, because the degree of the numerator is greater than the degree of the denominator.

Remark: In part (c), although there is no horizontal asymptote, the graph of the function for large values of x resembles the graph of "y=quotient" in the division of p(x) by q(x). In this case, it is y=2x-2, which is an oblique (slant) line, called oblique asymptote.

(d) $f(x) = \dfrac{4x}{2x^2 + 1}$

The degree of the numerator, 1, is less than the degree of the denominator, 2. Thus, the graph of f has the x-axis as a horizontal asymptote [see figure 3.26(a)] below. The equation of the horizontal asymptote is y=0

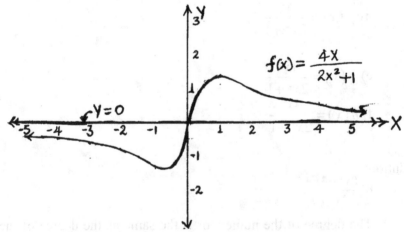

fig 3.26 (a) The horizontal asymptote of the graph is y=0
(the x-axis)

e) $g(x) = \dfrac{4x^2}{2x^2 + 1}$

The degree of the numerator, 2, is equal to the degree of the denominator, 2. The leading coefficients of the numerator and denominator, 4 and 2, are used to obtain the equation of the horizontal asymptote. The equation of the horizontal asymptote is $y = \dfrac{4}{2}$ or y=2 [see figure (3.26) (b)].

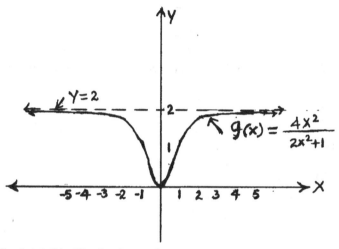

fig 3.26 (b). The horizontal asymptote of the graph is y=2

f) $h(x) = \dfrac{4x^3}{2x^2 + 1}$

The degree of the numerator, 3, is greater than the degree of the denominator, 2. Thus the graph h has no horizontal asymptote. [see figure 3.26 (c)]

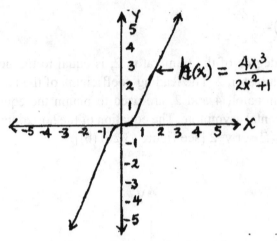

Fig 3.26 (c) The graph has no horizontal asymptote.

Exercise 3.10
Find the horizontal asymptote, if any, of the graph of each rational functions.

a) $g(x) = \dfrac{9x^2 - 2x}{x - 2}$

b) $f(x) = \dfrac{9}{3x + 7}$

c) $f(x) = \dfrac{3x + 13}{3x^2 + 2x - 5}$

d) $f(x) = \dfrac{9x^2}{3x^2 + 1}$

e) $g(x) = \dfrac{9x}{3x^2 + 1}$

f) $h(x) = \dfrac{8x^2}{4x^2 + 1}$

g) $g(x) = \dfrac{9x}{3x^2 + 1}$

h) $h(x) = \dfrac{9x^3}{3x^2 + 1}$

More over

3.14.3 Graph of $y = \dfrac{a}{x-h} + b$

The graph of $y = \dfrac{a}{x-h} + b$ has the following characteristics:

- If $|a| > 1$, the graph is a vertical stretch of the graph of $y = \dfrac{1}{x}$.

  If $0 < |a| < 1$, the graph is a vertical shrink of the graph of $y = \dfrac{1}{x}$

  , if $a < 0$, the graph is a reflection in the x-axis of the graph of $y = \dfrac{1}{x}$.

- The horizontal asymptote is y=b. The vertical asymptote is x=h.
- The domain of the function is all real numbers except x=h. The range is all real numbers except y=b.

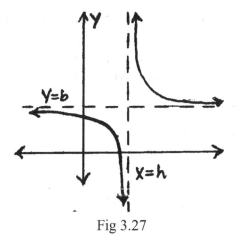

Fig 3.27

Example 4) Draw the graph of $f(x) = \dfrac{5}{x+3} - 2$.

Solution:-

The vertical asymptote of the graph is x=-3 and the horizontal asymptote is y=-2

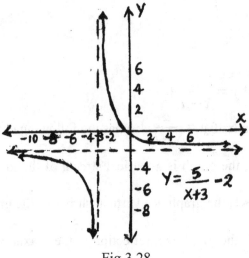

$$Y = \frac{5}{X+3} - 2$$

Fig 3.28

3.15 Oblique (Slant) Asymptotes

Examine the graph of

$$f(x) = \frac{x^2 + 1}{x - 1}$$

Shown in figure (3.29). Note that the degree of the numerator, 2, is greater than the degree of the denominator, 1. Thus, the graph of this function has no horizontal asymptote. However, the graph has an oblique or a slant asymptote, y=x+1.

The graph of a rational function has an oblique or a slant asymptote if the degree of the numerator is one more than the degree of the denominator. The equation of the slant asymptote can be found by division. For example, to find an oblique or a slant asymptote for the graph of $f(x) = \dfrac{x^2 + 1}{x - 1}$, divide x²-1 by x-1

$$\begin{array}{r} 1x+1+\dfrac{2}{x-1} \\ x-1\overline{)x^2+1} \\ (-)\underline{x^2-x} \\ x+1 \\ (-)\underline{x-1} \\ 2 \end{array}$$ remainder

Observe that

$$f(x)=\frac{x^2+1}{x-1}=x+1+\frac{2}{x-1}$$

The equation of an oblique or a slant asymptote is y=x+1.

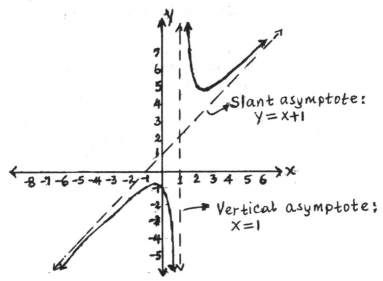

Fig 3.29 The graph of $f(x)=\dfrac{x^2+1}{x-1}$ with an oblique asymptote

As |x|→∞ the value of $\dfrac{2}{x-1}$ is approximately 0. Thus, when |x| is large, the function is very close to y=x+1+0. This means that as x→∞⁺ or as x→-∞, the graph of f gets closer and closer to the line whose equation is y=x+1. The line y=x+1 is an oblique or a slant asymptote of the graph.

In general, if $f(x) = \dfrac{p(x)}{q(x)}$, p and q have no common factors, and the degree of p is one greater than the degree of q, find the slant asymptote by dividing q(x) into p(x). The division will take the form:

$$\frac{p(x)}{q(x)} = mx + b + \frac{\text{remainder}}{q(x)}$$

The equation of an oblique asymptote is obtained by dropping the term with the remainder. Thus, the equation of an oblique asymptote is y=mx+b.

Example 5)   Find an oblique asymptote of $f(x) = \dfrac{x^2 - 4x - 5}{x - 3}$

Solution:-   Because the degree of the numerator, 2, is exactly one more than the degree of the denominator, 1. And x-3 is not a factor of x²-4x-5, the graph of f has an oblique asymptote. To find the equation of an oblique asymptote, divide x-3 into x²-4x-5:

$$\begin{array}{r} 1x - 1 - \dfrac{8}{x-3} \\ x-3\overline{)x^2 - 4x - 5} \\ (-)x^2 - 3x \\ \hline -x - 5 \\ (-)\underline{-x + 3} \\ -8 \leftarrow \text{remainder} \end{array}$$

Drop the remainder term and you will have the equation of an oblique asymptote.

The equation of an oblique asymptote is y=x-1. Using our strategy for graphing rational functions, the graph of $f(x) = \dfrac{x^2 - 4x - 5}{x - 3}$ is shown below in the figure (3.30).

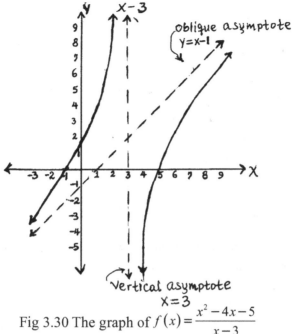

Fig 3.30 The graph of $f(x) = \dfrac{x^2 - 4x - 5}{x - 3}$

EXERCISE 3.11

1)   Find the vertical and horizontal asymptote of rational functions. If any.

a) $f(x) = \dfrac{-3}{5x - 3}$

b) $h(x) = \dfrac{7x - 3}{x + 5}$

c) $f(x) = \dfrac{5x^2 + x - 7}{x^2 + 9}$

d) $g(x) = \dfrac{2x^2 + 7x - 4}{2x^2 - 5x + 2}$

e) $f(x) = \dfrac{5x + 9}{x^2 + x + 1}$

f) $g(x) = \dfrac{x^2 + 2x + 1}{x^2 - 1}$

g) $f(x) = \dfrac{3x+13}{3x^2+2x-5}$

h) $g(x) = \dfrac{3x+4}{4x-5}$

i) $f(x) = \dfrac{3x^2-5}{x^2-x-2}$

j) $h(x) = \dfrac{7x^2+3}{x^2+x-6}$

2)   Find an oblique asymptote and vertical asymptote of the following rational functions.

a) $f(x) = \dfrac{2x^2-5x+7}{x-2}$

b) $g(x) = \dfrac{3x^2-7x+6}{x-2}$

c) $f(x) = \dfrac{5x^3+2x+1}{x^2-4}$

d) $f(x) = \dfrac{2x^2-7x-7}{x-5}$

e) $g(x) = \dfrac{3x^2+2}{x-1}$

3.16   Long Division of Polynomials

Polynomial Long Division is an algorithm for dividing a polynomial by an other polynomial of the same or arithmetic technique called Long division. It can be done easily by hand, because it separates an otherwise complex division problem into smaller ones.

Example 1)

    Find

$$\frac{x^3 - 12x^2 - 42}{x - 3}$$

This problem is written like this:-

$$\frac{x^3 - 12x^2 + 0x - 42}{x - 3}$$

The quotient and remainder can then be determined as follows.

1)   Divide the first term of the numerator by the highest term of the denominator (meaning the one with the highest power of x, which in this case is x). Place the result above the bar $(x^3 \div x = x^2)$.

$$x-3{\overline{\smash{\big)}\,x^3 - 12x^2 + 0x - 42}} \quad \overset{x^2}{}$$

2)   Multiply the denominator by the result just obtained (the first term of the eventual quotient). Write the under the first two terms of the numerator $(x^2 \cdot (x-3) = x^3 - 3x^2)$.

$$x-3{\overline{\smash{\big)}\,x^3 - 12x^2 + 0x - 42}} \quad \overset{x^2}{}$$
$$x^3 - 3x^2$$

3)   Subtract the product just obtained from the appropriate terms of the original numerator (being careful that subtracting something having a minus sign is equivalent to adding something having a plus sign), and write the result underneath. $[(x^3-12x^2)-(x^3-3x^2)=-12x^2+3x^2=-9x^2]$ Then, "bring down" the next term from the numerator.

$$x-3{\overline{\smash{\big)}\,x^3 - 12x^2 + 0x - 42}} \quad \overset{x^2}{}$$
$$\underline{x^3 - 3x^2}$$
$$-9x^2 + 0x$$

4)   Repeat the previous three steps, except this time use the two terms that have just been written as the numerator.

$$
\begin{array}{r}
x^2 - 9x \\
x-3{\overline{\smash{\big)}\,x^3 - 12x^2 + 0x - 42}} \\
\underline{x^3 - 3x^2} \\
-9x^2 + 0x \\
\underline{(-)\underline{-9x^2 + 27x}} \\
-27x - 42
\end{array}
$$

5)   Repeat Step 4. This time, there is nothing to "pull down".

$$
\begin{array}{r}
x^2 - 9x - 27 \\
x-3{\overline{\smash{\big)}\,x^3 - 12x^2 + 0x - 42}} \\
\underline{(-)x^3 - 3x^2} \\
-9x^2 + 0x \\
\underline{(-)\underline{-9x^2 + 27x}} \\
-27x - 42 \\
\underline{(-)\underline{-27x + 81}} \\
-123
\end{array}
$$

The polynomial above the bar is the quotient, and the number left over (-123) is the remainder.

$$
\frac{x^3 - 12x^2 - 42}{x - 3} = \underbrace{x^2 - 9x - 27}_{q(x)} - \underbrace{\frac{123}{\underset{g(x)}{\underbrace{x - 3}}}}_{r(x)}
$$

3.16.1 Division transformation.

    Polynomial division allows for a polynomial to be written in a divisor-quotient form which is often advantageous. Consider polynomials p(x), D(x) where degree (D)<degree (p). Then, for some quotient polynomial Q(x) and remainder polynomial R(x) with degree (R)<degree (D).

$$\frac{P(x)}{D(x)} = Q(x) + \frac{R(x)}{D(x)} \Rightarrow P(x) = D(x) \cdot Q(x) + R(x)$$

This rearrangement is known as the division transformation, and derives from arithmetical identity.

Dividend = Divisor x quotient + remainder.

Example 2) Divide $2x^2+3x+4$ by $x+1$

$$x+1\overline{)2x^2 + 3x + 4}^{\,2x}$$

 - Divide $2x^2$ by x

$$\underline{(-1)2x^2 + 2x}$$
$$x + 4$$

 - Subtract $2x^2+2x$ from $2x^2+3x$
 - Bring down 4

Now we divide x by x to obtain 1, multiply 1 and the divisor, and subtract.

$$x+1\overline{)2x^2 + 3x + 4}^{\,2x+1}$$

 - Divide: $\dfrac{x}{x} = 1$

$$\underline{2x^2 + 2x}$$
$$x + 4$$
$$\underline{(-)x + 1}$$
$$3 \quad \text{~Remainder}$$

 - Subtract: x+1 from x+4

Thus, $\dfrac{2x^2 + 3x + 4}{x + 2} = 2x + 1 + \dfrac{3}{x + 1}$ Remainder above divisor

Multiplying both sides of this equation by x+1 results in the following equation.

$2x^2+3x+4 =$	$(x+1)$	$(2x+1) +$	3
Dividend	Divisor	Quotient	Remainder

Example 3)   Divide $x^2+10x+21$ by $x+3$

Solution $x+3\overline{)x^2+10x+21}$ Arrange the terms of the dividend and the divisor in decreasing power of x.

$$\begin{array}{r} x \\ x+3\overline{)x^2+10x+21} \end{array}$$

$$\begin{array}{r} x \\ x+3\overline{)x^2+10x+21} \end{array}$$ Divide x^2 by x

$$\begin{array}{r} x^2+3x \\ \hline x \\ x+3\overline{)x^2+10x+21} \end{array}$$ Multiply each term in the divisor by x.

$$\begin{array}{r} (-)x^2+3x \\ \hline 7x \\ x \\ x+3\overline{)x^2+10x+21} \end{array}$$ Subtract x^2+3x from x^2+10x

$$\begin{array}{r} x^2+3x \\ 7x+21 \quad x+7 \\ x+3\overline{)x^2+10x+21} \end{array}$$ Bring down 21 from its original position.

$$\begin{array}{r} x^2+3x \\ 7x+21 \end{array}$$ Divide $(7x+21)$ by x

$$\begin{array}{r} x+7 \\ x+3\overline{)x^2+10x+21} \\ x^2+3x \\ 7x+21 \\ (-)7x+21 \\ \hline 0 \end{array}$$ Multiply $(x+3)$ by 7.

0 ◄——— Remainder

The quotient is x+7. Because the remainder is 0. We can conclude that x+3 is a factor of $x^2+10x+21$:

and $\dfrac{x^2+10x+21}{x+3} = x+7$

3.17 Division Algorithm

Polynomial long division is checked by multiplying the divisor with the quotient and then adding the remainder. This should give the dividend. The process illustrates the Division Algorithm.

The Division Algorithm

If f(x) and d(x) are polynomials, with d(x)≠0, and the degree of d(x) is less than or equal to the degree of f(x), then there exist unique polynomials q(x) and r(x) such that:-

$$f(x) = d(x) \bullet q(x) + r(x)$$
Dividend Divisor Quotient Remainder

Example 5)   Divide $6x^4+5x^3+3x-5$ by $3x^2-2x$

Soln:-

$$3x^2-2x \overline{)6x^4+5x^3+0x^2+3x-5} \quad \frac{2x^2+3x+2}{}$$

$$\frac{(-)6x^4-4x^3}{}$$

$$9x^3+0x^3$$

$$\frac{(-)9x^3-6x^2}{}$$

$$6x^2+3x$$

$$\frac{(-)6x^2-4x}{}$$

$$7x-5 \quad \longleftarrow \text{Remainder}$$

The process of division is finished because the degree of 7x-5, which is 1, is less than the degree of the divisor $3x^2-2x$, which is 2. The answer is

$$\frac{6x^4+5x^3+3x-5}{3x^2-2x} = 2x^2+3x+2+\frac{7x-5}{3x^2-2x}.$$

3.18   Dividing Polynomials Using Synthetic Division

Synthetic division is a method of performing polynomial long division, with less writing and fewer calculations. It is mostly taught for division by binomial of the form

x-a,

but the method generalizes to division by any monic-polynomial. This method can be used instead of long-division on integers by considering 10=x and only substituting 10 back in at the end.

The most useful aspects of synthetic division are that it allows one to calculate without writing variables and uses fewer calculations. As well, it takes significantly less space than long division. Most importantly, the subtraction in long division are converted to additions by switching the signs at the very beginning preventing sign errors.

Let's compare the two methods showing

$$\begin{array}{r} 3x^2 + 4x + 7 \quad \longleftarrow \text{Quotient} \\ \text{Divisor} \longrightarrow x-2\overline{)3x^2 - 2x^2 - x + 2} \quad \longleftarrow \text{Dividend} \\ \underline{(-)3x^2 - 6x^2} \\ 4x^2 - x + 2 \\ \underline{(-)4x^2 - 8x} \\ 7x + 2 \\ \underline{(-)7x - 14} \\ 16 \quad \longleftarrow \text{Remainder} \end{array}$$

3.18.1 Long Division

Synthetic Division

$$\begin{array}{r|rrrr} 2 & 3 & -2 & -1 & 2 \\ & & 6 & 8 & 14 \\ \hline & 3 & 4 & 7 & 16 \end{array}$$

Notice the relationship between the polynomials in the long division process and the numbers that appears in synthetic division.

Coefficients of the Dividend

The divisor x-2

$$\begin{array}{r|rrrr} 2 & 3 & -2 & -1 & 2 \\ & & 6 & 8 & 14 \\ \hline & 3 & 4 & 7 & 16 \end{array}$$

Remainder

Coefficients of the Quotient

Example 1)   Use synthetic division to divide 5x³-6x²+7x-3 by x-3.

$x-3{\overline{\smash{\big)}\,5x^3-6x^2+7x-3}}$ (Arrange the power in descending power with coefficient for any missing term)

3 | 5 -6 7 -3   Write 3 for the divisor to the right and write the coefficients of the dividend.

3 | 5 -6 7 -3
————————————
    5   Bring down 5.

3 | 5 -6 7 -3
      15
————————————
    5   Multiply by 3: 3x5=15

3 | 5 -6 7 3
      15
————————————
    5   9   Add

3 | 5 -6 7 3
      15 27
————————————
    5   9 34   Multiply by 3: 3x9=27 and odd 7+27=34

3 | 5 -6 7 3
      15 27 102
————————————
    5   9 34 105   Multiply by 3: 3x34=102 and add 3+102=105

Use the numbers in the last row to write the quotient, plus the remainder above the divisor.
written form

$$5 \qquad 9 \qquad 34 \qquad 105$$

The last row of the synthetic division:

$$\begin{array}{r} 5x^2 + 9x + 34 + \dfrac{105}{x-3} \\ x-3 \overline{\smash{\big)}\, 5x^3 - 6x^2 + 7x - 3} \end{array}$$

The degree of the first term of the quotient is one less than the degree of the first term of the dividend. The final value in this row is the remainder.

Example 2)   Use the synthetic division to divide 4x³+5x+6 by x+2

Solution:- The divisor must be written in the form x-a. Thus, we write x+2 as x-(-2). This means x=-2, we have to write a 0 coefficient for the missing x²-term in the dividend, we can write the division as follows:

$$x-(-2) \overline{\smash{\big)}\, 4x^3 + 0x^2 + 5x + 6}$$

Now we can use synthetic division.

$$-2 \,\big|\, \begin{array}{cccc} 4 & 0 & 5 & 6 \end{array}$$

$$-2 \,\big|\, \begin{array}{cccc} 4 & 0 & 5 & 6 \end{array}$$ Bring down 4

$$\begin{array}{c} \underline{} \\ 4 \end{array}$$

$$-2 \,\big|\, \begin{array}{cccc} 4 & 0 & 5 & 6 \\ & -8 \end{array}$$ Multiply: -2x4=-8

$$-2 \,\big|\, \begin{array}{cccc} 4 & 0 & 5 & 6 \\ & -8 \end{array}$$ Add: 0+(-8)=-8

$$\begin{array}{c} \underline{} \\ 4 \end{array}$$

```
-2 | 4   0   5   6
         -8  16
   ─────────────────
     4  -8
```

```
-2 | 4   0   5   6
         -8  16
   ─────────────────
     4  -8  21
```

```
-2 | 4   0   5   6
         -8  16  -42
   ─────────────────
     4  -8  21
```

3.19   The Remainder Theorem

Let's consider the Division Algorithm when the dividend, $f(x)$, is divided by x-c. In this case, the remainder must be a constant because its degree is less than one, the degree of x-c.

$f(x) = (x - c) q(x) + r$
Dividend   Divisor   quotient   The remainder, r, is a constant when dividing by x-c.

Now let's evaluate f at c,
$f(c)=(c-c)q(c)+r$ Find f(c) by letting x=c in $f(x)=(x-c)q(x)+r$.
This will give an expression for r.
$f(c)=0 \cdot q(c)+r$ c-c=0
$f(c)=r$ $0 \cdot q(c)=0$ and 0+r=r.

What does this last equation mean? If a polynomial is divided by x-c, the remainder is the value of the polynomial at c. This result is called the Remainder Theorem.

The Remainder Theorem
If the polynomial f(x) is divided by x-c, then the remainder is f(c).

Example 7) Using the remainder theorem evaluate a polynomial function: $f(x)=x^3-4x^2+5x+3$, use the Remainder Theorem to find f(2).

Solution:- By the Remainder Theorem, if f(x) is divided by x-2, then the remainder is f(2). We'll use synthetic division to divide.

$$2 \begin{array}{|cccc} 1 & -4 & 5 & 3 \\ & 2 & -4 & 2 \\ \hline 1 & -2 & 1 & 5 \end{array}$$ ← Remainder

The remainder, 5, is the value of f(2). Thus, f(2)=5. We can verify that this is correct by evaluating f(2) directly. Using $f(x)=x^3-4x^2+5x+3$, we obtain

$$f(2)=2^3-4\cdot2^2+5\cdot2+3=8-16+10+3=5.$$

3.20 The Factor Theorem

Let's look again at the Division Algorith when the divisor is of the form x-c.

$$f(x) \quad = \quad (x-c) \quad q(x) \quad + \quad r$$

Dividend   Divisor   Quotient   Constant remainder

By the Remainder Theorem, the remainder r is f(c), so we can substitute f(c) for r:

$$f(x)=(x-c)q(x)+f(c).$$

Notice that if f(c)=0, then

$$f(x)=(x-c)q(x)$$

So that x-c is a factor of f(x). This means that for the polynomial function f(x), if f(c)=0, then x-c is a factor of f(x).

Let's reverse directions and see what happens if x-c is a factor of f(x). This means that

$$f(x)=(x-c)q(x).$$

If we replace x in f(x)=(x-c)q(x) with c, we obtain

$$f(c)=(c-c)q(c)=0 \cdot q(c)=0.$$

Thus, if x-c is a factor of f(x), then f(c)=0.

We have proved a result known as the Factor Theorem.

The Factor Theorem

Let f(x) be a polynomial.

a.   If f(c)=0, then x-c is a factor of f(x).

b.   If x-c is a factor of f(x) then f(c)=0

Note:- The example that follows shows how the Factor Theorem can be used to solve a polynomial equation.

Example 8) Solve the equation: $2x^3-3x^2-11x+6=0$ using the factor theorem. If 3 is a zero of $f(x)=2x^3-3x^2-11x+6$.

Solution: We have given that 3 is a zero of $f(x)=2x^3-3x^2-11x+6$. This means that f(3)=0. Because f(3)=0, the Factor Theorem tells us that x-3 is a factor of f(x). We will use synthetic division to divide f(x) by x-3.

$$
\begin{array}{r|rrrr}
3 & 2 & -3 & -11 & 6 \\
  &   & 6 & 9 & -6 \\
\hline
  & 2 & 3 & -2 & 0
\end{array}
$$

$$
\begin{array}{r}
2x^2+3x-2 \\
x-3\overline{)2x^3-3x^2-11x+6}
\end{array}
$$

The remainder, 0, verifies that x-3 is a factor of $2x^3-3x^2-11x+6$.

Equivalently, $2x^3-3x^2-11x+6=(x-3)(2x^2+3x-2)$

Now we can solve the polynomial equation.

$2x^3-3x^2-11x+6=0$ -   This is the given equation

$(x-3)(2x^2+3x-2)=0$ -   Factor using the result from the synthetic division.

$(x-3)(2x-1)(x+2)=0$ - Factor the trinomial

$x-3=0$ or $2x-1=0$ or $x+2=0$ Set each factor equal to 0.

$x=3$ $x=\dfrac{1}{2}$ $x=-2$ Solve for x.

The solution set is $\left\{-2, \dfrac{1}{2}, 3\right\}$.

Based on the factor theorem, the following statements are useful in solving polynomial equations:

1.   If f(x) is divided by x-c and the remainder is zero, then c is a zero of f and c is a root of the polynomial equation f(x)=0.
2.   If f(x) is divided by x-c and the remainder is zero, then x-c is a factor of f(x).

EXERCISE 3.12

Divide using long division. State the quotient, q(x), and the remainder, r(x)

1) $(3x^2+5x+7)\div(x^2+3x)$
2) $(x^3-1)\div x-3$
3) $(x^3+5x^2+7x+2)\div(x+2)$
4) $(x^5-3)\div(x^3-x^2)$
5) $(2x^3-x^2-4x+5)\div(x^2+1)$

EXERCISE·3.13

Divide using synthetic division.
6) $(2x^2+x-10)\div(x-2)$
7) $(3x^2+7x-20)\div(x+5)$
8) $(5x^3-6x^2+3x+11)\div(x+2)$
9) $(5x^3+3x-2)\div(x+3)$
10) $(4x^2+x+10)\div(x+1)$

Exercise 3.14

In Exercise 11-27, use the Remainder Theorem to find the indicated function value.

11. $f(x)=3x^3-2x+4$, $f(2)$
12. $f(x)=4x^3-3x^2+3x-1$, $f(-1)$

13. $f(x)=6x^5-2x^3+3x^2+x-1$, $f(0)$
14. $f(x)=x^2-6x+6$, $f(3)$
15. $f(x)=x^3-2x^2-5x+6$, $f(½)$
16. $f(x)=3x^4-3x^2-x+3$, $f(-2)$
17. $f(x)=2x^4-5x^3-x^2+3x+1$, $f(-⅓)$
18. $f(x)=3x^3+9x^2+5$, $f(3)$
19. $f(x)=-4x^3-3x^2+2x-1$, $f(2)$
20. $f(x)=2x^4+5x^2+6x+1$, $f(-3)$

21. Find the value of k such that when $5x^2+7x+k$ is divided by $x+1$, the remainder is -3.

22. When $100x^2+5kx+2$ is divided by $x+1$, the remainder is 4. What is the value of k?

23. Find the value of m such that $\dfrac{5x^3+m}{x+1}$, the remainder is -1.

24. If $f(x)=1000x^3-200x+300$, then $f(-1)$ is _____

25. If $f(x)=400x^3+300x^2-500$, then $f(-1)+f(-2)=$_____

26. If $f(x)=2kx^3+6x+9$ is divided by $x-2$ the remainder is 1, what is the value of k?

27. Find the value of r such that $\dfrac{x^3-10x+r}{x-2}$ is -1

3.21 Simplifying Rational Expression

We have seen that the sum, difference, or product of polynomial is again a polynomial. However, the quotient of two polynomials, called a rational expression, may not be a polynomial. In this section we will see how to simplify a rational expressions.

3.21.1 Rational Expression

An algebraic expression $\dfrac{p(x)}{q(x)}$, where p(x) and q(x) are polynomial expressions and q(x) is a non-zero polynomial, is called a Rational Expression.

For example,

$\dfrac{x^2-9}{x^2-6x+9}$ is a rational expression and $\dfrac{\sqrt{x^2-9}}{x^2-6x+9}$

is not a rational expression (why?).
A rational expression is undefined when the denominator is 0. A number that makes a rational expression undefined is called an excluded value. A rational expression is in simplest form if the numerator and denominator have no factor in common other than 1.

3.21.2 A• Reduction of a rational Expression

As in the case of a rational number, a rational expression is also reduced to its simplest form by canceling factor(s) common to the numerator and the denominator.

A rational expressions $\dfrac{x\cdot y}{z\cdot y}=\dfrac{x}{z}$, for y≠0

Note: The process of removing common factors from the numerator and denominator is called reduction of a rational expression.

For example, $\dfrac{x^2-5x+6}{x^2-3x}$

Here, $\dfrac{x^2-5x+6}{x^2-3x}=\dfrac{(x-2)(x-3)}{x(x-3)}$, the denominator is zero,

when x=0 and x=3, these values 0 and 3 are called Excluded Values. In the above rational expression, we have to state that x≠3. Before we cancel the factor (x-3).
Thus,

$$\frac{x^2 - 5x + 6}{x^2 - 3x} = \frac{(x-2)(x-3)}{x(x-3)}$$

$$= \frac{(x-2)}{x} \quad \text{for x} \neq 3$$

Example 1)   Find the excluded values, if any, of the following expression.

a) $\dfrac{2x-1}{-3x}$

b) $\dfrac{x}{x^2 - 4}$

c) $\dfrac{x^2}{x^2 + 8}$

d) $\dfrac{x+2}{x^2 - 9x + 20}$

Solution:-

a) $\dfrac{2x-1}{-3x}$

The expression $\dfrac{2x-1}{-3x}$ is undefined when -3x=0, or x=0. The excluded value is 0.

b)   The expression $\dfrac{x}{x^2 - 4}$ is undefined when x²-4=0, or x=2 or x=-2. The excluded values are -2 and 2

c)   The expression $\dfrac{x^2}{x^2 + 8}$ is undefined when x²+8=0 the graph of f(x)=x²+8 doesn't cross the x-axis. So, the quadratic equation has no real roots. There are no excluded values.

Example 2)   Reduce $\dfrac{x^2+9x+18}{x+3}$ to its simplified form

Solution: $\dfrac{x^2+9x+18}{x+3}=\dfrac{(x+3)(x+6)}{x+3}$ Factor the numerator

$\dfrac{(x-3)(x+6)}{(x+3)}$ Divide out common factors

$= x+6$ Simplify

The excluded value is -3. (x=-3)

Example 3)   Reduce $\dfrac{x^2-5x+6}{x^4-4x^2}$ to its simplified form.

Solution:

$\dfrac{x^2-5x+6}{x^4-4x^2}=\dfrac{(x-2)(x-3)}{x^2(x^2-4)}=\dfrac{(x-2)(x-3)}{x^2(x-2)(x+2)}$ Factor numerator and denominator.

$=\dfrac{(x-2)(x-3)}{x^2(x-2)(x+2)}$

$=\dfrac{(x-3)}{x^2(x+2)}$ Divide out common factors.

Simplify

The excluded values are 0, 2 and -2. (x≠0, 2, -2)

Example 4)   Simplify $\dfrac{2x}{x-3}$ to its simplest form.

Solution:-

The expression $\dfrac{2x}{x-3}$ is already in simplest form. The excluded

value is 3. (x≠3).

Example 5)   Reduce $\dfrac{15(5x+3)^2(2x+5)^2-4(5x+3)^3(2x+5)}{(2x+5)^4}$ to its simplest form.

Solution:-

$$\dfrac{15(5x+3)^2(2x+5)^2-4(5x+3)^3(2x+5)}{(2x+5)^4}$$

$$= \frac{(5x+3)^2 (2x+5) \left[15(2x+5) - 4(5x+3)\right]}{(2x+5)(2x+5)^3}$$

Factor the numerator and denominator

$$= \frac{(5x+3)^2 (2x+5) \left[15(2x+5) - 4(5x+3)\right]}{(2x+5)(2x+5)^3}$$

Divide out common factors.

$$= \frac{(5x+3)^2 (10x+63)}{(2x+5)^3}$$

The excluded value is $-\frac{5}{2}. \left(x \neq -\frac{5}{2}\right)$

Example 6) Simplify $\dfrac{x^3 + x^2}{x+1}$ to its simplest form
Solution

$$\frac{x^3 + x^2}{x+1} = \frac{x^2(x+1)}{x+1}$$ Factor the numerator.

$$= \frac{x^2(x+1)}{x+1}$$ Divide out the common factor, x+1

$$= x^2, \text{ for } x \neq -1$$

Example 7) Simplify $\dfrac{x^2 + 6x + 5}{x^2 - 25}$ to its simplest form.

Solution:-

$$\frac{x^2 + 6x + 5}{x^2 - 25} = \frac{(x+5)(x+1)}{(x+5)(x-5)}$$ Factor the numerator and denominator, (x+5)(x-5), x≠-5 and x≠5.

$$= \frac{(x+5)(x+1)}{(x+5)(x-5)}$$ Divide out the common factor, x+5.

$$= \frac{x+1}{x-5}, \text{ x≠-5, x≠5}$$

Example 8)   Simplify $\dfrac{x^3+3x^2}{x+3}$
Solution

$\dfrac{x^3+3x^2}{x+3}=\dfrac{x^2(x+3)}{x+3}$ Factor the numerator, because the denominator is (x+3), x≠-3.

$=\dfrac{x^2(x+3)}{(x+3)}$ Divide out the common factor, x+3.

$= x^2$, x≠-3

Example 9)   Simplify: $\dfrac{5x+10}{3x^2+6x}$

Solution

$\dfrac{5x+10}{3x^2+6x}=\dfrac{5(x+2)}{3x(x+2)}$ Factor the numerator and denominator because the denominator is 3x(x+2), x≠0 and x≠-2.

$=\dfrac{5(x+2)}{3x(x+2)}$ Divide out the common factor, x+2.

$\dfrac{5}{3x}$, x≠0, x≠-2.

Example 10)   Simplify: $\dfrac{9x^2+6xy-3y^2}{12x^2-12y^2}$

Solution

$\dfrac{9x^2+6xy-3y^2}{12x^2-12y^2}=\dfrac{3(x+y)(3x-y)}{3(4)(x+y)(x-y)}$ Factoring the numerator and the denominator, because the denominator is 3(4)(x+y)(x-y); x+y≠0 and x-y≠0

$=\dfrac{3(x+y)(3x-y)}{3(4)(x+y)(x-y)}$ Divide out the common factor, 3(x+y)

$=\dfrac{3x-y}{4(x-y)}$, x+y≠0, x-y≠0

Example 11)   Simplify: $\dfrac{x^2-1}{2x^2-x-1}$

Solution

$\dfrac{x^2-1}{2x^2-x-1}=\dfrac{(x-1)(x+1)}{(2x+1)(x-1)}$

Factoring the numerator and the denominator, because the denominator is (2x+1)(x-1); x≠-½ and x≠1

$=\dfrac{(x-1)(x+1)}{(2x+1)(x-1)}$

Divide out the common factor, (x-1)

$=\dfrac{x+1}{2x+1}\ ,\ x\neq\dfrac{-1}{2},x\neq1$

Example 12) Write and simplify a rational expression for the ratio of the perimeter to the area of the given figures below.

a) Rectangle   Figure (b) Triangle   c) Square

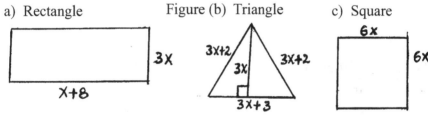

Solutions

a)   Area of a Rectangle = L•W
          = (x+8)(3x)
          A= 3x²+24x

          Perimeter of Rectangle = 2L+2W
          = 2(x+8+3x)
          = 2(4x+8)
          P = 8x+16

$$\frac{\text{Perimeter of Rectangle}}{\text{Area of Rectangle}}$$

$$= \frac{8x+16}{3x^2+24x}, \quad x>0$$

b)   Area of triangle $= \frac{1}{2}$ bh

$$= \frac{1}{2}(3x+3)(3x)$$

$$A = \frac{1}{2}(9x^2+9x)$$

Perimeter of triangle $= S_1+S_2+S_3$
P=S+S+S$_3$
=2S+S$_3$
= 2(3x+2)+3x+2
= 6x+4+3x+2
P = 9x+6

$$\frac{\text{Perimeter of triangle}}{\text{Area of triangle}}$$

$$= \frac{9x+6}{\frac{1}{2}(9x^2+9x)}$$

$$= \frac{2(9x+6)}{9x^2+9x}$$

$$= \frac{6(3x+2)}{9x(x+1)}$$

$$= \frac{2(3x+2)}{3x(x+1)}, \quad x>0, \text{ (x is a positive number)}$$

In this case the triangle is Isosceles, so the Perimeter is equal to:
P=S+S+S$_3$

c)   Area of square = S²
     = (6x)²
     A = 36x²

Perimeter of square = 4S
     = 4(6x)
     = 24x

$$\frac{\text{Perimeter of square}}{\text{Area of square}}$$

$$= \frac{24x}{36x^2}, \quad x > 0$$

$$= \frac{2x}{3x}, \quad x > 0$$

Example 13) The expression $\dfrac{m}{3x+3}$ simplifies to $\dfrac{x-4}{3}$. What is the value of m?

Solution

$\dfrac{m}{3x+3}$ Simplifies to $\dfrac{x-4}{3}$ implies

$$\frac{m}{3x+3} = \frac{x-4}{3}$$

⇒   3m=(x-4)(3x+3)   Cross multiplication

⇒   3m=3x²-9x-12

⇒ $$m = \frac{3x^2 - 9x - 12}{3}$$

⇒ $$m = \frac{3\left(x^2 - 3x - 4\right)}{3}$$

m=x²-3x-4

Exercises 3.15

1)   Find the excluded values, if any of the following

a) $\dfrac{x^2 - 5x + 7}{3x - 2}$

b) $\dfrac{x^2 - 5}{x^2 + 8}$

c) $\dfrac{x^3 - 1}{x^3 + 1}$

d) $\dfrac{4x^2 - 12x}{2x^2 - 5x - 3}$

e) $\dfrac{x^2 + 7x + 10}{2x^3 - 8x}$

f) $\dfrac{y^2 - 6y + 9}{y^2 - 9}$

g) $\dfrac{2x^2 + 5x}{4x^2 + 8x - 5}$

2)   Simplify each rational expression to its simplest form.

a) $\dfrac{6x^2 + x - 2}{3x^2 + 2x}$

b) $\dfrac{3x^2 - 13x + 4}{3x^2 + 11x - 4}$

c) $\dfrac{6 + 7x - 3x^2}{x^2 - 2x - 3}$

d) $\dfrac{x^2 - 4}{x^2 - 4x + 4}$

e) $\dfrac{2x^3 + 3x^2}{2x^2 + 5x + 3}$

f) $\dfrac{h^2}{h^2 + 6h + 9}$

g) $\dfrac{3x^2 - 13x + 4}{x^2 - 16}$

3)   Write and simplify a rational expression for the ratio of the perimeter to the area of the given figures below.

a) Rectangle   Figure (b) Triangle

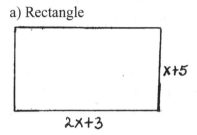

4)   The expression $\dfrac{x^2 - 4}{m}$ simplifies to $\dfrac{x+2}{x-1}$. What is the value of m?

3.22

A)   Multiplying Rational Expressions

The product of two rational expressions is the product of their numerators divided by the product of their denominators. Here is a step-by-step procedure for multiplying rational expressions:

Multiplying Rational Expressions

1.   Factor all numerators and denominators completely.
2.   Divide numerators and denominators by common factors.
3.   Multiply the remaining factors in the numerators and multiply the remaining factors in denominators.

Example 1)   Multiply and simplify: $\dfrac{x-2}{x-1} \cdot \dfrac{x^2-1}{2x-4}$

Solution

$\dfrac{x-2}{x-1} \cdot \dfrac{x^2-1}{2x-4}$ This is the given multiplication problem.

$= \dfrac{x-2}{x-1} \cdot \dfrac{(x+1)(x-1)}{2(x-2)}$ Factor as many numerators and denominators as possible. Because the denominator has factors of x-1 and x-2, x≠1 and x≠2.

$$= \frac{x-2}{x-1} \cdot \frac{(x+1)(x-1)}{2(x-2)}$$

Divide numerators and denominators by common factors.

$$= \frac{x+1}{2} , x \neq 1, x \neq 2$$

Multiply the remaining factors in the numerators and denominators.

3.23 B) Dividing Rational Expressions

The quotient of two rational expressions is the product of the first expression and the multiplicative inverse, or reciprocal, of the second expression. The reciprocal is found by interchanging the numerator and denominator. Thus,

We find the quotient of two rational expressions by inverting the divisor and multiplying.

Example 2) Divide and simplify: $\dfrac{x^2-2x-8}{x^2-9} \div \dfrac{x-4}{x+3}$

Solution

$$\frac{x^2-2x-8}{x^2-9} \div \frac{x-4}{x+3}$$

This is the given division problem.

$$= \frac{x^2-2x-8}{x^2-9} \cdot \frac{x+3}{x-4}$$

Invert the divisor and multiply.

$$= \frac{(x-4)(x+2)}{(x+3)(x-3)} \cdot \frac{(x+3)}{(x-4)}$$

Factor as many numerators and denominators as possible. For nonzero denominators, $x \neq -3$, $x \neq 3$, and $x \neq 4$.

$$= \frac{(x-4)(x+2)}{(x+3)(x-3)} \cdot \frac{(x+3)}{(x-4)}$$

Divide numerators and denominators by common factors.

$$= \frac{x+2}{x-3}, x \neq -3, x \neq 3, x \neq 4$$

Multiply the remaining factors in the numerators and denominators.

Example 3) Divide and simplify: $\dfrac{x^2-25}{2x-2} \div \dfrac{x^2+10x+25}{x^2+4x-5}$

Solution:

$$\frac{x^2-25}{2x-2} \div \frac{x^2+10x+25}{x^2+4x-5}$$

This is the given division problem.

$$= \frac{x^2 - 25}{2x - 2} \cdot \frac{x^2 + 4x - 5}{x^2 + 10x + 25}$$

Invert the divisor and multiply.

$$\frac{(x+5)(x-5)}{2(x-1)} \cdot \frac{(x-1)(x+5)}{(x+5)(x+5)}$$

Factor as many numerators and denominators as possible. For nonzero denominators, x≠1, and x≠-5.

$$= \frac{(x+5)(x-5)}{2(x-1)} \cdot \frac{(x-1)(x+5)}{(x+5)(x+5)}$$

Divide numerators and denominators by common factors.

$$= \frac{x-5}{2}, x \neq -5, x \neq 1$$

Multiply the remaining factors in the numerators and denominators.

Example 4) Divide and simplify: $\dfrac{x^3 - 25x}{4x^2} \cdot \dfrac{2x^2 - 2}{x^2 - 6x + 5} \div \dfrac{x^2 + 5x}{7x + 7}$

Solution:

$$\frac{x^3 - 25x}{4x^2} \cdot \frac{2x^2 - 2}{x^2 - 6x + 5} \div \frac{x^2 + 5x}{7x + 7}$$

This is the given division problem.

$$= \frac{x^3 - 25x}{4x^2} \cdot \frac{2x^2 - 2}{x^2 - 6x + 5} \cdot \frac{7x + 7}{x^2 + 5x}$$

Invert the divisor and multiply.

$$= \frac{x(x+5)(x-5)}{4x^2} \cdot \frac{2(x+1)(x-1)}{(x-1)(x-5)} \cdot \frac{7(x+1)}{x(x+5)}$$

Factor as many numerators and denominators as possible. For nonzero denominators, x≠0, x≠1, x≠5 and x≠-5.

$$= \frac{x(x+5)(x-5)}{4x^2} \cdot \frac{2(x+1)(x-1)}{(x-1)(x-5)} \cdot \frac{7(x+1)}{x(x+5)}$$

Divide numerator and denominators by common factors

$$= \frac{(x+1) \cdot 7(x+1)}{2x^2}$$

$$= \frac{7x^2 + 14x + 7}{2x^2}, x \neq 0, x \neq 1, x \neq 5, x \neq -5$$

Multiply the remaining factors in the numerators and denominators.

3.24 C. Addition and Subtraction of rational expressions

When rational expressions have the same denominator, we can add or subtract them by adding or subtracting the numerators and retaining the common denominator. If the denominators are not the same. We then find equivalent expressions with the same denominator and add.

Example 5)   Add $\dfrac{2x^2+5x-3}{x^2+8}+\dfrac{3x^2-2x+2}{x^2+8}$

Solution

$$\frac{2x^2+5x-3}{x^2+8}+\frac{3x^2-2x+2}{x^2+8}=\frac{5x^2+3x-1}{x^2+8}$$

In the following example, one denominator is the opposite of the other. We find a common denominator by multiplying by $-\frac{1}{-1}$.

Example 6)   Add: $\dfrac{2x^2+6}{x-y}+\dfrac{x^2-2}{y-x}$

Solution

$$\frac{2x^2+6}{x-y}+\frac{x^2-2}{y-x}=\frac{2x^2+6}{x-y}+\frac{-1}{-1}\cdot\frac{x^2-2}{y-x}$$

We multiply by 1 using $-\frac{1}{-1}$ to convert the second denominator to its opposite.

$$=\frac{2x^2+6}{x-y}+\frac{-1(x^2-2)}{-1(y-x)}$$

$$=\frac{2x^2+6}{x-y}+\frac{2-x^2}{x-y},\ -1(y-x)=-y+x=(x-y)$$

$$=\frac{x^2+8}{x-y}$$

Example 7)   Subtract $\dfrac{7x+4}{x^4+1}-\dfrac{3x+2}{x^4+1}$
Solution:

$$\frac{7x+4}{x^4+1}-\frac{3x+2}{x^4+1}=\frac{(7x+4)-(3x+2)}{x^4+1}=\frac{4x+2}{x^4+1}$$

When denominators are different, but not opposites, we find a common denominator by factoring the denominators. Then we multiply each term by 1 in such a way as to get the common denominator in each expression.

Example 8)   Add: $\dfrac{1}{2x}+\dfrac{2x}{x^2-1}+\dfrac{4}{x+1}$

Solution: We first find the least common multiple (LCM) of the denominators, also referred to as the least common denominator (LCD). To find the LCD, we first factor each denominator.

2x=2x

x²-1=(x+1)(x-1)   The LCD is 2x(x+1)(x-1).

x+1=x+1

We consider how often each factor occurs in each factorization. We make up a product using each factor the greatest number of times that it occurs - in each factorization. We use 2 as a factor once, x as a factor once, x+1 as a factor once even though it occurs in two of the factorizations, and x-1 as a factor once. The LCD is 2x(x+1)(x-1). Now we multiply each rational expression by 1- in such a way to get the LCD:

$$\dfrac{1}{2x}+\dfrac{2x}{x^2-1}+\dfrac{4}{x+1}$$

$$=\dfrac{1}{2x}\cdot\dfrac{(x+1)(x-1)}{(x+1)(x-1)}+\dfrac{2x}{(x+1)(x-1)}\cdot\dfrac{2x}{2x}+\dfrac{4}{x+1}\cdot\dfrac{2x(x-1)}{2x(x-1)}$$

$$=\dfrac{1(x+1)(x-1)}{2x(x+1)(x-1)}+\dfrac{2x(2x)}{(x+1)(x-1)(2x)}+\dfrac{4(2x)(x-1)}{(x+1)(2x)(x-1)}$$

$$=\dfrac{(x+1)(x-1)+4x^2+8x(x-1)}{2x(x+1)(x-1)}$$

$$=\dfrac{13x^2-8x-1}{2x(x+1)(x-1)}$$

Example 9)   Simplify: $\dfrac{2}{x^2+x-6}-\dfrac{1}{x^2-x-2}$

Solution:

$\dfrac{2}{x^2+x-6}-\dfrac{1}{x^2-x-2}$

$=\dfrac{2}{(x+3)(x-2)}-\dfrac{1}{(x-2)(x+1)}$ Factor each denominator.

$=\dfrac{2}{(x+3)(x-2)}-\dfrac{1}{(x-2)(x+1)}$ Find the least common denominator (LCD). The LCD = product of all different factors; each factor is raised to its highest exponents.

LCD=(x+3)(x-2)(x+1)

$=\dfrac{2(x+1)}{(x+3)(x-2)(x+1)}-\dfrac{1(x+3)}{(x-2)(x+1)(x+3)}$ Compare each denominator with the LCD and multiply the numerator and denominator by the missing factors (s).

$=\dfrac{2x+2-x-3}{(x+3)(x-2)(x+1)}$ Write the common numerator and add or subtract the numerators.

$=\dfrac{x-1}{(x+3)(x-2)(x+1)}$ Reduce the rational expression, if possible.

Example 10)   Add and simplify: $\dfrac{x-2}{x^2+4x+4}+\dfrac{x-1}{x^2+5x+6}$

Solution

$\dfrac{x-2}{x^2+4x+4}+\dfrac{x-1}{x^2+5x+6}$

$=\dfrac{x-2}{(x+2)^2}+\dfrac{x-1}{(x+2)(x+3)}$ Factoring each denominator.

The LCD is (x+2)²+(x+3)

$$= \frac{(x-2)(x+3)}{(x+2)^2(x+3)} + \frac{(x-1)(x+2)}{(x+2)^2(x+3)}$$

Comparing each denominator with LCD and multiplying the denominator and the denominator by the missing factor(s).

$$= \frac{(x-2)(x+3)(x-1)(x+2)}{(x+2)^2(x+3)} = \frac{2x^2+2x-8}{(x+2)^2(x+3)}$$

Adding the rational expressions with like denominators.

$$= \frac{2(x^2+x-4)}{(x+2)^2(x+3)}$$

Factoring the numerator.

Example 11) Write the sum: $\frac{1}{y^2+y} + \frac{2y}{y^2-1}$ in lowest terms.

Solution: Since $y^2+y=y(y+1)$ and $y^2-1=(y+1)(y-1)$, the least common denominator (LCD) is $y(y+1)(y-1)$. Thus, we proceed as follows:

$$\frac{1}{y^2+y} + \frac{2y}{y^2-1} = \frac{1}{y(y+1)} + \frac{2y}{(y+1)(y-1)}$$

Factoring the denominators

$$= \frac{1}{y(y+1)} \cdot \frac{y-1}{y-1} + \frac{2y}{(y+1)(y-1)} \cdot \frac{y}{y}$$

Obtain the common denominator.

$$= \frac{y-1}{y(y+1)(y-1)} + \frac{2y^2}{y(y+1)(y-1)}$$

$$\frac{2y^2+y-1}{y(y+1)(y-1)}$$

Adding numerators.

To write this expression in lowest terms, we must factor the numerator and divide out any common factors.

$$\frac{2y^2+y-1}{y(y+1)(y-1)} = \frac{(2y-1)(y+1)}{y(y+1)(y-1)}$$

$$= \frac{2y-1}{y(y-1)}$$

3.25 D. Complex Fractions

Some times the numerator and denominator are not polynomials but rational expressions. Such expressions can be simplified as discussed in the following examples:-

Example 12)   Simplify: $\dfrac{\dfrac{1}{x} - \dfrac{1}{y}}{\dfrac{1}{x} + \dfrac{1}{y}}$

Solution

Method 1:   We first simplify the fractions in the main numerator and the main denominator.

Main numerator: $\dfrac{1}{x} - \dfrac{1}{y} = \dfrac{y}{xy} - \dfrac{x}{xy}$

$$= \dfrac{y - x}{xy}$$

Main denominator: $\dfrac{1}{x} + \dfrac{1}{y} = \dfrac{y}{xy} + \dfrac{x}{xy}$

$$= \dfrac{y + x}{xy}$$

After replacing the main numerator and denominator with their simplified expressions, we invert and multiply.

$\dfrac{\dfrac{1}{x} - \dfrac{1}{y}}{\dfrac{1}{x} + \dfrac{1}{y}} = \dfrac{\dfrac{y - x}{xy}}{\dfrac{y + x}{xy}}$ Replacing the main numerator and denominator.

$= \dfrac{y - x}{xy} \cdot \dfrac{xy}{y + x}$ Inverting and multiplying

$= \dfrac{y - x}{y + x}$ Simplifying

Method 2:   We can clear all fractions in one step simply by multiplying the main numerator and main denominator by the least common multiple of all denominators in the expression. In this case, the least common multiple is xy, and so we proceed as follows:

$$\frac{\frac{1}{x}-\frac{1}{y}}{\frac{1}{x}+\frac{1}{y}} = \frac{\left(\frac{1}{x}-\frac{1}{y}\right)xy}{\left(\frac{1}{x}+\frac{1}{y}\right)xy} \qquad \text{Multiplying numerator and denominator by xy.}$$

Distributing

$$= \frac{\frac{1}{x}xy-\frac{1}{y}xy}{\frac{1}{x}xy+\frac{1}{y}xy}$$

Simplifying

$$= \frac{y-x}{y+x}$$

Example 13)   Simplify: $\dfrac{\dfrac{1}{(x+y)^2}-\dfrac{1}{x^2}}{2x+a}$.

Solution:

The least common multiple of the denominators $(x+a)^2$, and x^2 is $x^2(x+a)^2$

$$= \frac{\dfrac{1x^2(x+a)^2}{(x+a)^2}-\dfrac{1x^2(x+a)^2}{x^2}}{(2x+a)x^2(x+a)^2} = \frac{x^2-(x+a)^2}{x^2(2x+a)(x+a)^2}$$

We multiply each term in the numerator and the denominator by $x^2(x+a)^2$ and cancel the common factors.

$$= \frac{x^2-(x+a)^2}{x^2(2x+a)(x+a)^2} = \frac{x^2-(x^2+2ax+a^2)}{x^2(2x+a)(x+a)^2}$$

We simplify the numerator $(x+a)^2=x^2+2ax+a^2$

$$= \frac{-2ax - a^2}{x^2(2x+a)(x+a)^2}$$

$$= \frac{-a(2x+a)}{x^2(2x+a)(x+a)^2}$$

$$= \frac{-a(2x+a)}{x^2(2x+a)(x+a)^2}$$

$$= \frac{-a}{x^2(x+a)^2}$$ Reducing the rational expression to the simplest form.

Example 14)   Simplify: $\dfrac{x^{-3} - y^{-3}}{x^{-1} - y^{-1}}$

     Solution   We first note that

$$\frac{x^{-3} - y^{-3}}{x^{-1} - y^{-1}} = \frac{\dfrac{1}{x^3} - \dfrac{1}{y^3}}{\dfrac{1}{x} - \dfrac{1}{y}}$$

Method 1.   The denominators within the complex rational expression are x, y, x^3 and y^3. The LCM of these expressions is x^3y^3. We multiply by 1 using $x^3y^3 \Big/ x^3y^3$:

$$\frac{x^{-3} - y^{-3}}{x^{-1} - y^{-1}} = \frac{\dfrac{1}{x^3} - \dfrac{1}{y^3}}{\dfrac{1}{x} - \dfrac{1}{y}} = \frac{\dfrac{1}{x^3} - \dfrac{1}{y^3}}{\dfrac{1}{x} - \dfrac{1}{y}} \cdot \frac{x^3y^3}{x^3y^3} = \frac{\left(\dfrac{1}{x^3} - \dfrac{1}{y^3}\right)x^3y^3}{\left(\dfrac{1}{x} - \dfrac{1}{y}\right)x^3y^3}$$

Then
$$= \frac{\dfrac{1}{x^3}(x^3y^3) - \dfrac{1}{y^3}(x^3y^3)}{\dfrac{1}{x}(x^3y^3) - \dfrac{1}{y}(x^3y^3)} = \frac{y^3 - x^3}{x^2y^3 - x^3y^2}$$

$$= \frac{(y-x)(y^2 + xy + x^2)}{x^2y^2(y-x)}$$

$$= \frac{(y^2 + xy + x^2)}{x^2y^2}.$$

Method 2. We carry out the subtractions in the numerator and the denominator separately to obtain a single rational expression for both the numerator and the denominator. Then we divide:

$$\frac{x^{-3}-y^{-3}}{x^{-1}-y^{-1}}=\frac{\dfrac{1}{x^3}-\dfrac{1}{y^3}}{\dfrac{1}{x}-\dfrac{1}{y}}=\frac{\dfrac{1}{x^3}\cdot\dfrac{y^3}{y^3}-\dfrac{1}{y^3}\cdot\dfrac{x^3}{x^3}}{\dfrac{1}{x}\cdot\dfrac{y}{y}-\dfrac{1}{y}\cdot\dfrac{x}{x}}=\frac{\dfrac{y^3}{x^3y^3}-\dfrac{x^3}{x^3y^3}}{\dfrac{y}{xy}-\dfrac{x}{xy}}=\frac{\dfrac{y^3-x^3}{x^3y^3}}{\dfrac{y-x}{xy}}$$

$$=\frac{y^3-x^3}{x^3y^3}\cdot\frac{xy}{y-x}=\frac{(y-x)(y^2+xy+x^2)xy}{xy(x^2y^2)(y-x)}$$

$$=\frac{y^2+xy+x^2}{x^2y^2}$$

Example 15) Simplify $\dfrac{\dfrac{1}{a+b}-\dfrac{1}{a}}{b}$.

Solution: We will use the method of multiplying each of the three terms, $\dfrac{1}{a+b}$, $\dfrac{1}{a}$ and b by the least common denominator. The least common denominator is a(a+b).

$$\dfrac{\dfrac{1}{a+b}-\dfrac{1}{a}}{b}$$

$$=\frac{\left(\dfrac{1}{a+b}-\dfrac{1}{a}\right)a(a+b)}{ba(a+b)}$$ Multiply the numerator and denominator by a(a+b), b≠0, a≠0, a≠-b.

$$=\frac{\dfrac{1}{a+b}\cdot a(a+b)-\dfrac{1}{a}\cdot a(a+b)}{ba(a+b)}$$ Use the distributive property in the numerator.

$$= \frac{a-(a+b)}{ba(a+b)} \quad \text{Simplify:} \quad \frac{1}{a+b}a(a+b)=a \quad \text{and}$$

$$\frac{1}{a} \cdot a(a+b)=a+b$$

$$= \frac{a-(a-b)}{ba(a+b)} \quad \text{Subtract in the numerator.}$$

$$= \frac{-b}{ba(a+b)} \quad \text{Simplify: a-a-b=-b.}$$

$$= -\frac{1}{a(a+b)} \quad , b \neq 0, \text{ a} \neq 0, \text{ a} \neq \text{-b} \quad \text{Divide the numerator and}$$
denominator by b.

3.26 E.) Evaluating a Rational Expression

To evaluate a rational expression $\dfrac{p(x)}{q(x)}$, for x=a, we substitute a for x in the expression and simplify.

Example 16)   Evaluate $\dfrac{x^2-6x+3}{x+5}$ for;

   a) x=6
   b) x=2

Solutions:

a) For $x=6$: $\dfrac{x^2-6x+3}{x+5}=\dfrac{(6)^2-6(6)+3}{6+5}=\dfrac{36-36+5}{11}=\dfrac{5}{11}$.

b) For $x=2$: $\dfrac{x^2-6x+3}{x+5}=\dfrac{(2)^2-6(2)+3}{2+5}=\dfrac{4-12+3}{7}=\dfrac{-5}{7}$

Example 17)   Evaluate $\dfrac{2x-1}{30-x}$ for
   a) -3
   b) 30

Solutions:
a) For $x=-3$: $\dfrac{2x-1}{30-x}=\dfrac{2(-3)-1}{30-(-3)}=\dfrac{-6-1}{30+3}=\dfrac{-7}{33}$

b) For $x=30$: $\dfrac{2x-1}{30-x}=\dfrac{2(30)-1}{30-30}=\dfrac{60-1}{0}=\dfrac{59}{0}$

This is undefined because we cannot divide by zero.

Note:- When on substituting a for x in a rational expression $\dfrac{p(x)}{q(x)}$, the result is of the form $\dfrac{k}{0}$, where k≠0, we say that the expression.

$\dfrac{p(x)}{q(x)}$ is not defined (or is undefined) for x=a.

This happens when p(a)≠0 but q(a)=0.

If p(a) and q(a) are both zero, then $\dfrac{p(x)}{q(x)}$ is said to be indeterminate at x=a

Example 18)   Evaluate $\dfrac{x^2-2x-3}{2x-6}$ for x=3

Solution:

For x=3

$$\frac{x^2-2x-3}{2x-6}=\frac{(3)^2-2(3)-3}{2(3)-6}=\frac{9-6-3}{6-6}=\frac{9-9}{6-6}=\frac{0}{0}$$

Thus; $\dfrac{x^2-2x-3}{2x-6}$ is indeterminate for x=3.

Example 19)   Evaluate $\dfrac{x^2-4}{2x-4}$ for x=2

Solution:

For x=2

$$\frac{x^2-4}{2x-4}=\frac{(2)^2-4}{2(2)-4}=\frac{4-4}{4-4}=\frac{0}{0}$$

We say that $\dfrac{x^2-4}{2x-4}$ is indeterminate for x=2

Example 20)   Approximate the value of $\dfrac{6x^2-7x+9}{2x^3+8x^2+1}$ when x is very large or x→∞.

Solution: By factoring out the variable part of the leading terms from the numerator and the denominator gives:

$$\frac{x^2\left(6-\dfrac{7}{x}+\dfrac{9}{x^2}\right)}{x^3\left(2+\dfrac{8}{x}+\dfrac{1}{x^3}\right)}=\frac{6-\dfrac{7}{x}+\dfrac{9}{x^2}}{x\left(2+\dfrac{8}{x}+\dfrac{1}{x^3}\right)}=\left(\dfrac{1}{x}\right)\frac{6-\dfrac{7}{x}+\dfrac{9}{x^2}}{2+\dfrac{8}{x}+\dfrac{1}{x^3}}$$

Since x is large (x→∞), so $\dfrac{1}{x}$, $\dfrac{1}{x^2}$, and $\dfrac{1}{x^3}$ are approximately zero. This gives

$$\left(\frac{1}{x}\right)\frac{6-\dfrac{7}{x}+\dfrac{9}{x^2}}{2+\dfrac{8}{x}+\dfrac{1}{x^3}}=(0)\frac{6-0+0}{2+0+0}=(0)\left(\frac{6}{2}\right)=0$$

Thus for large x-values the value of a rational expression $\dfrac{p(x)}{q(x)}$

, where the degree of p(x)<degree of q(x), is approximately equal to zero.

EXERCISES 3.16

1) Multiply and simplify each of the following rational expression.

a) $\dfrac{x^3-8}{x^2-4}\cdot\dfrac{x+2}{3x}$

b) $\dfrac{3x+6}{x-4}\cdot\dfrac{x-5}{4x-6}$

c) $\dfrac{x^2-2x+1}{2x+2}\cdot\dfrac{x^2-1}{x^2-4}$

d) $\dfrac{x^2-9}{x^2}\cdot\dfrac{x^2-3x}{x^2+x-12}$

e) $\dfrac{x^2+5x+6}{x^2+x-6}\cdot\dfrac{x^2-9}{x^2-x-6}$

2) Divide and simplify each of the following rational expression.

a) $\dfrac{x^3-3x}{3x}\cdot\dfrac{2x^2-2}{x^2-6x+5}\div\dfrac{x^2+5x}{7x+7}$

b) $\dfrac{4x^2+10}{x-3}\div\dfrac{6x^2+15}{x^2-9}$

c) $\dfrac{x^2+x}{x^2+4x+4}\div\dfrac{x^2-1}{x^2-4}$

d) $\dfrac{x^2-4}{x-2}\div\dfrac{x+2}{4(x-1)}$

e) $\dfrac{x^2+x-12}{x^2+x-30}\cdot\dfrac{x^2+5x+6}{x^2-2x-3}\div\dfrac{x+3}{x^2+7x+6}$

3) Add and simplify each of the following rational expression

a) $\dfrac{2x}{x+2}+\dfrac{x+2}{x-2}+\dfrac{x-1}{x-2}$

b) $\dfrac{x^2-1}{x^2-5x+4}+\dfrac{x-1}{x-4}+\dfrac{x^2-1}{x-4}$

c) $\dfrac{2x-1}{5x+2}+\dfrac{4x}{16x^2-4}$

d) $\dfrac{4x+1}{6x+5}+\dfrac{8x+9}{6x+5}$

e) $\dfrac{3x}{x^2+3x-10}+\dfrac{2x}{x^2+x-6}$

4) Subtract and simplify each of the following rational expression.

a) $\dfrac{5x}{x-5}-\dfrac{4x}{x-5}$

b) $\dfrac{4x^2 + x - 6}{x^2 + 3x + 2} - \dfrac{3x}{x+1}$

c) $\dfrac{x}{x^2 - 2x - 24} - \dfrac{x}{x^2 - 7x + 6}$

d) $\dfrac{3x+1}{x^2 - 8x + 16} - \dfrac{2x}{x-4}$

e) $\dfrac{5}{x-5} - \dfrac{4}{x^2 - 5x} - \dfrac{5x+2}{x^2}$

5)　　Simplify each complex rational expression

a) $\dfrac{\dfrac{3}{x-2} - \dfrac{4}{x+2}}{\dfrac{7}{x^2 - 4}}$

b) $\dfrac{\dfrac{x}{x-2} + 1}{\dfrac{3}{x^2 - 4} + 1}$

c) $\dfrac{1 - \dfrac{8}{x} + \dfrac{16}{x^2}}{1 - \dfrac{5}{x} + \dfrac{4}{x^2}}$

d) $\dfrac{\dfrac{2}{x-1} + \dfrac{1}{x+3}}{\dfrac{3x+5}{x-1}}$

e) $\dfrac{\dfrac{x}{3} - 1}{x - 3}$

f) $\dfrac{x-3}{x-\dfrac{3}{x-2}}$

6)   Evaluate each of the following rational expression,

a) $\dfrac{x^2-4x+1}{x^4+2x+1}$, for x=-2

b) $\dfrac{2x^2+5x+3}{3x^2-2x+1}$, for $x=\dfrac{1}{2}$

c) $\dfrac{x^2+6x+5}{2x^2+5x+3}$, for x=-1

d) $\dfrac{x^3-4x+6}{x^3+8}$, for x=-2

e) $\dfrac{5x+8}{5x^2+6x+2}$, for x=3

3.27A) Rational Equations

An equation that contains rational expression is called a Rational Equation.

More about Rational Equation
- Rational Expression: A rational expression is an expression of the form $P\!\!\Big/\!\!_q$ where p and q are nonzero polynomials.
- Rational Function: A function written, as a quotient of polynomials is a rational function.

That is, if p(x) and q(x) are polynomial functions and q(x)≠0,

then $f(x)=\dfrac{p(x)}{q(x)}$ is called rational function.

Examples of Rational Equation

- The following are examples of rational equation.

$$\frac{3x}{4}+\frac{2}{x+3}=\frac{1}{2}, \quad \frac{5x+3}{2}=\frac{3x+1}{x-1}, \quad \frac{3x+2}{x^2-4}=\frac{x}{x-3}$$

Finding the solutions of a rational equations often involves multiplying by expressions with variables, as in the following example. We multiply by the LCM of all the denominators.

Example 1)　　Solve $\frac{x-3}{x-5}=\frac{5}{x-5}$

Solution: The LCM of the denominators is x-5. We multiply by LCM:

$$(x-5)\cdot\frac{x-3}{x-5}=(x-5)\cdot\frac{5}{(x-5)} \quad \text{Multiplying by x-5}$$

$$= \text{x-3=5} \quad\quad \text{Simplifying}$$

x=8

The possible solution is 8. We check

$$\frac{x-3}{x-5}=\frac{5}{x-5}$$

$$\frac{8-3}{8-5}=\frac{5}{8-5}$$

$$\frac{5}{3}=\frac{5}{3}$$

The solution set is {8}

Example 2)　　Solve: $\frac{x^2}{x-2}=\frac{4}{x-2}$

Solution: The LCM of the denominators is x-2. We multiply by LCM:

$$(x-2)\cdot\frac{x^2}{x-2}=(x-2)\cdot\frac{4}{x-2} \quad \text{Multiplying by x-2}$$

$x^2=4$ Simplifying

$x^2-4=0$

$(x+2)(x-2)=0$

$x+2=0$ or $x-2=0$ Principle of zero products.

$x=-2$ or $x=2$.

The possible solutions are 2 and -2. We must check, since we have multiplied by an expression with a variable.

for x=2   For x=-2

$$\frac{x^2}{x-2}=\frac{4}{x-2} \qquad\qquad \frac{x^2}{x-2}=\frac{4}{x-2}$$

$$\frac{2^2}{2-2}=\frac{4}{2-2} \qquad\qquad \frac{(-2)^2}{-2-2}=\frac{4}{-2-2}$$

$$\frac{4}{0}=\frac{4}{0} \qquad\qquad\qquad \frac{4}{-4}=\frac{4}{-4}$$

We can not divide by zero, so the solution set is {-2}

Example 3)   Solve $\dfrac{x}{x-3}=\dfrac{3}{x-3}+9$.

Solution: We must avoid any values of the variable x that make a denominator zero.

$$\frac{x}{x-3}=\frac{3}{x-3}+9, x\neq 3$$

> This denominators are zero if x=3.

We see that x cannot equal 3. With denominators of x-3, x-3, and 1, the least common denominator is x-3. We multiply both sides of the equation by x-3. We also write the restriction that x cannot equal 3 to the right of the equation.

$$\frac{x}{x-3}=\frac{3}{x-3}+9, x\neq 3$$
                                This is the given equation.

$$(x-3)\cdot\frac{x}{x-3}=(x-3)\left(\frac{3}{x-3}+9\right)$$

Multiply both sides by x-3.

$$(x-3)\cdot\frac{x}{(x-3)}=(x-3)\cdot\frac{3}{(x-3)}+(x-3)\cdot9$$

Use the distributive property.

$$(x-3)\cdot\frac{x}{(x-3)}=(x-3)\cdot\frac{3}{(x-3)}+9(x-3)$$

Divide out common factors in two of the multiplications.

x=3+9(x-3)   Simplify.

The resulting equation is cleared of fractions. We now solve for x.

x=3+9x-27   Use the distributive property.
x=9x-24   Combine numerical terms.
x-9x=9x-24-9x   Subtract 9x from both sides.
-8x=-24   Simplify.

$$\frac{-8x}{8}=\frac{-24}{-8}$$ Divide both sides by -8.

x=3   Simplify.

The proposed solution, 3, is not a solution, because of the restriction that x≠3. There is no solution to this equation. The solution set for this equation contains no elements. The solution set is { }, the empty set.

Example 4)   Solve: $\dfrac{x}{x-2}=\dfrac{2}{x-2}-\dfrac{2}{3}$, for x ≠ 2

Solution: The LCM of the denominators is 3(x-2). We multiply both sides of the equations by the LCM.

Multiply by both sides by 3(x-2)

$$3(x-2)\cdot\frac{x}{x-2}=3(x-2)\cdot\frac{2}{x-2}-\frac{2}{3}\cdot3(x-2)$$

3x=6-(2x-4)

3x=6-2x+4   Use the distributive property.
3x+2x=10
5x=10 Divide both sides by 5.

$$x = \frac{10}{5}$$

x=2

Since the domain is x ≠ 2, the solution set is empty set.

The general procedure for solving rational equations involves multiplying on both sides by the LCM of all the denominators. This procedure is called clearing of fractions.

Example 5)   Solve: $\dfrac{14}{x+2} - \dfrac{1}{x-4}$

Solution: We note at the outset that -2 and 4 are not meaningful replacements. We multiply by the LCM of all the denominators: (x+2)(x-4).

$$(x+2)(x-4)\cdot\left[\frac{14}{x+2} - \frac{1}{x-4}\right] = (x+2)(x-4)\cdot 1$$

Using the distributive law.

$$(x+2)(x-4)\cdot\frac{14}{x+2} - (x+2)(x-4)\cdot\frac{1}{x-4} = (x+2)(x-4)\cdot 1$$

14(x-4)-(x+2)=(x+2)(x-4)   Simplifying
14x-56-x-2=x²-2x-8
13x-58=x²-2x-8
0=x²-15x+50
x²-15x+50=0
(x-10)(x-5)=0
x-10=0 or x-5=0   Principle of zero products.
x=10 or x=5

The possible solutions are 10 and 5. These check, so the solution set is {10, 5}.

Example 6)   Solve: $\dfrac{x+2}{2} + \dfrac{3x+1}{5} = \dfrac{x-2}{4}$

Solution: The LCM of the denominator is 20. We multiply both sides of the equation by the LCM.

$$20 \cdot \dfrac{x+2}{2} + 20 \cdot \dfrac{3x+1}{5} = 20 \cdot \dfrac{x-2}{4}$$

10(x+2)+4(3x+1)=5(x-2)
10x+20+12x+4=5x-10

10x+12x-5x=-10-20-4	Collect like terms
17x=-34	Divide both sides by 17
x=-2	Simplify.

The solution set is {-2.}

3.28B) Equivalent Equations
    Equations that have the same solution set are called Equivalent equations.

Example 1)   Determine whether the equations 2x=6, 2x+4=10, and -5x=-15 are equivalent.

Solution

2x=6	2x+4=10	-5x=-15
x=3	x=3	x=3
Solution set: {3}	Solution set: {3}	Solution set: {3}

Each equation has the same solution set as the others, {3}. Thus all three equations are equivalent.

Example 2)   Determine whether the equations x=2 and x²=4 are equivalent.

Solution:

x=2   x²=4

Solution set: {2}   Solution set: {-2, 2}

The solution sets are not the same, so the equations are not equivalent.

Example 3)   Determine whether the equations 5x=8x and $\frac{5}{x} = \frac{8}{x}$ are equivalent.

Solution:

5x=8x $$\frac{5}{x} = \frac{8}{x}$$

Solution set: {0}   The solution set is φ, the empty set (no solution, since division by 0 is not defined).

The empty set φ and the set containing the number 0, {0}. There is one element in {0}, the number 0. The solution sets are not the same, so the equations are not equivalent.

Exercise 3.17

1.   Solve each of the following

a) $\dfrac{3x}{x+2} + \dfrac{6}{x} = \dfrac{12}{x^2 + 2x}$

b) $\dfrac{3}{x+2} + \dfrac{2}{x-2} = \dfrac{4x-4}{x^2-4}$

c) $\dfrac{4}{y} - \dfrac{4}{y-6} = \dfrac{24}{6y-y^2}$

d) $\dfrac{1}{x} + \dfrac{1}{2x} + \dfrac{1}{3x} = 5$

e) $\dfrac{5t}{t-4} - \dfrac{20}{t} = \dfrac{80}{t^2-4t}$

f) $\dfrac{y^2 - 1}{3} = \dfrac{y - 1}{4}$

2.   Determine whether the equations of the pair are equivalent.

a)   x=4

       x²=16

b)   3y+5=6

       12y+6=42

c) $\dfrac{(x-3)(x+9)}{x-3} = x+9$

       x+9=x+9

d)   x²=-4x

       x=-4

e)   2x+1=-3,

       8x+4=-12

3.29 Average Rate of Change

Definition: A function is a process by which every input is associated with exactly one output. When create a process (or series of steps) to do a certain task we are often creating a function. If we want to use it over and over again then to make our lives easier we give it a name. It helps us remember the name when it has something to do with the process that is being described.

The average rate of change function describes the average rate at which one quantity is changing with respect to something else changing.

You are already familiar with some average rate of change calculations:

(a) Mile per gallon- calculated by dividing the number of miles by the number of gallons used.

(b) Cost per killowatt- Calculated by dividing the cost of electricity by the number of killowatts used.

(c) Miles per hour- Calculated by dividing the number of miles traveled by the number of hours it takes to travel them.

In general, an average rate, a change of function is a process that calculates the amount of change in one item divided by the corresponding amount of change in another. Using function notation, we can define the average rate of change of a function f from a to x as

$$A(x) = \frac{f(x) - f(a)}{x - a}$$

- A is the name of this average rate of change function.
- x-a, represents the change in the input of the function f.
- f(x)-f(a) represents the change in the function f as the input changes from a to x.

You might have noticed that the average rate of change function looks a lot like the formula for the slope of a line. In fact, if you take any two distinct point on a curve, (x_1, y_1) and (x_2, y_2), the slope of the line connecting the points will be the average rate of change from x_1 to x_2.

Example 1: Find the slope of the line going through the curve $f(x) = \frac{1}{4}x^2 - 2$ as x changes from 4 to 0.

Solution:
   Step 1: f(4)=2 and f(0)=-2

   Step 2: use the slope formula to create the ratio.
   $$\frac{f(0) - f(4)}{0 - 4}$$

   Step 3: Simplify
   $$\frac{f(0) - f(4)}{0 - 4} = \frac{-2 - 2}{0 - 4} = 1$$

Step 4: so the slope of the line going through the curve $f(x) = \frac{1}{4}x^2 - 2$ as x changes from 4 to 0 is 1.

Example 2)   Find the average rate of change of $f(x) = \frac{1}{4}x^2 - 2$ from 4 to 0.

Solution:- Since the average rate of change of a function is the slope of the associated line we have already done the work in the last problem.

That is, the average rate of change of $f(x) = \frac{1}{4}x^2 - 2$ from 4 to 0 is 1.

That is, over the interval [0, 4], for every 1 unit change in x, there is a 1 unit change in the value of the function.

Here is a graph of the function, the two points used, and the line connecting those two points.

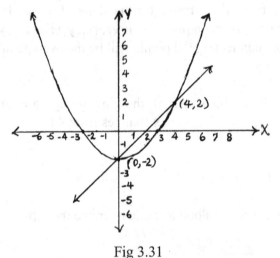

Fig 3.31

Now suppose you need to find series of slopes of lines that go through the curve and the point (4, f(4)) but the other point keeps moving. We will call the second point (x, f(x)). It will be useful to have a process (function) that will do just that for us. The average rate of change function also determines slope so that process is what we will use.

Example 3) Find the average rate of change of $f(x)=\dfrac{1}{4}x^2-2$ - from
4 to x.

Solution

Step 1: f(4)=2 and $f(x)=\dfrac{1}{4}x^2-2$.
Step 2: Use the average rate of change formula to define A(x) and simplify.

$$A(x)=\frac{f(x)-f(4)}{x-4}$$

$$A(x)=\frac{\left(\dfrac{1}{4}x^2-2\right)-2}{x-4}$$

$$=\frac{\dfrac{1}{4}x^2-4}{x-4}$$

$$=\frac{x^2-16}{4(x-4)}$$

$$A(x)=\frac{(x+4)(x-4)}{4(x-4)},x\neq 4$$

$$A(x)=\frac{x+4}{4},x\neq 4$$

Example 4) Use the result of example 3 to find the average rate of change of $f(x)=\dfrac{1}{4}x^2-2$ from 4 to 7.

Solution:- The average rate function of change of $f(x)=\dfrac{1}{4}x^2-2$ from

4 to x is $A(x)=\dfrac{x+4}{4}$, x≠4

So, the average rate of change of $f(x)=\dfrac{1}{4}x^2-2$ from 4 to 7 is

$$A(7)=\frac{7+4}{4}=11\big/\!\!\big/_{4}$$

Example 5) Use the result of example 3 to find the average rate of change of $f(x) = \frac{1}{4}x^2 - 2$ from 4 to 0.

Solution:- The average rate of change of $f(x) = \frac{1}{4}x^2 - 2$ from 4 to 0 is

$$A(0) = \frac{0+4}{4} = 1$$

Exercise 3.18

1) Find the average rate of change of $f(x) = \frac{1}{8}x^3 - 2x$ from x_1 to x_2.
   a) $x_1 = 5, x_2 = 1$
   b) $x_1 = 0, x_2 = 3$

2) Find the average rate of change of $g(x) = x^2 - 1$ from x_1 to x_2.

   a) $x_1 = 4, x_2 = \frac{1}{2}$

   b) $x_1 = -2, x_2 = 0$

3.30 Number Sequences
    In mathematics, the term number sequence is a list of numbers arranged in an ordered. Consider the following examples:-

A: 2, 4, 6, 8, 10
B: -1, 2, 5, 8, 11, ...

C: 1, $\frac{1}{3}$, $\frac{1}{9}$, $\frac{1}{27}$, ...

A   is a sequence of numbers, because it has a pattern. Each number is obtained by adding 2 to the preceding number.

B   is a sequence of numbers, because it has a pattern. Each number is obtained by adding 3 to the preceding number.

C   is a sequence of numbers, because it has a pattern. Each number is obtained multiplying by $\frac{1}{3}$ to the preceding number.

The numbers that make up a sequence are called terms. For example, the terms of sequence A are 2, 4, 6, 8, and 10. Sequences like A that have finite numbers of terms are called finite sequence. Where as sequences B and C that have infinite numbers of terms are called infinite sequences. Terms of a sequence are named as, the first term, the second term, the third term, ..., as a_1, a_2, a_3, a_4, etc. In general n^{th} term of a sequence will be named as a_n.

Example 1) Find the first five terms of each of the following sequence whose n^{th} term, or general term, is given:

a) $a_n = 2n-1$

b) $a_n = \dfrac{1}{2^{n+1}}$

solution:

a) $a_n = 2n-1$
$a_1 = 2\times1-1=1$
$a_2 = 2\cdot1-1=4-1=3$
$a_3 = 2\cdot3-1=6-1=5$
$a_4 = 2\cdot4-1=8-1=7$
$a_5 = 2\cdot5-1=10-1=9$

so, the first five terms are 1, 3, 5, 7, 9

b) $a_n = \dfrac{1}{2^{n+1}}$

$a_1 = \dfrac{1}{2^{1+1}} = \dfrac{1}{2^2} = \dfrac{1}{4}$

$a_2 = \dfrac{1}{2^{2+1}} = \dfrac{1}{2^3} = \dfrac{1}{8}$

$a_3 = \dfrac{1}{2^{3+1}} = \dfrac{1}{2^4} = \dfrac{1}{16}$

$a_4 = \dfrac{1}{2^{4+1}} = \dfrac{1}{2^5} = \dfrac{1}{32}$

$a_5 = \dfrac{1}{2^{5+1}} = \dfrac{1}{2^6} = \dfrac{1}{64}$

So, the first five terms of a sequence are:- $\dfrac{1}{4}$, $\dfrac{1}{8}$, $\dfrac{1}{16}$, $\dfrac{1}{32}$, $\dfrac{1}{64}$

Example 2) Write the first five terms of the sequence with the given n^{th} term.

a) $fn = \dfrac{2n}{n+1}$

b) $a_n = \dfrac{(-1)^n}{3^n - 1}$

solution:-

a) $fn = \dfrac{2n}{n+1}$

In fn, we substitute n=1, n=2, n=3, n=4 and n=5

$$f1 = \dfrac{2 \cdot 1}{1+1} = 1$$

$$f2 = \dfrac{2 \cdot 2}{2+1} = \dfrac{4}{3}$$

$$f3 = \dfrac{2 \cdot 3}{3+1} = \dfrac{6}{4} = \dfrac{3}{4}$$

$$f4 = \dfrac{2 \cdot 4}{4+1} = \dfrac{8}{5}$$

$$f5 = \dfrac{2 \cdot 5}{5+1} = \dfrac{10}{6} = \dfrac{5}{3}$$

b) $a_n = \dfrac{(-1)^n}{3^n - 1}$

$$a_1 = \frac{(-1)^1}{3^1-1} = \frac{-1}{3-1} = \frac{-1}{2}$$

$$a_2 = \frac{(-1)^2}{3^2-1} = \frac{-1}{9-1} = \frac{1}{8}$$

$$a_3 = \frac{(-1)^3}{3^3-1} = \frac{-1}{27-1} = \frac{-1}{26}$$

$$a_4 = \frac{(-1)^4}{3^4-1} = \frac{1}{81-1} = \frac{1}{80}$$

$$a_5 = \frac{(-1)^5}{3^5-1} = \frac{-1}{243-1} = \frac{-1}{242}$$

3.31 SEQUENCE AS A FUNCTION
3.31.1 Definition of sequence
A sequence is a function whose domain is the set of positive integers.

1)　Domain of a finite sequence is the set of positive integers: {1, 2, 3, ..., n} for some positive integer n.

2)　Domain of an infinite sequence is the set of positive integers: {1, 2, 3, 4, ...}

Example 3)　Find the nth term of the following sequence
　　a) 2, 4, 6, 8, ...

　　b) $1, \dfrac{1}{2}, \dfrac{1}{4}, \dfrac{1}{8}, ...$

　　c) -3, -1, 1, 3, 5, 7, ...

Solution

a)　2, 4, 6, 8, ...

If we see the pattern, each term of the sequence is obtained by multiplying the preceding term by 2. That is, $a_1=2(1)$, $a_2=2(2)$, $a_3=2(3)$, $a_4=2(4)$, and so on. Thus, we would then write the sequence with general term as $a_n=2n$.

b) $1, \dfrac{1}{2}, \dfrac{1}{4}, \dfrac{1}{8}, \cdots$

If we see the patterns, each denominator is a power of 2. Thus, we can rewrite the sequence as

$$\dfrac{1}{2^0}, \dfrac{1}{2^1}, \dfrac{1}{2^2}, \dfrac{1}{2^3}, \dfrac{1}{2^4}, \cdots$$

Since the power of 2 in the denominator is one less than the position of the term. The nth term is:- $a_n = \dfrac{1}{2^{n-1}}$

c)   -3, -1, 1, 3, 5, 7, ...

As we see the pattern, we obtain that the difference between each two successive terms is 2. Hence, these terms can be rewritten as:-

$a_1 = -3 = -3+2(0) = -3+2(1-1)$
$a_2 = -1 = -3+2(1) = -3+2(2-1)$
$a_3 = 1 = -3+2(2) = -3+2(3-1)$
$a_4 = 3 = -3+2(3) = -3+2(4-1)$

Thus,
$a_1 = -3+2(1-1)$
$a_2 = -3+2(2-1)$
$a_3 = -3+2(3-1)$
$a_4 = -3+2(4-1)$
$\vdots \quad \vdots$
$a_n = -3+2(n-1)$

Hence, the nth term is $a_n = -3+2(n-1)$

We can rewrite the sequence in the form of function as:-
$\qquad f(n) = -3+2(n-1).$
$\qquad f(n) = 2n-5.$

3.32 Arithmetic Sequence (Arithmetic Progression)

A sequence in which each term after the first differs from the preceding term by a constant number is called an arithmetic sequence or arithmetic progression. The difference between consecutive terms is called the common difference of the sequence. The common difference is usually represented by the letter d.

The common difference, d, is obtained by subtracting any term from the term that directly follows it. consider the following sequence:-

$$a_1, a_2, a_3, a_4, a_5, ..., a_n, ...$$

Thus, the above sequence is an arithmetic sequence if $d=(a_2-a_1)=(a_3-a_2)=(a_4-a_3)=...=(a_{n+1}-a_n)=...$

Examples

a)   2, 5, 8, 11, 14, ... is an arithmetic sequence.
$d=5-2=8-5=11-8=...$
The common difference is 3

b)   -3, 3, 9, 15, 21, ... is an arithmetic sequence.
$d=3-(-3), 9-3, 15-9, ...$
The common difference is 6.

c)   -8, -6, -4, -2, 0, 2, 4, ... is an arithmetic sequence
$d=-6-(-8), -4-(-6), -2-(-4), ...$
The common difference is 2.

3.33 The General Term of an Arithmetic Sequence

Consider an arithmetic sequence that have first term a_1 and common difference d. It is possible to find a formula for the general term, a_n. Let's begin by writing the first five terms. The first term is a_1. The second term is a_1+d. The third term is a_1+2d, the fourth term is a_1+3d. and fifth term is a_1+4d Thus, we start with a_1 and add d to each successive term. The first five terms are written below:-

$a_1,$

$a_2 = a_1 + d$

$a_3 = a_2 + d = (a_1 + d) + d = a_1 + 2d$

$a_4 = a_3 + d = (a_1 + 2d) + d = a_1 + 3d$

$a_5 = a_4 + d = (a_1 + 3d) + d = a_1 + 4d$

As we see above the coefficient of d in each case is 1 less than the number of the term, n. or the subscript of a denoting the term number.

Generalizing we obtain the following.

The nth term of an arithmetic sequence with first term a_1 and common difference d is

$$a_n = a_1 + (n-1)d, \text{ for } n \geq 1.$$

Example 4)   Find the 13th term of the arithmetic sequence 3, 8, 13, ...

Solution:- In this case $a_1 = 3$, $d = 5$, and $n = 13$, then using the formula $a_n = a_1 + (n-1)d$, we can find

$\qquad a_{13} = 3 + (13-1) \cdot 5$
$\qquad = 3 + 12 \cdot 5$
$\qquad a_{13} = 63$

Therefore, the 13th term is 63

Example 5)   Find the 100th term of the arithmetic sequence 7, 16, 25, 34, ...

Solution: In this case, $a_1 = 7$, $d = 9$ and $n = 100$, using the formula $a_n = a_1 + (n-1) \cdot d$, we can find the required term.

Thus,

$\qquad a_n = a_1 + (n-1)d$

$a_{100}=7+(100-1)\cdot9$
$= 7+99\cdot9$
$a_{100}=898$

Therefore, the 100th term is 898.

Example 6) Find the tenth term of the arithmetic sequence whose first term is 6 and whose common difference is -6.

Solution: In this case to find the tenth term, a_{10}, we just replace n in the formula by 10, a_1 by 6, and d by -6. Thus,
$a_n=a_1+(n-1)\cdot d$
$a_{10}=6+(10-1)(-6)$
$=6+9(-6)$
$=6+(-54)$
$a_{10}=-48$
Therefore, the tenth term is -48.

Example 7) In the arithmetic sequence -2, 5, 12, 19, ... which term is 306?

Solution:- In this case, $a_1=-2$, $d=7$ and $a_n=306$. Thus, we are asked to find n. So we substitute into the formula of General Term of an arithmetic sequence.
$a_n=a_1+(n-1)d$
$306=-2+(n-1)\cdot7$
$306=-2+7n-7$
$306=-9+7n$
$315=7n$
$45=n$
The 45th term is 306.

Example 8) The 6th and 20th terms of an arithmetic sequence are 10 and 52, respectively. Find the 136th term.

Solution: Substitute n=6 and $a_6=10$ into the nth-term formula and we obtain

$$a_n = a_1 + (n-1)d$$
$$a_6 = a_1 + (6-1)d$$

(1)
$$10 = a_1 + 5d$$

Again, substitute $n=20$ and $a_{20}=52$ into the nth-term formula and we obtain

$$a_n = a_1 + (n-1)d$$
$$a_{20} = a_1 + (20-1)d$$

(2)
$$52 = a_1 + 19d$$

Subtracting equation (1) from equation (2) gives

$$52 = a_1 + 19d$$
$$\underline{-10 = a_1 + 5d}$$
$$42 = 14d$$
$$d = \frac{42}{14} = 3$$

Substituting $d=3$ into equation (1) yields

$$10 = a_1 + 5(3)$$
$$10 = a_1 + 5$$
$$10 - 5 = a_1$$
$$-5 = a_1$$
$$a_1 = 5$$

We obtained $a_1 = -5$ and $d=3$, we apply the nth-term formula with $n=136$.
$$a_n = a_1 + (n-1)d$$

$$a_{136} = -5 + (136-1) \cdot 3$$
$$= -5 + 135 \cdot 3$$
$$= -5 + 405$$
$$a_{136} = 400$$

The 136th term is 400.

3.34 Arithmetic Means

If we have two numbers, we can insert numbers between them and the new set of numbers form an arithmetic sequence. The numbers that we inserted between the two numbers are called Arithmetic Means. (AM) between the two numbers.

3.34.1 Let us see two cases of Arithmetic Means. (AM)

1)   Finding single Arithmetic means between a and b.
        Let's consider the three numbers a, m, and b. that forms an arithmetic sequence.

        a, m, b. is an arithmetic sequence, then
        m-a=b-m
        m+m=a+b
        2m=a+b

        $$M = \frac{a+b}{2}$$

Example 1)   Find the arithmetic means of -3 and 7.

        Solution:- Arithmetic means of -3 and $7 = \frac{-3+7}{2} = 2.$

Example 2)   Find the arithmetic means of 12 and 56.

        Solution:- Arithmetic means of 12 and $56 = \frac{12+56}{2} = 34$.

2)   Finding more than one arithmetic means between two numbers.

        Let a, A_1, A_2, A_3, A_4, A_5, and b form an arithmetic sequence. a is the first term and b is the seventh term. Hence, by the formula for general term of arithmetic sequence:-
        b=a+(7-1)d
        b=a+6d
        6d=b-a
        $$d = \frac{b-a}{6}$$

As we see above the value of the common difference is

$d = \dfrac{b-a}{6}$. Since a and b are known. It is necessary to find

the arithmetic means using the first term a, and the common difference d.

$A_1 = a + d$

$A_1 = a + \dfrac{b-a}{6}$

$A_1 = \dfrac{5a+b}{6}$

$A_2 = a + 2d$

$\quad = a + 2\left(\dfrac{b-a}{6}\right)$

$\quad = a + \dfrac{b-a}{3}$

$A_2 = \dfrac{2a+b}{3}$

$A_3 = a + 3d$

$\quad = a + 3\left(\dfrac{b-a}{6}\right)$

$\quad = a + \dfrac{b-a}{3}$

$A_3 = \dfrac{a+b}{2}$

$A_4 = a + 4d$

$$= a + 4\left(\frac{b-a}{6}\right)$$

$$= a + 2\left(\frac{b-a}{3}\right)$$

$$= a + \frac{2b - 2a}{3}$$

$$= \frac{3a + 2b - 2a}{3}$$

$$A_4 = \frac{a + 2b}{3}$$

$A_5 = a + 5d$

$$= a + 5\left(\frac{b-a}{6}\right)$$

$$= a + \frac{5b - 5a}{6}$$

$$= \frac{6a + 5b - 5a}{6}$$

$$A_5 = \frac{a + 5b}{6}$$

Example 3)   Insert five arithmetic means between 2 and 32.

Solution:- Let the five arithmetic means between 2 and 32 be x, y, z, w and P, (in the order written). Thus, 2, x, y, z, w, P, 32 is an arithmetic sequence with:-

$a_1=2$, $a_2=x$, $a_3=y$, $a_4=z$, $a_5=w$, $a_6=P$ and $a_7=32$

To find a_2, a_3, a_4, a_5 and a_6, we need to know the common difference d.

Thus, $a_7=32$, $a_1=2$, n=7

$a_n=a_1+(n-1)d$

$$32=2+(7-1)d \qquad \text{Using formula for general term.}$$
$$32=2+6d$$
$$6d=32-2$$
$$6d=30 \qquad \text{solving for d.}$$
$$d=5$$

Using the formula for general term we get:-

$a_2=x, a_1=2, n=2$
$a_n=a_1+(n-1)d$
$a_2=a_1+(2-1)(5)$
$x=2+5$
$x=7$

$a_3=y, a_1=2, n=2$
$y=a_3=2+(3-1) \cdot 5$
$=2+10$
$y=12$

$a_4=z,$
$z=a_4=2+(4-1) \cdot 5$
$z=2+15$
$z=17$

$a_5=w$
$w=a_5=2+(5-1)5$
$w=2+20$
$w=22$

$a_6=p$
$p=a_6=2+(6-1) \cdot 5$
$p=2+25$
$p=27$

Hence, the five arithmetic means between 2 and 32 are 7, 12, 17, 22 and 27.

3.35 Geometric Sequences

We have seen an arithmetic sequence to be a sequence in which its consecutive terms have a common difference, d. A geometric sequence is a sequence in which its consecutive terms have a common ratio.

3.35.1 Definition:- Geometric sequences

A sequence a_1, a_2, a_3, a_4, ..., is a geometric sequence, if there exists a constant number r, called common ratio, such that:-

$$\frac{a_2}{a_1} = r \quad, \frac{a_3}{a_2} = r \quad, \frac{a_4}{a_3} = r \quad,..., \frac{a_{n+1}}{a_n} = r \quad, \text{ for any n} \geq 1.$$

or

$$a_2 = a_1 r, \ a_3 = a_2 r, \ a_4 = a_3 r, \ ..., \ a_{n+1} = a_n r, \text{ for any n} \geq 1.$$

A geometric sequence is also called a geometric progression.

Example 1) Find the common ratio r for each of the following geometric progression.

a) 2, 8, 32, 128, ...

b) -3, 9, -27, 81, -243, ...

c) $1, \frac{1}{4}, \frac{1}{16}, \frac{1}{64}, \frac{1}{256}, ...$

solution:-

a) 2, 8, 32, 128, ...

$$r = \frac{8}{2} = 4$$

b) -3, 9, -27, 81, -243, ...

$$r = \frac{9}{-3} = -3$$

c) $1, \frac{1}{4}, \frac{1}{16}, \frac{1}{64}, \frac{1}{256}, ...$

$$r = \frac{\frac{1}{4}}{1} = \frac{1}{4}$$

3.35.2 n^{th} Term of Geometric Sequence

Like the arithmetic sequence, a geometric sequence is given by a recursive formula. To find a precise formula for the nth term of a geometric sequence. Let a_1 be the 1st term, and let r be the common ratio. We just start with the first term a_1 and use the fact that each successive term is found by multiplying the preceding one by the common ratio r.

a_1

$a_2 = a_1 r,$

$a_3 = a_2 r = (a_1 r)r = a_1 r^2,$

$a_4 = a_3 r = (a_1 r^2)r = a_1 r^3,$

$a_5 = a_4 r = (a_1 r^3)r = a_1 r^4,$

Thus, we see that the exponent is 1 less than the number of term. (the subscript of the term).

Generalizing, we have the following formula.

The nth term of a geometric sequence with first term a_1 and common ratio r is given by

$$\boxed{a_n = a_1 r^{n-1}, \text{ for any } n \geq 1}$$

Example 2)   Using the nth term formula for geometric sequence. Find the 36th term of the geometric sequence having common ratio 3 and first term $a_1 = 27$.

Solution
Given, r=3, $a_1 = 27$, and n=36, $a_n = ?$

$a_n = a_1 r^{n-1}$

$a_n = 27 \cdot 3^{36-1}$

$= 3^3 \cdot 3^{35}$

$= 3^{35+3}$

$a_n = 3^{38}$

Example 3)   Find the 20th term of the geometric progression with common ratio $r = \dfrac{1}{5}$ and first term $a_1 = 4$

Solution:- Given that: $a_1 = 4$, $r = \dfrac{1}{5}$ and n=20, $a_{20} = ?$

$a_n = a_1 r^{n-1}$, for $n \geq 1$.

$$a_n = 4 \cdot \left(\frac{1}{5}\right)^{20-1}$$

$$a_n = 4 \cdot \left(\frac{1}{5}\right)^{19}$$

Example 4)   Find the first term of a geometric progression with the 7th term 2,187 and common ratio r=3.

Solution:- Given that: $a_7 = 2,187$, r=3 and n=7, $a_1 = ?$

$a_n = a_1 r^{n-1}$, for $n \geq 1$

$2,187 = a_1 \cdot 3^{7-1}$

$3^7 = a_1 \cdot 3^6$

$\dfrac{3^7}{3^6} = a_1$

$a_1 = 3^{7-6} = 3$

Therefore, the first term is 3.

Exercises 3.19

1)   Find the first six terms of each of the following sequence whose nth term, or general term is given:

a) $a_n = 3n^2 + 1$

b) $x_n = \dfrac{n-3}{n+3}$

c) $S_n = \left(2 + \dfrac{2}{n}\right)^n$

d) $a_n = \sqrt{2n+1}$

e) $r_n = \dfrac{1}{2n}$

2)   Find the nth term of a sequence whose few terms are given.

a) 3, 7, 11, 15, 19, 23, ...

b) 10, 6, 2, -2, -6, -10, -14, ...

c) $3, 2, \dfrac{4}{3}, \dfrac{8}{9}, \ldots$

d) $\dfrac{2}{3}, \dfrac{3}{4}, \dfrac{4}{5}, \dfrac{5}{6}, \ldots$

e) 0.3, 0.03, 0.003, 0.0003, ...

3)   Find the 45th term of the arithmetic sequence 7, 11, 15, 19, 23, ...

4)   Find the 64th term of the arithmetic sequence -9, -6, -3, 0, 3, 6, 9, ...

5)   The 24th and 7th terms of an arithmetic sequence are 49 and -19, respectively. Find the 10th term.

6)   Find the 100th term of the arithmetic sequence whose first term is -8 and whose common difference is -6.

7)   In the arithmetic sequence -11, -7, -3, 1, ..., which term is 381?

8) Find the arithmetic means of -8 and 60.

9) Find the arithmetic means of 17 and 24

10) Insert four arithmetic means between -1 and -16.

11) Insert five arithmetic means between 7 and 31.

12) Find the common ratio r for each of the following geometric progression.

a) -3, 1, $\dfrac{-1}{3}$, $\dfrac{1}{9}$, $\dfrac{-1}{27}$, ...

b) 5, 25, 125, ...

c) 7, -7, 7, -7, 7, ...

d) 4, 16, 64, ...

e) 1, $\dfrac{1}{5}$, $\dfrac{1}{25}$, $\dfrac{1}{125}$, ...

13) Using the nth term formula geometric sequence find the 34th term of the geometric sequence having common ratio 4 and first term $a_1 = -2$

14) Find the first term of a geometric sequence with the 10th term 1024 and common ratio r=2.

CHAPTER FOUR

4. Geometry: Reasoning and Proof, Lines, and Congruent Triangles.

Geometry is a branch of mathematics that deals with the properties, measurement, and relationships between points, lines, surfaces, and solids. The word, geometry, comes from the Greek word geometria. The litral translation of geometria is "to measure the earth."

The ancient Greeks developed geometry from their efforts to measure distances and calculate areas. The two main areas of geometry are plane geometry and solid geometry. Plane geometry deals with two dimensional (flat or planar) shapes like circles, lines, and triangles. Three dimensional shapes like cubes, spheres, and cones are studied in solid geometry.

4.1   Point, Line and Plane

In Geometry, points, lines and planes are basic terms they are considered undefined terms since they are explained only interms of examples and descriptions. All other geometrical terms can be defined using these basic terms. Even though, the terms points, line and plane are undefined terms in geometry. Still we can give descriptions about these undefined terms.

4.1.1   Point:- An element in geometry having definite position, but no length, width or thickness. No one has seen a point, we can only think about a point. A point is represented by a dot. This dot is designed by a capital letter.

•M   •A
Point M   Point A

4.1.2   Line:- A line is a set of continuous points that extend indefinitely
in either direction.
A line has only length, but has no width or thickness. A line
may be straight or curved or both. If the tip of your pen or
pencil is moving in the same direction you can form straight
line.

A line can be named by any two points on the line and drawing
a line over the letters $\overrightarrow{BC}$. (The letters that we used to name are
capital letters)

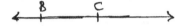

This is line BC ... (B and C are points on the line). A line may
be named by using a single small letter. (line l).
- Notation $\overrightarrow{BC}$ (read as line BC).
- A straight line is unlimited in length. (It will extend in both
directions indefinitely).

4.1.3   Plane:- A plane is a set of points that forms a flat surface that
has no depth and that extends indefinitely in all directions.
A plane is usually represented as a closed four-sided figure and
is named by placing a capital letter at one of the corners.

Fig 4.1

4.2   Line Segment and Ray

A line segment is a part of a line that has two end points. The two given points are the end points of the segment.

Notation:- $\overline{XY}$ (read as line segment xy.)

xy (distance between points x and y or it is length of the line segment xy.)

Note:- A line has no end point, but a line segment has two end points.

A ray is the union of half line and its end point. The end point is the vertex of the ray.

Ray XA is the union of half line XA and its end point X.

4.3   Angles:-

Definition and Notations

4.3.1   Definition:- An angle is the figure formed by two rays sharing a common endpoint, called the vertex of the angle.

Each ray is called side or arm of the angle.

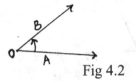

Fig 4.2

This is angle AOB

- $\overline{OA}$ and $\overline{OB}$ are sides or arms of the angle.
- O is the vertex of the angle.

- $\overline{OA}$ is initial ray and $\overline{OB}$ is terminal ray.

4.4   Naming an Angle

Angles are named in several ways.

- By naming the vertex of the angle (only if there is only one angle formed at the vertex; the name must be non-ambiguous) $\angle B$.
- By naming a point on each side of the angle with the vertex in between $\angle ABC$
- By placing a small number on the interior of the angle near the vertex.

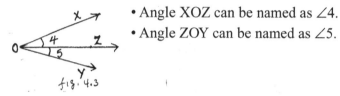

- Angle XOZ can be named as $\angle 4$.
- Angle ZOY can be named as $\angle 5$.

Fig 4.3

4.5   Classification of Angles by Degree Measure

4.5.1   Acute Angle

- An angle is said to be acute if it measures between 0 and 90 degrees, exclusive.

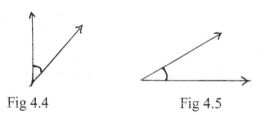

Fig 4.4   Fig 4.5

4.5.2   Right Angle

- An angle is said to be right if it measures 90 degrees.
- Notice the small box placed in the corner of a right angle, unless the box is present it is not assumed the angle is 90 degrees.
- All right angles are congruent.

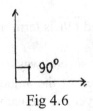

Fig 4.6

4.5.3 Obtuse Angle
- An angle is said to be obtuse if it measures between 90 and 180 degrees, exclusive.

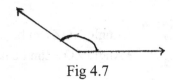

Fig 4.7

4.5.4 Straight Angle
- An angle that measure 180°.

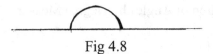

Fig 4.8

4.5.5 Reflex Angle
- An angle whose measure is more than 180° and less than 360°.

Fig 4.9

4.6 Relation between Angles

4.6.1 Adjacent Angles
- Adjacent angles are angles with a common vertex and a common sides.
- Adjacent angles have no interior points in common.

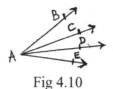

Fig 4.10

Example 1)   ∠BAC and ∠CAD are adjacent angles, because A is
common vertex and AC is common side.
∠BAD and ∠CAE are not adjacent angles because they
have common interior points.

4.6.2   Adjacent Supplementary Angles.
Here ∠1 and ∠2 are adjacent supplementary angles.

∠1+∠2=180°

Fig 4.11

Two adjacent angles on a straight line are always supplementary.

4.7   Definition: If a ray CD meets straight line ACB so as to make
two adjacent angles equal, each angle is called a right angle.

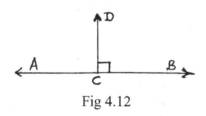

Fig 4.12

In the above figure $\overline{CD}$ is said to be at right angle to $\overline{AB}$ or
perpendicular to $\overline{AB}$ and C is called the foot of the perpendicular
from D to $\overline{AB}$.

4.8.   Complementary Angles
•   Complementary angles are two angles whose sum is 90°.
•   Complementary angles may or may not be adjacent.
Example 2) 30° and 60° are complementary.
30°+60°=90°, (30° is complement of 60°)

4.9   Supplementary Angles
- Two angles are said to be supplementary if their sum is 180°.
- Supplementary angles need not be adjacent.
- If supplementary angles are adjacent, then the sides they do not share form a line.

Example 3) 130° and 50° are supplementary angles.
130°+50°=180°, (50° is supplement of 130°)

4.10. Congruent Angles
- Angles that have the same measures are called congruent angles.
  If m(∠A)=m(∠B), then ∠A≡∠B

4.11. Angles at a point

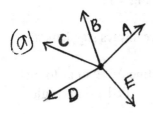

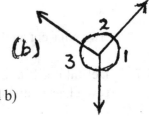

Fig 4.13 a) and b)

Such angles which have the same vertex are called angles at a point. If any number of straight lines are drawn from a given point, the sum of all the successive adjacent angles so formed is equal to four right angles (360°).

Example:- The sum of the angles marked 1, 2, 3, 4 and 5 is 360°.

Fig 4.14

4.12   Vertical Angles
- Angles with a common vertex whose sides form opposite rays are called vertical angles.
- Vertical angles are congruent.

4.13   Angle Bisector and Perpendicular Bisector

4.13.1 A) ANGLE BISECTOR

Ray OY divides angle XOZ into two-equal parts and measure of angle XOY is equal to measure of angle YOZ. Hence $\overline{OY}$ is called angle bisector.

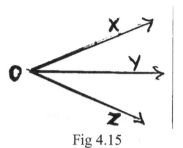

Fig 4.15

An angle bisector will divide an angle into two equal parts. This construction can be done using a compass and a straight edge. We don't need protractor.

4.13.2 B) Perpendicular Bisector.

If R is mid point of $\overline{AB}$, any line through R will bisect line segment $\overline{AB}$.

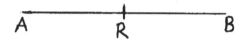

Perpendicular bisector of $\overline{AB}$ is a line through R which is perpendicular to $\overline{AB}$.

4.14   Classifying Triangles

When three noncollinear points are connected by segments, a the figure which is formed is called a triangle. The three segments are the sides of the triangle. Each pair of segments forms an angle of the triangle. The vertex of each angle is a vertex of the triangle.

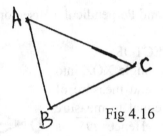

Fig 4.16

The symbol for a triangle is Δ.

The figure in the above shows ΔABC. We read as triangle ABC. The sides of ΔABC are $\overline{AB}$, $\overline{BC}$, and $\overline{AC}$. Its angles are ∠A, ∠B and ∠C. And its vertices are A, B, and C.

In the above ΔABC, the side opposite ∠A is $\overline{BC}$, the side opposite ∠B is $\overline{AC}$ and the side opposite ∠C is $\overline{AB}$. also, the angle opposite $\overline{BC}$ is ∠A, the angle opposite $\overline{AB}$ is ∠C and the angle opposite $\overline{AC}$ is ∠B.

Example 1)

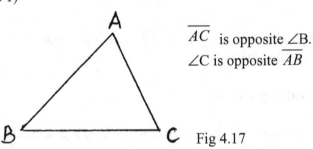

$\overline{AC}$ is opposite ∠B.

∠C is opposite $\overline{AB}$

Fig 4.17

4.14.1 A) Classifying Triangles Based on Sides.

In a triangle we compare three of the sides as: Three of the sides may be congruent, only two of the sides may be congruent or none of the sides may be congruent. Hence we have the following classifications:-

4.14.1.1

1) Scalene Triangle- A triangle that has no congruent sides.

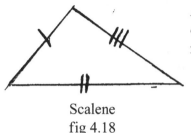

As you see from the picture the number of slashes are different showing that none of the sides are congruent.

Scalene
fig 4.18

4.14.1.2

2) Isosceles Triangle: A triangle that has two congruent sides.

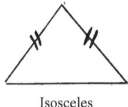

As you see from the picture the two sides the same number of slashes, showing that the two sides are congruent.

Isosceles
fig 4.19

4.14.1.3

3) Equilateral Triangle: A triangle that has three congruent sides.

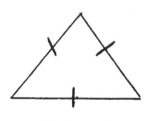

As you see from the picture, all sides have the same number of slash, showing that all three sides are congruent to each other.

Equilateral
fig 4.20

Note: An equilateral triangle is isosceles but an isosceles triangle may not be equilateral.

4.14.2 B) Classifying Triangles Based on Angles.

In a triangle, the measure of each angle may be less than 90°, one of the angle may be 90° or one of the angle may be greater than 90°. Hence we have the following classifications.

4.14.2.1

1) Acute Triangle: A triangle that has three acute angles.
    (An angle m is acute if 0°<m(<m)<90°).

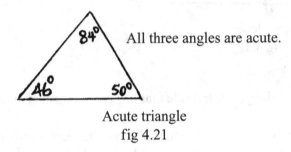

All three angles are acute.

Acute triangle
fig 4.21

4.14.2.2

2) Right Triangle: A triangle that has a right angle.
    (An angle R is right if m(R)=90°)

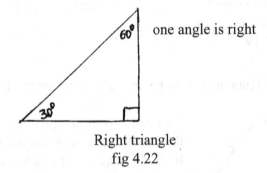

one angle is right

Right triangle
fig 4.22

Note: In a right triangle, two of the angles are acute.

4.14.2.3

(3) Obtuse Triangle: A triangle that has an obtuse angle.
        (An angle k is obtuse if 90°<m(<k)<180°).
Note: In an obtuse triangle, two of the angles are acute.

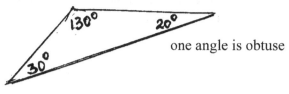

one angle is obtuse

Obtuse triangle
fig 4.23

4.14.2.4

(4) Equiangular or equilateral triangle:- A triangle in which all three angles are congruent to each other.

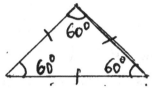

All three angles are congruent to each other.
Equiangular triangle
fig 4.24

- An equilateral triangle is a regular polygon.

4.15   Distance and Midpoint Formulas

4.15.1 The Distance Formula

Distance is a numerical description of how far an object are. In physics or every day discussion, distance may refer to a physical length or an estimation based on other criteria. In mathematics, a distance function or metric is a generalization of the describing what it means for elements of some space to be "close to" or "far away from" each other. In most cases, "Distance from A to B" is interchangeable with "distance between B and A."

The distance 'd' between the points A=(X_1, Y_1) and B=(X_2, Y_2) is given by the formula:

$$d = \sqrt{\left(x_2 - x_1\right)^2 + \left(y_2 - y_1\right)^2}$$

4.16   Pythagoras' Theorem:- If a right angled triangle has legs of length a and b, hypotenuse of length c, then, $c^2 = a^2 + b^2$.

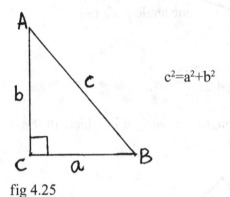

$$c^2 = a^2 + b^2$$

fig 4.25

The above distance formula $d = \sqrt{(x_2 - x_1)^2 + (y_2 - y_1)^2}$ is an application of Pythagoras' theorem for right angled triangles.

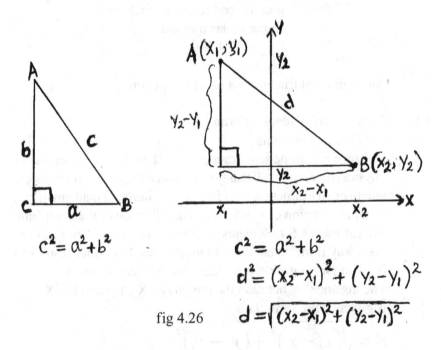

$$c^2 = a^2 + b^2$$

$$c^2 = a^2 + b^2$$

$$d^2 = (x_2 - x_1)^2 + (y_2 - y_1)^2$$

fig 4.26 $$d = \sqrt{(x_2 - x_1)^2 + (y_2 - y_1)^2}$$

Example 1)   Using the Pythagoras' formula find the length of the legs of each of the following right angled triangle

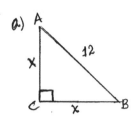 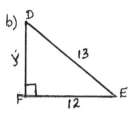 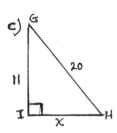

fig 4.27 a, b, and c

Solution:- a) In right triangle ABC

$$x^2+x^2=(12)^2$$
$$2x^2=144$$
$$x^2=72$$
$$x = 6\sqrt{2}$$

b) In right triangle DEF

$$y^2+(12)^2=(13)^2$$
$$y^2+144=169$$
$$y^2=25$$
$$y=5$$

c) In right triangle GHI

$$(11)^2+x^2=(20)^2$$
$$121+x^2=400$$
$$x^2=400-121$$
$$x^2=279$$
$$x = \sqrt{279}$$

Example 2)   Find the distance between the points:-

a) A(3, 5), B(-2, 6)
b) $P_1(6, 0)$, $P_2(5, 4)$
c) C(4,7), D(8,4)
d) A(-1, -3), B(4, 6)

Solution

a) A(3, 5), B(-2, 6)

$$d = \sqrt{\left(x_2 - x_1\right)^2 + \left(y_2 - y_1\right)^2}$$
$$= \sqrt{\left(-2-3\right)^2 + \left(6-5\right)^2}$$
$$= \sqrt{25+1}$$
$$d = \sqrt{26}$$
$$d = \sqrt{26}$$

b) $P_1(6, 0)$, $P_2(5, 4)$

$$d = \sqrt{\left(x_2 - x_1\right)^2 + \left(y_2 - y_1\right)^2}$$
$$d = \sqrt{\left(5-6\right)^2 + \left(4-0\right)^2}$$
$$d = \sqrt{1+16}$$
$$d = \sqrt{17}$$

c) $C(4,7)$, $D(8,4)$

$$d = \sqrt{\left(x_2 - x_1\right)^2 + \left(y_2 - y_1\right)^2}$$
$$d = \sqrt{\left(8-4\right)^2 + \left(4-7\right)^2}$$
$$d = \sqrt{16+9} = \sqrt{25}$$
$$d = 5$$

d) $A(-1, -3)$, $B(4, 6)$

$$d = \sqrt{\left(x_2 - x_1\right)^2 + \left(y_2 - y_1\right)^2}$$
$$d = \sqrt{\left(4-(-1)\right)^2 + \left(6-(-3)\right)^2}$$
$$d = \sqrt{25+81}$$
$$d = \sqrt{106}$$

Example 3)   The distance between (-4, 2) and (5, x) is $\sqrt{97}$ units. Find the possible value of x.

Solution

Using distance formula with $d = \sqrt{97}$. Let $(x_1, y_1) = (-4, 2)$ and $(x_2, y_2) = (5, x)$. Then

$$d = \sqrt{(x_2 - x_1)^2 + (y_2 - y_1)^2} \qquad \text{Distance formula}$$

$$\sqrt{97} = \sqrt{(5 - (-4))^2 + (x - 2)^2}$$

$97 = 81 + x^2 - 4x + 4$ Square both sides.

$x^2 - 4x + 85 - 97 = 0$

$x^2 - 4x - 12 = 0$ Writing in standard form.

$(x+2)(x-6) = 0$ Factor

$x+2 = 0$ or $x-6 = 0$ Zero product property.

$x = -2$ or $x = 6$ Solve for x.

The possible value of x is -2 or 6.

4.17   The Midpoint Formula

The midpoint is the middle point of a line segment. It is equidistant from both end points.

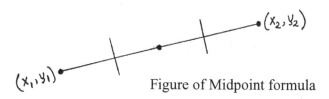

Figure of Midpoint formula

The midpoint of the segment (x_1, y_1) to (x_2, y_2) is given by $\left(\dfrac{x_1 + x_2}{2}, \dfrac{y_1 + y_2}{2} \right)$.

The formula for determining the midpoint of a segment in the plane, with endpoints (x_1, y_1) and (x_2, y_2) is:

$$\left(\frac{x_1 + x_2}{2}, \frac{y_1 + y_2}{2} \right)$$

Example 1) What is the midpoint of the line segment with endpoints
(8, -1) and (4, 3)?

Solution:

Let $(x_1, y_1)=(8, -1)$ and $(x_2, y_2)=(4, 3)$.

$$\text{Midpoint} = \left(\frac{x_1+x_2}{2}, \frac{y_1+y_2}{2}\right) = \left(\frac{8+4}{2}, \frac{-1+3}{2}\right) = (6,1)$$

Example 2) Find the midpoint of the line segment with endpoints (0, 0) and (-2, 8).

$$\text{Midpoint} = \left(\frac{x_1+x_2}{2}, \frac{y_1+y_2}{2}\right) = \left(\frac{0+(-2)}{2}, \frac{0+8}{2}\right) = (-1,4)$$

Example 3) Find the values of x and y, if the midpoint of the line segment (x, 6) and (-5, y) is (4, 2).

Solution:

Let $(x_1, y_1)=(x, 6)$ and $(x_2, y_2)=(-5, y)$, then

$$\left(\frac{x_1+x_2}{2}, \frac{y_1+y_2}{2}\right) = (4,2)$$

$$\frac{x_1+x_2}{2} = 4, \quad \frac{y_1+y_2}{2} = 2$$

$$\frac{x+(-5)}{2} = 4, \quad \frac{6+y}{2} = 2$$

$$x-5=8, \quad 6+y=4$$

$$x=13 \quad y=-2$$

The value of x is 13 and the value of y is -2.

Exercise 4.1

1)   Using the Pythagoras' formula, find the length of the hypotenuse of the following right angled triangle.

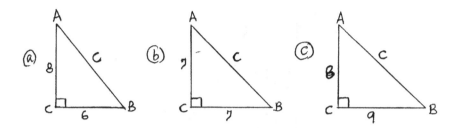

2)   Using the Pythagoras' formula, find the length of the legs of the following right angled triangle.

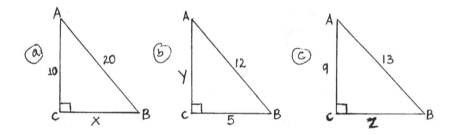

3)   Find the distance between the points:
       a) (2, 7) and (4, -3)
       b) (1, 0) and (5, 4)
       c) (6, 6) and (4, 2)
       d) (4, -4) and (2, -5)
       e) (1, -1) and (7, 4)

4)   The distance between (4, 2) and (x, -5) is $5\sqrt{2}$ units. Find the possible value of x.

5) The distance between (3, 4) and (2, b) is $\sqrt{10}$ units. Find the possible value of b.

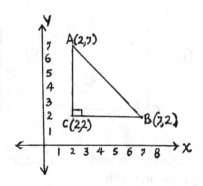

6) In the figure shown above. If the distance between consecutive grid lines is 5 miles.
a) Find the area of △ABC.
b) Find the perimeter of △ABC.
c) What is the name of the triangle?

7) Find the midpoint of the line segment with the given end points.
a) (-1, -2), (-3, -4)
b) (2, 7), (8, -6)
c) (0, 0), (8, 9)
d) (-7, -4), (6, 2)
e) (-5, 1), (6, 9)
f) (4, 2), (-1, 5)

8) Find the values of a and b, if the midpoint of the line segment (a, 4) and (6, b) is (5, 4).

9) Find the values of x and y, if the midpoint of the line segment (x, 2y) and (3x, y) is (-5, 4).

4.18 What is a Conjecture?

4.18.1 A conjecture is a proposition that is unproven but is believed to be true and has not been disproven. (in mathematics, a conjecture is an unproven proposition or theorem that appears correct.)

4.18.2 Inductive reasoning is a making of conclusion based on patterns in a specific cases that we observe. The conclusion that we reached is called conjecture.

4.18.3 A Counter Example is a specific case for which a conjecture is false. In mathematics, counterexamples are often used to prove the boundaries of possible theorems. By using counterexamples to show that certain conjectures are false.

The following are examples of conjecture.

- "The degree measure of the straight angle is 180°."
- "If two parallel lines are crossed by a transversal, then vertically opposite angles are congruent."
- "Opposite sides of a parallelogram are congruent."
- Consider the following sequence of numbers:- 1, 2, 4, 8, 16, 32,

Conjecture: $y=2^x$

Let us verify the conjecture
When x=0, $y=2^0=1$
When x=1, $y=2^1=2$
When x=2, $y=2^2=4$
When x=3, $y=2^3=8$
When x=4, $y=2^4=16$
When x=5, $y=2^5=32$
What is the next number in the above given series?
When x=6, $y=2^6=64$.
Therefore, the next number is 64.

Consider the following patterns of numbers.

Given the sequence: ¼, 1, 4, 16, 64, ---

Conjecture: $y=4^{x-1}$

Let us verify the conjecture:-

When x=0, $y=4^{0-1}=4^{-1}=¼$
When x=1, $y=4^{1-1}=4^{0}=1$
When x=2, $y=4^{2-1}=4^{1}=4$
When x=3, $y=4^{3-1}=4^{2}=16$
When x=4, $y=4^{4-1}=4^{3}=64$

What is the next number in the given series.
When x=5, $y=4^{5-1}=4^{4}=256$
Therefore, the next number is 256.

More Examples:-

Look at the following visual patterns and then use the inductive reasoning to reach to the conclusion and sketching the next figure.

1)   Sketch the next figure in the following patterns.

 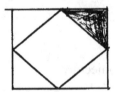 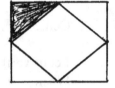

Fig 4.28

If you observe the patterns carefully, you may reach to the conclusion to sketch the next figure shown below.

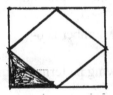

Fig 4.28.1

2)   Sketch the next figure in the following patterns.

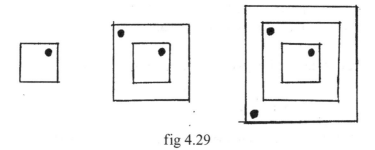

fig 4.29

If you observe the patterns carefully, you may reach to the conclusion to sketch the next figure shown below.

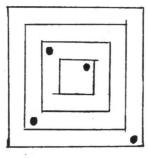

fig 4.29.1

Examples of Counterexample

1)   A counterexample to the statement "All prime numbers are odd numbers," is the number 2, as it is a prime number but is not an odd number.

2)   A counterexample to the statement "All natural numbers are either prime or composite." is the number 1. As it is neither prime nor composite.

3)   A counter example to the statement "Every set is uncountable." would be the set containing 1, 2, 3 {1, 2, 3},as it is countable set. That is, it has three elements.

Exercise 4.2

1) Consider the following sequence of numbers:-

1, 3, 9, 27, 81, ...

What is the next number in the above given series?

2) Consider the following sequence of numbers:-

$$1, \frac{1}{2}, \frac{1}{4}, \frac{1}{8}, \frac{1}{16}, \dots$$

What is the next number in the above given series?

3) Sketch the next figure in the following patterns.

4) Sketch the next figure in the following patterns.

 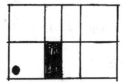 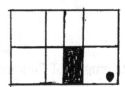

5) Show the conjecture is false by finding a counterexample.

a) Every number is either positive or negative.

b) The sum of two irrational number is always irrational number.

c) The quotient of any two numbers is defined.

d) The product of every two number is always positive.

e) The area of a rectangle can never be an odd number.

4.19 LOGIC

4.19.1 Introduction

Logic involves the systematic study of valid methods of argument and inference. It can be seen as a subset of philosophy or mathematics, and provides the foundation of each discipline.

4.19.2 History

Although many ancient civilizations developed systems of argumentation and studied logical paradoxes, the history of modern logic is typically traced back to ancient Greece and the writings of Aristotle (384-322 BC). His formulation of so-called Aristotelian logic was the dominant form of formal logic until the late 18th and early 19th centuries, during which the development of symbolic logic and mathematical logic introduced new paradoxes and more powerful techniques to deal with them. Mathematical logic is the science of thinking and reasoning correctly. Reasoning is the process of drawing conclusions from known facts. When we give reason, we use sentences which we know to be either true or false. In Mathematics we are much interested on declarative or assertive types of sentences. English sentences that tell about people, things and ideas are called Declarative Sentences. You can also recall English sentences such as commands and interrogative sentences.

4.19.3 Propositional Logic

Propositional Logic is the branch of logic that studies ways of joining and/or modifying entire propositions, statements or sentences to form more complicated propositions, statements or sentences, as well as the logical relationships and properties that are derived from these methods of combining or altering statements. In propositional logic, the simplest statements are considered as indivisible units, and hence, propositional logic doesn't study those logical properties and relations that depends upon parts of statements that are not themselves statements on their own, such as the subject and predicate of a statement. The most thoroughly researched branch of propositional logic is classical truth-functional propositional logic, which studies logical operators and connectives that are used to produce complex statements whose truth-value depends entirely on the truth-values of the simpler statements making them up, and in which it is assumed that every statement is either true or false and not both. However, there are other forms of propositional logic in which other truth-values are considered, or in which there is consideration of connectives that are used to produce statements whose truth-values depend not simply on the truth-values of the parts, but additional things such as their necessity, possibility or relatedness to one another.

4.20   A statement or proposition can be defined as a declarative sentence, or part of sentence, that is capable of having a truth value, such as being true or false but not both. Consider the following examples:

- Addis Ababa is the capital city of Ethiopia.
- Barack Obama is the 44th president of the United States.
- The integer next to nine is ten.
- x+0=x, for all real number x.

In the above statements, we can say each of the statement is true or we can say the truth value of of each statement is true (T).

Sometimes, a statement can contain one or more other statements as parts. Consider the following example:-

Either 3 is a prime number or 3 is an even number.

While the above compound sentence is itself a statement, because it is true, the two parts, "3 is a prime number" and "3 is an even number", are themselves statements, because the first is true and the second is false.

Consider the following sentences:-

a) Look at the lion.
b) Have you done your home work?
c) I wish I will be a pilot.

Are the above sentences (a, b, and c) true or false? It is not possible to assign a true or false value to these sentences. So these sentences are not statements or propositions.

Example 1)   Consider the following sentences

1) Cats are not mammals.
2) 6 is the square root of 36.
3) y+0=y, for all real number y.
4) -5 is an integer.

All the above sentences are propositions.
      - The truth value of number 1 is false. (F)
      - The truth values of number 2, 3 and 4 are true (T).

Notation:- We use small letters p, q, r, s, ... etc to denote proposition.

Example 1)   q ≡ Addis Ababa is the capital city of Ethiopia
      Here q stands for the proposition, "Addis Ababa is the capital city of Ethiopia"
      What is the truth value of q?
      Answer:- The truth value of q is true (T).

Exercise 4.3

1)   Identify each of the following sentences as proposition or not proposition.
      a) -3 is the cube root of -27.
      b) How are you today?
      c) Will there be life in the moon?
      d) x+1=6
      e) The integer next to six is seven.

2)   Find the truth value of the following propositions.
      a) 9 is an odd number.
      b) All rectangles are square.
      c) Opposite sides of a parallelogram are congruent.
      d) Nyala is endemic animal to Ethiopia.
      e) 4+6=12

4.21 LOGICAL CONNECTIVES

In logic, a logical connective is a symbol or word used to connect two or more sentences in a grammatically valid way, such that the compound sentence produced has a truth value dependent on the respective truth values of the original sentences.

4.21.1 Definition: Compound or Complex Proposition

If two or more propositions are combined by one or more propositional connectives, the resulting proposition is called a compound or complex proposition.

Examples) All of the following propositions are compound or complex propositions. (They are not simple statements or propositions.)

A) A triangle is equilateral if and only if it is equiangular.
B) -5 is an integer and Ethiopia is in Asia.
C) 2 is a prime number or 2 is an even number.
D) Every triangle is not equilateral.
E) If today is Friday, then tomorrow is Saturday.

The table below gives the five propositional (logical connectives) with their symbols and the corresponding compound propositions.

	Connective	Name	Symbol	Complex Proposition	Read as
1	not	negation	$\neg$	$\neg P$	Negation of P. (not P)
2	and	conjunction	$\wedge$	$P \wedge q$	P and q
3	or	disjunction	$\vee$	$P \vee q$	P or q
4	If..., then	implication	$\Rightarrow$	$P \Rightarrow q$	If P, then q. (P implies q)
5	If and only if	bi-implication	$\Leftrightarrow$	$P \Leftrightarrow q$	P if and only if q

Example) Let $P \equiv$ A triangle is equilateral.
$q \equiv$ It is equiangular.

See the following compound proposition below.
- A triangle is equilateral and it is equiangular. (P∧q).
- A triangle is not equilateral. (¬P)
- A triangle is equilateral or it is equiangular. (P∨q)
- If a triangle is equilateral, then it is equiangular. (P⟹q)
- A triangle is equilateral if and only if it is equiangular. (P⟺q)

Exercise 4.4

1) Identify the following sentences as simple proposition or compound proposition.
A) If it is raining, then I am indoors.
B) Today is not Monday.
C) A triangle is equilateral if and only if the three sides are congruent.
D) Jack is my brother.
E) A parallelogram is a square or a rectangle.

2) Identify the following compound propositions as negation, conjunction, disjunction, implication or bi-implication.
a) 8 is an even number or 8 is a composite number.
b) If all angles of a triangle are congruent, then it is an equilateral triangle.
c) 13 is not a prime number.
d) A rectangle is a square if and only if all sides are congruent.
e) Every isosceles triangle is equilateral or, every square is a rectangle.

4.22   Truth Tables and Propositional Connectives

Definition:- The table in which we assign the truth-values of a proposition that result from all the possible combinations of the truth values of its components is called Truth table.

In this section we will consider each propositional connective and see the rules on how to find the truth values.

4.22.1 NEGATION
Negation is denoted by "¬".

Given a proposition p, its logical negation is denoted by ¬P.

1) ¬P (negation P) is true if and only if P is false. (If P is true, then the negation P is false and if p is false, then negation P is true.)

Note:- Negation is different from the other basic propositional connectives since it acts only in one proposition.

4.22.1.1 TRUTH TABLE FOR NEGATION

P	¬P
T	F
F	T

The table above is easy to understand, if P is true, its negation is false. If P is false, then the negation of P is true.

Example 1)   Let P≡ All parallelograms are rectangles.
(P is the same as all parallelograms are rectangles.)

What is the negation of P or ¬P?
¬P ≡ Not all parallelograms are rectangle.
Here, the truth value of ¬P is true.
Thus, P is F and ¬P is T.

Example 2)   Let q ≡ 5 is an even number.
¬q ≡ 5 is not an even number.
Truth value of q is F and truth value of ¬q is T.

4.22.1.2 LAW OF DOUBLE NEGATION
For any proposition P, the statement ¬(¬P) has the same truth-value as P.   .
¬(¬P) ≡ P

4.22.2 Conjunction
When two propositions are connected by "and", the resulting complex proposition is conjunction.
"And" is symbolized by "∧".
The conjunction of two proposition P and q is denoted as "P∧q."

4.22.2.1 Rule for conjunction of two propositions:
- P∧q is true if and only if P is true and q is true. The truth value of two propositions connected by the propositional connective "and" will be true if and only if both propositions are true.

4.22.2.2 Truth table for conjunction

P	q	P∧q
T	T	T
T	F	F
F	T	F
F	F	F

Example 1) Let P≡ 2 is a prime number....(It is simple proposition)
             q≡ 0 is a positive number (It is simple proposition)

  * What are truth values of P and q?
  a) The truth value of P is true (T).
  b) The truth value of q is false (F).

  * What is the compound proposition (P∧q)?
  P∧q≡ 2 is a prime and 0 is a positive number.

  * What is the truth value of (P∧q)?
  P is true and q is false, therefore P∧ is false (F).

Example 2) Let P≡ Addis Ababa is the capital city of Ethiopia.
             q≡ The degree measure of a circle is 360°.

  * What are the truth values of P and q?

a) The truth value of P is true (T).
b) The truth value of q is true (T)

What is the truth value of (P∧q)?
P is true and q is true, therefore, P∧q is true (T).

Example 3)   Let P≡ A square is a rectangle.
q≡ A square is a parallelogram.

A) A square is both a rectangle and a parallelogram.
Symbolic form: P∧q .... truth value T.

B) A square is a rectangle but not a parallelogram.
Symbolic form: P∧¬q .... truth value F.

C) A square is not a rectangle but a parallelogram.
Symbolic form: ¬P∧q ... truth value F.

4.22.3 Disjunction

When two propositions are joined by the connective "or", the resulting complex proposition is called Logical Disjunction. "or" is symbolized by "∨". Hence the disjunction of P and q is written as P∨q (read as P or q).

4.22.3.1 Rule for determining the truth value of P∨q is:

- P∨q is true if and only if at least one of the two propositions is true. Or P∨q is false if and only if both propositions are false and is true in all the remaining three cases.

4.22.3.2 Truth Table for Disjunction

P	q	P∨q
T	T	T
T	F	T
F	T	T
F	F	F

Note:- The use of "or" in propositional logic is different from the normal use in English language. In normal use of the sentence "They will go to USA this week or next week." It means they will go to USA in one of the weeks but not both weeks. Such use of "or" is the exclusive case. (Only one of the two alternatives.) But, in propositional logic "or" is inclusive. (Both cases can hold.). That means they may go to USA, in one or both weeks.

Example: Let P≡ Every square is a parallelogram.
             (The truth value is T)
             q≡ 2 is an odd number.
             (The truth value is F)

Observe and determine the truth value of the following logical disjunction.

A) P∨q≡ Every square is a parallelogram, or, 2 is an odd number.
        (P is true and q is false and truth value of P∨q is true)
B) ¬P∨q≡ Not every square is a parallelogram, or, 2 is an odd number.
        (¬P is false and q is also false and hence the truth value of ¬P∨q is false.)
             Note:- (F∨F) ≡ F
C) P∨¬q ≡ Every square is a parallelogram, or 2 is not an odd number.
(P is true and ¬P is also true and hence the truth value of P∧¬q is true.)
D) ¬P∨¬q ≡ Not every square is a parallelogram, or, 2 is not an odd number.
        (¬P is false and ¬q is true and hence the truth value of ¬P∨¬q is true.)

Example 2)   Consider the proposition r and s.
             $r \equiv \sqrt{3}$ is a rational number.
             $s \equiv -5$ is an element of the set of integers.

A) $rVs \equiv \sqrt{3}$ is a rational number, or -5 is an element of the set of integers.

    (r is false and s is true and hence the truth value of rvs is true.)

B) $\neg rVs \equiv \sqrt{3}$ is not a rational number or, -5 is an element of the set of integers.)

    (¬r is true and s is true and hence the truth value of ¬rvs is true.)

C) $rV\neg s \equiv \sqrt{3}$ is a rational number or, -5 is not an element of the set of integers.

    (r is false and ¬s is false and hence the truth value of rV¬s is false.)

D) $\neg rV\neg s \equiv \sqrt{3}$ is not a rational number or, -5 is not an element of the set of integers.

    (¬r is true and ¬s is false and hence the truth value of ¬rv¬s is true.).

Example: Let P≡The degree measure of each angles of an equilateral triangle is 60°.

    What is the truth value of:

a) ¬P∨P   b)¬P∧P

a) ¬P∨P≡¬P is false and P is true and hence the truth value of ¬P∨P is true.

b) ¬P∧P≡¬P is false and P is true and hence the truth value of ¬P∧P is false.

    Note:- ¬P∨P is always true... (It is called Law of Excluded Middle)

¬P∧P is always false... (It is called Law of Non-contradiction)

4.22.4 Implication

When two propositions are joined with the connective "implies", the proposition formed is called a logical implication, or conditional statement.

"implies" is denoted by "⇒"

P implies q is symbolized as P⇒q.

P⇒q is read as
- If P, then q.
- P implies q.
- P only if q.

In P⇒q ... (P is called antecedent or hypothesis and q is called consequent or conclusion of the conditional statement.)
In the conditional statement "P⇒q";

P is called the hypothesis or sufficient condition for q and q is called the conclusion or necessary condition for P.

4.22.4.1 Rule of Implication

P⇒q is false if and only if P is true and q is false.

4.22.4.2 Truth table for logical implication

P	q	P⇒q
T	T	T
T	F	F
F	T	T
F	F	T

Example 1) Let r≡13 is the square root of 169.

     s≡ -5<-6

Observe the truth value of each of the following logical implication.

a) r⇒s ≡ r is true and s is false and hence the truth value of r⇒s is false.

b) s⇒r ≡ s is false and r is true and hence the truth value of s⇒r is true.

c) ¬r⇒s ≡ ¬r is false and s is also false and hence the truth value of ¬r⇒s is true.

d) r⇒¬s ≡ ¬r is true and ¬s is also true and hence the truth value of r⇒¬s is true.

e) ¬r⇒¬s ≡ ¬r is false and ¬s is true and hence the truth value of ¬r ≡ ¬s is true.

Example 2)   Consider the following conditional statement. "If 2 is an odd number, then a rectangle is a square."
Let P≡ 2 is an odd number.
q≡ A rectangle is a square.

Here the antecedent P is false and the consequent q is false. P⇒q is true. (P is false and q is also false)

4.23   Converse, Inverse and Contrapositive.
Given a conditional statement P⇒q.

i) q⇒P is called the converse of P⇒q.
ii) ¬P⇒¬q is called the inverse of P⇒q
iii) ¬q⇒¬P is called the contrapositive of P⇒q.

4.23.1 Converse:- is a conditional statement written by switching the hypothesis and the conclusion. That is, the original statement of P then q will be if q then P.

Use the truth table to show the converse of the conditional statement P⇒q.

P	q	P⇒q	q⇒P
T	T	T	T
T	F	F	T
F	T	T	F
F	F	T	T

Example 1) Consider the following conditional statement.
    "If 5 is an integer, then it is an odd number."
Converse:- If 5 is an odd number, then it is an integer.

Example 2) Consider the following conditional statement.
    "If a figure is a square, then it is a rectangle."
    Converse:- If a figure is a rectangle, then it is a square.

Example 3) Consider the following conditional statements.
    "If two triangles are congruent, then they are similar."
    Converse:- If two triangles are similar, then they are congruent.

Example 4) "If you drink too much alcohol, then you will get sick."
    Converse:- If you get sick, then you drink too much alcohol.

4.23.2 Inverse:- is a conditional statement written by making negative both the hypothesis and conclusion. That is the original statement if P then q will be if not P then not q.

Use truth table to show the inverse of the conditional statement $P \Rightarrow q$.

P	q	$\neg P$	$\neg q$	$P \Rightarrow q$	$\neg P \Rightarrow \neg q$
T	T	F	F	T	T
T	F	F	T	F	T
F	T	T	F	T	F
F	F	T	T	T	T

Example 1) Consider the following conditional statement.
    If two lines are parallel and crossed by a transversal line, then corresponding angles are congruent.

Inverse:- If two parallel lines are not crossed by transversal lines, then corresponding angles are not congruent.

Example 2) If -2 is an integer then 6 is the square root of 36.
Inverse:- If -2 not an integer then 6 is not the square root of 36.

Example 3) If all cats are mammals, then a rectangle is a square.
Inverse:- If all cats are not mammals, then a rectangle is not a square.

Example 4) If zero is positive, then a triangle has four sides.
Inverse: If zero is not a positive, then a triangle doesn't have four sides.

4.23.3 Contrapositive:- is a conditional statement written by switching the hypothesis and the conclusion and then making negative both the hypothesis and the conclusion. That is, the original, if P, then q will be if not q, the not P.

Use truth table to show the contrapositive of the conditional statement $P \Rightarrow q$.

P	q	¬q	¬P	$P \Rightarrow q$	$¬q \Rightarrow ¬P$
T	T	F	F	T	T
T	F	T	F	F	F
F	T	F	T	T	T
F	F	T	T	T	T

Note:- The contrapositive of any statement has the same truth-value as the original statement. You can easily observe this from the truth table shown in the above.

Example 1) If two straight lines are parallel, then they do not meet.
Contrapositive:- If two straight lines meet, then they are not parallel.

Example 2) If Addis Ababa is a capital city of Ethiopia, then a square is a rectangle.

Contrapositive:- If a square is not a rectangle, then Addis Ababa is not a capital city of Ethiopia.

Example 3) If two angles are complementary, then their sum is 90°.
Contrapositive:- If the sum of two angles is not 90°, then they are not complementary.

Example 4) If today is Monday, then tomorrow will be Tuesday.
Contrapositive:- If tomorrow will not be Tuesday, then today is not Monday.

Example 5) If x+3≥5, then x+6<7.
Contrapositive:- If x+6≥7, then x+3<5.

4.24.BI - Implication:-
When two propositions are connected by a bi-implication, the new proposition that is formed is called a logical bi-implication or a bi-conditional.

A bi-implication is denoted by "⇔".
The bi-implication of propositions P and q is written as: P⇔q.... and read as:-
* P if and only if q.
* P is necessary and sufficient condition for q.
* q is necessary and sufficient condition for P.

Note:- It is usual to use "iff" as a short form or abbreviation for if and only if.
P iff q ... (read as P if and only if q.)

Example 1) Consider the proposition P and q where
Let P≡ 4 is an even number.... (truth value T)
q≡ 18 is a multiple of 6.... (truth value T)

a) P⇔q ≡ 4 is an even number if and only if 18 is a multiple of 6.
(The truth value of P⇔q is true.)
b) What is the truth value of ¬P⇔q?
Answer: ¬P is F and q is T and hence truth value of ¬P⇔q is F.

c) What is the truth value of $\neg P \Leftrightarrow \neg q$?

Answer:- The truth value of $\neg P$ is F and truth value of $\neg q$ is also F and hence truth value of $\neg P \Leftrightarrow q\neg$ is T.

Rule:- The truth value of $P \Leftrightarrow q$ is true (T) if and only if P and q have the same truth values.

4.24.1 Truth table for Bi-implication

P	q	$P \Leftrightarrow q$
T	T	T
T	F	F
F	T	F
F	F	T

Remark:- A bi-conditional is the conjunction of two conditional statements where the hypothesis and conclusion of the first conditional have been switched in the second. Therefore, a bi-conditional statement $P \Leftrightarrow q$ is the same as $(P \Rightarrow q) \wedge (q \Rightarrow P)$. We can also construct the truth table of $P \Leftrightarrow q$ by constructing the table of $(P \Rightarrow q) \wedge (q \Rightarrow P)$.

See the table for $(P \Rightarrow q) \wedge (q \Rightarrow P)$

p	q	$P \Rightarrow q$	$q \Rightarrow P$	$(P \Rightarrow q) \wedge (q \Rightarrow P)$
T	T	T	T	T
T	F	F	T	F
F	T	T	F	F
F	F	T	T	T

Observe the last column of the two truth table in the above, they have the same truth value, i.e.,

$$P \Leftrightarrow q \equiv (P \Rightarrow q) \wedge (q \Rightarrow P)$$

Remark: The five compound propositions we considered so far are called basic compound propositions. We can form more complex propositions using the rules of these basic compound propositions.

Exercise 4.5

1)   Let r and s be propositions where
       r≡ Water is denser than ice.
       s≡ Mercury is liquid metal.

       Express each of the following in symbolic form.

       A) Water is denser than ice and mercury is liquid metal.
       B) Water is not denser than ice and mercury is liquid metal.
       C) Water is not denser than ice or mercury is not liquid metal.
       D) If water is denser than ice, then mercury is liquid metal.
       E) Water is denser than ice only if mercury is liquid metal.
       F) Water is denser than ice if and only if mercury is not liquid metal.
       G) Neither water is denser than ice nor mercury is liquid metal.
       H) Water is denser than ice, but mercury is not liquid metal.

2)   Let P≡ A rectangle has three sides.
       q≡ Tuesday is the day after Monday.

       Translate each of the following into ordinary English statement.
       a) ¬P
       b) P∧q
       c) P∨q
       d) P⇒q
       e) P⇔q
       f) ¬P⇒q
       g) ¬P⇔¬q
       h) ¬P⇒¬q
       i) ¬q

3)   Determine the truth values of the complex-propositions above. (Question number 2)

4)   Let P≡ Dogs are mammals.
         q≡ All squares are rectangle.
         r≡ 2+1<13
         s≡ 121 is a prime number, and

Determine the truth values of each of the following statements.
         a) ¬s∨r
         b) (P∨¬q)⟺s
         c) (r∧s)⟹q
         d) (r∨s)⟺(P∧q)
         e) ¬r⟹(¬q∨P)
         f) r∨(P∧s)
         g) ¬s⟹(¬P∧q)
         h) r⟺s
         i) (¬r∨q)∧P
         j) (P∧r)⟺(q∧s)

5)   Given that the truth value of P is T, truth value of q is T, truth value of r is F and truth value of s is F, determine the truth value of the following.
         a) (P∧q)∨¬s
         b) (P∨¬s)∧r
         c) r⟹¬q
         d) (r⟹s)⟺P
         e) (¬r∧¬q)⟺P
         f) s⟺r
         g) ¬P⟹(¬P∧r)
         h) (r∧q)⟹¬P
         i) (r∧s)⟺(P∧q)
         j) (P∧q)⟹¬(r∧s)
         k) [(r∧s)∨(P∧s)]⟹q

6) Determine the converse, inverse and contrapositive of each of the following.

a) If 2 is a prime, then Addis Ababa is a capital city of Ethiopia.

b) If -5 is an integer, then all birds have three legs.

c) If $\sqrt{3}$ is rational, then 0 is a negative number

7) Determine the truth values of all the statements given in number 6 above and also determine the truth value of the converse, inverse and contrapositive of the statements given in number 6 above.

8) Let r be "It is hot." and let s be "It is not raining." Translate each of the following into ordinary English statement.

  a) r∧s

  b) ¬r

  c) ¬s

  d) s⇔r

  e) r⇔¬s

  f) s∨¬r

  g) ¬s⇒¬r

4.25 Equivalent Compound Propositions

4.25.1 Definition:- Two compound propositions involving the same component propositions are said to be logically equivalent if and only if they have the same truth values or identical truth tables.

Example 1) Construct truth tables for

  a) ¬(P∨q)

  b) ¬P∧¬q

Solution a)

P	q	Pvq	¬(Pvq)
T	T	T	F
T	F	T	F
F	T	T	F
F	F	F	T

b)

P	q	¬P	¬q	¬P∧¬q
T	T	F	F	F
T	F	F	T	F
F	T	T	F	F
F	F	T	T	T

* Observe the truth table of ¬(Pvq) and (¬P∧¬q) ... Observe the last columns of the above two tables in A and B.)

What do you observe? The truth values are the same. We call it, they are equivalent compound proposition. Equivalence is denoted by "≡"

Thus, ¬(Pvq) ≡ ¬P∧¬q

Example 2:   Construct truth table for
        A) ¬(¬Pvq)
        B) P∧¬q

ANSWER: A)

P	q	¬P	¬Pvq	¬(¬Pvq)
T	T	F	T	F
T	F	F	F	T
F	T	T	T	F
F	F	T	T	F

B)

P	q	¬q	P∧¬q
T	T	F	F
T	F	T	T
F	T	F	F
F	F	T	F

As we observe the above two truth tables, the last columns are identical. Therefore:-
¬(¬P∨q) and (P∧¬q) are equivalent compound-
Propositions or ¬(¬P∨q) ≡ (P∧¬q)

4.25.2 De Morgan's Law
A) ¬(P∧q) ≡ ¬P∨¬q
B) ¬(P∨q) ≡ ¬P∧¬q

4.26 Tautology and Contradiction

Example 1)   Construct the truth table of (P⇒q)⇔(¬P∨q).

P	q	¬P	P⇒q	¬P∨q	(P⇒q)⇔(¬P∨q)
T	T	F	T	T	T
T	F	F	F	F	T
F	T	T	T	T	T
F	F	T	T	T	T

Note:- The truth value of the compound proposition in the above example is T in every combination. Such propositions are called tautologies.

4.26.1 Definition:- A compound propositions are said to be a tautology if and only if it attains a truth value T for every possible combinations of assignments of the truth values for the component propositions.

Example 2)   Construct a truth table for the complex proposition
$(P \land q) \Rightarrow (P \lor q)$.

P	q	P∧q	P∨q	$(P \land q) \Rightarrow (P \lor q)$
T	T	T	T	T
T	F	F	T	T
F	T	F	T	T
F	F	F	F	T

Do you observe that $(P \land q) \Rightarrow (P \lor q)$ is a tautology?

4.27 Note:- The negation of a tautology is a contradiction.

4.27.1 Definition:- A proposition is said to be a contradiction if and
only if it attains a truth value F for every combination of truth
values of the component proposition.

P	q	P∧q	P∨q	¬(P∨q)	(P∧q)∧¬(P∨q)
T	T	T	T	F	F
T	F	F	T	F	F
F	T	F	T	F	F
F	F	F	F	T	F

Observe the last column, you can see that the complex
proposition is a contradiction.

More on truth values and truth tables.

As we were able to form compound propositions from two
component propositions, we can do the same using three or
more component propositions.

Example 1)   Construct truth table to show $(P \land q) \land \neg(P \lor q)$ is a contradiction.
Answer: $P \Leftrightarrow q$ is F, means P and q have opposite truth values.
Since $P \Rightarrow q$ is T, it is not the case P is T and q is F. So q is T.
Or using the table:

P	q	P⇒q	P⇔q
T	T	T	T
T	F	F	F
F	T	T	F
F	F	T	T

We can see from the table that P⇔q is F and P⇒q is T in the third row. There P is F and q is T.

Example 2)   Construct the truth tables of each of P∧(q∨r) and (P∧q)∨(P∧r).

Answer:- Truth table for P∧(q∨r)

There are three component propositions (p, q and r). We expect $2^3=8$, possible combination of T and F.

Truth table for P∧(q∨r)

P	q	r	q∨r	P∧(q∨r)
T	T	T	T	T
T	T	F	T	T
T	F	T	T	T
T	F	F	F	F
F	T	T	T	F
F	T	F	T	F
F	F	T	T	F
F	F	F	F	F

Truth table for (P∧q)∨(P∧r)

P	q	r	P∧q	P∧r	(P∧q)∨(P∧r)
T	T	T	T	T	T
T	T	F	T	F	T
T	F	T	F	T	T

T	F	F	F	F	F
F	T	T	F	F	F
F	T	F	F	F	F
F	F	T	F	F	F
F	F	F	F	F	F

From the above two truth tables, it is clear that

$P \wedge (q \vee r) \equiv (P \wedge q) \vee (P \wedge r)$ ... (They are equivalent). This equivalence shows that "$\wedge$" is distributive over "$\vee$." Similarly, we can show that "$\vee$" is distributive over "$\wedge$" or $P \vee (q \wedge r) \equiv (P \vee q) \wedge (P \vee r)$.

Note:- In general if a compound proposition involves n-component propositions the possible combinations of T and F are 2^n.

Exercise 4.6

1) Construct a truth table to show each of the following are equivalence.
   a) $\neg(P \wedge q) \equiv \neg P \vee \neg q$
   b) $P \vee (q \vee r) \equiv (P \vee q) \vee r$
   c) $P \wedge q \equiv q \wedge P$
   d) $\neg(P \vee q) \equiv \neg P \wedge \neg q$
   e) $P \Rightarrow q \equiv \neg(P \wedge \neg q) \equiv \neg P \vee q$

2) By constructing truth tables show that each of the following compound propositions are tautologies.
   a) $(P \Rightarrow q) \Leftrightarrow (\neg P \vee q)$
   b) $(P \vee q) \Leftrightarrow (q \vee P)$
   c) $q \Rightarrow (P \Rightarrow q)$

3) By constructing truth table show that the following complex propositions are contradiction.
   a) $P \Leftrightarrow \neg P$
   b) $(r \wedge s) \wedge \neg(r \vee s)$

4)   Determine the truth values of the following.
      a) ¬q, if P and P∧q are T.
      b) P, if q and ¬P⇒¬q are T.
      c) ¬q, if ¬(P∧q) and P are T.
      d) P, if P⇒q and ¬q are T.

4.28 Postulate and Theorem

An axiom or postulate is a proposition that is not proved or demonstrated but considered to be either self-evident, or subject to necessary decision. That is to say, an axiom is a logical statement that is assumed to be true. Therefore, its truth is taken for granted, and serves as a starting point for deducing and inferring other (theory dependent) truths.

(Postulate or axiom is a mathematical statement that is accepted without proof.)

Example-1)   Through two points there is exactly one line or two points determine a unique line.

              (This postulates is called unique line postulate)

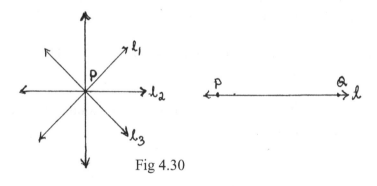

Fig 4.30

As you can easily understand through a single point it is possible to draw infinite lines. But through two points we can only draw one line. This is obvious fact, such obvious facts that can be accepted without proof are called postulates.

Example 2)   Angle addition postulate states that if a point M lies in the interior of ∠ABC, then ∠ABM+∠MBC=∠ABC

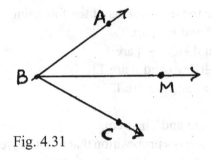

Fig. 4.31

4.28.1 Segment Addition Postulate

If M is between A and B, then AM+MB=AB

If AM+MB=AB, then M is between A and B.

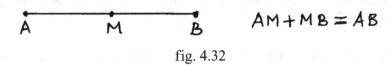

$$AM + MB = AB$$

fig. 4.32

4.28.2 Angle Addition Postulate

If P is in the interior of $\angle XYZ$; then the measure of $\angle XYZ$ is equal to the sum of the measures of $\angle XYP$ and $\angle PYZ$.

Example 3) If M is between A and B, and $\overline{AB} = 20cm$ and $\overline{AM} = 6cm$, then find $\overline{MB}$

Solution

Given: $\overline{AB} = 20cm$, $\overline{AM} = 6cm$ we are required to find the length of $\overline{MB}$

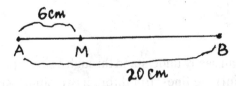

$\overline{AB} = \overline{AM} + \overline{MB}$ .....   Segment addition postulate.

$20cm = 6cm + \overline{MB}$ ....   Substitution

$\overline{MB} = 20cm - 6cm = 14cm$ ....   Subtraction

$\therefore$ The length of $\overline{MB}$ is 14cm.

Example 4)   Suppose ∠a is complementary to ∠b, ∠c is complementary to ∠b, and m(∠b)=43°.

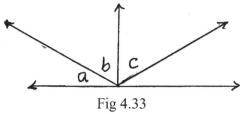

Fig 4.33

a) Find m(∠a)
b) Find m(∠b)
c) What can we say about m(∠a) and m(∠c)?

Solution
a) Given m(∠a)+m(∠b)=90°, m(∠b)+m(∠c)=90° and m(∠b)=43°

    a+b=90
    b+c=90

    ————

    a+43=90....   Substitution given that ∠b=43°
    a=90-43=47 ....   Subtraction
    m(∠a)=47°

    b) b+c=90
    43+c=90 .....   Substitution
    c=90-43=47 ....   Subtraction
    m(∠c)=47°

c)   From the above a and b we can say m(∠a) and m(∠c) are congruent. i.e., m(∠a)≅m(∠c)=47°

Examples 3) In the figures below, m(∠R)≅m(∠S),then

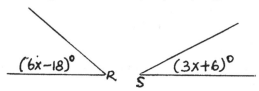

a) Solve for x

b) Find m(∠R) or m(∠S)

Solution:-

a) Since m(∠R)≅m(∠S)

6x-18=3x+6	Given
6x-3x=6+18	Collecting like terms
3x=24	Dividing
x=8	

b)
m(∠R)=(6x-18)° .... Given
= (6x8-18)°	Substitution
= (48-18)° ...	Subtraction
m(∠R)=30°	
m(∠R)≅m(∠S)=30°	

Definition: Theorem is a mathematical proposition that has been or is to be proved on the basis of explicit assumption or by using chain of steps.

To build our geometry we need set of postulates. From the set of postulates we can produce our first theorem and prove it. The set of postulates and proved theorem will give rise to another theorem.

We prove theorems to convince ourself or to convince others. In proving theorems we can use.

1)   THE PARAGRAPH FORM OF PROOF
We can prove the theorem by writing logical arguments until we reach the conclusion.

2)   TWO COLUMN PROOF
In providing a geometrical theorem by using two column proof one can follow the following procedures.

1) State:- the theorem.

2) Restatement:- re-write the theorem using notation on your figure if necessary.

3) Given:- write what is given or write your hypothesis.

4) To prove:- Write what you want to prove or write what your conclusion is.

5) Construction:- Do any construction if necessary to prove the theorem and state it.

	Statement	Reason
6) Proof	_____	_____
	_____	_____
	_____	_____

Write statements of the proof by giving reason for each step until you reach the conclusion required. Your reason may be given facts, definitions, postulates or previously proved theorems.

Example 4)   Parallel postulate:- (Euclid's fifth postulate)
Through a point not on a given line there is exactly one line parallel to the given line.

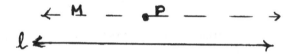

Restatement: Through point P there is exactly one line which is parallel to l.

Postulate: If a transversal cuts two parallel lines in a plane, then the corresponding angles are congruent.

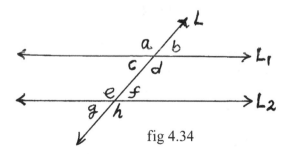

fig 4.34

If L_1 and L_2 are two parallel lines and L is a transversal, then the corresponding angles, that is, (a and e), (c and g), (b and f) and (d and h) are congruent.

Postulate: If two straight lines in a plane are cut by a transversal and if a pair of corresponding angles are congruent, then the first two straight lines are parallel.

Restatement:

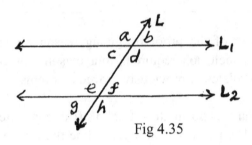

Fig 4.35

If a and e are congruent then L_1 is parallel to L_2.

Theorem 4-1: If two straight lines intersect, then vertically opposite angles are congruent.

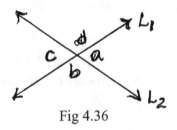

Fig 4.36

Given L_1 and L_2 are two lines intersecting at a point and forming vertically opposite angles b and d.

To prove $\angle b \cong \angle d$

Statement	Reason
Proof 1) $\angle a + \angle b = 180°$	Adjacent angles on a straight line.
2) $\angle d + \angle a = 180°$	Adjacent angles on a straight line.

3) $\angle a + \angle b \cong \angle d + \angle a$ .... From Steps 1 and 2.
4) $\angle b \cong \angle d$ ..... From Steps 3

In the same way c and a are congruent vertically opposite angles.

Theorem 4-2:If two parallel lines are cut by a transversal, then alternate interior angles are congruent.

Given: $l_1 // l_2$, $\angle 4$ and $\angle 5$ alternate interior angles and l a transversal.
To prove $\angle 4 \cong \angle 5$

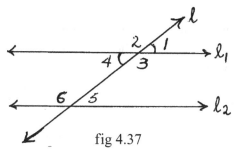

fig 4.37

Proof

	Statement	Reason
	1.) $l_1 // l_2$	Given
	2.) $\angle 4 \cong \angle 1$	Vertically opposite angles
	3) $\angle 1 \cong \angle 5$	Corresponding angles.
	4) $\angle 4 \cong \angle 5$	by steps 2 and 3 (transitivity)

Theorem 4-3:If a transversal is perpendicular to one of two parallel lines, then it is perpendicular to the other lines.

Restatement: If $l_1 // l_2$ and l a transversal where $l \perp l_2$, then $l \perp l_1$.

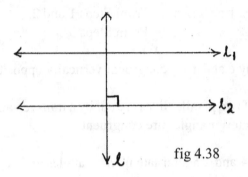

fig 4.38

Theorem 4-4:If two parallel lines are cut by a transversal, then interior angles on the same side of the transversal are supplementary.

Restatement: If $l_1 // l_2$ then a+b=180° or c+d=180°.

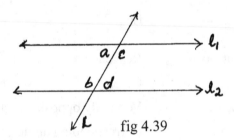

fig 4.39

Theorem 4-5:In a plane if two lines are both parallel to the same line, then they are parallel to each other.

Restatement: If $l_1 // l$ and $l_2 // l$ then $l_1 // l_2$

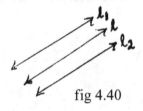

fig 4.40

Example 1) If l_1 and l_2 are parallel lines crossed by a transversal l, find the values of x and y.

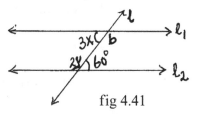

fig 4.41

Solution
1) $3x=60°$ ...... (Alternate interior angles are congruent)
$x=20°$

2) b and 60° are supplementary angles.
(Interior angles on the same side of a transversal)
$b+60°=180°$
$b=120°$

$b=2y$ ..... (Alternate interior angles)
$120°=2y$
$\therefore y=60°$

Example 2) If $m(\angle AOB)=2x$, $m(\angle BOC)=(x+36)$, and $m(\angle AOC)$ is a right angle, then find x.

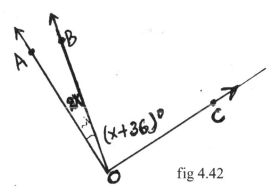

fig 4.42

Solution:- Since $\angle AOC$ is a right angle, $m(\angle AOC)=90°$.

m(∠AOB)+m(∠BOC)=m(∠AOC)   Angle addition postulate

2x+(x+36)=90 ...   Substitute the values of the angles.

3x+36=90 ...   Collecting like terms.

3x=90-36=54 ....   Subtraction

x=18° ....   Division

Example 3) Find m(∠BOC), if m(∠AOC) is right angle.

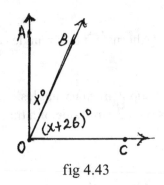

fig 4.43

Solution:- m(∠AOB+m(∠BOC)=90°... Definition of complementary
                                   angles.

x°+(x+26)°=90° ...   Substitution

2x+26=90 ...   Combining like terms

2x=90-26=64 ...   Subtraction

x=32° ...   Division.

Now to find m(∠BOC), we just substitute the value x into the expression (x+26)°

i.e, m(∠BOC)=(x+26)°

= 32°+26°

m(∠BOC)=58°

Example 4) If l is transversal and l_1 and l_2 are parallel lines then find
            the values of x, y and z

A) X=_____

B) Y=_____

C) Z=_____

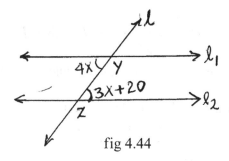

fig 4.44

Solution

A) 4x=3x+20 ....   Alternate interior angles
    4x-3x=20 ....   Collecting like terms.
    x=20

B) 4x+y=180° .....   Straight angle.
    From A above the value of x is 20, then

    4(20)+y=180 ....   Substitution
    80+y=180
    y=180-80=100 ....   Subtraction
    ∴ y=100°

C) 3x+20+z=180° ....   Straight angle.
    From A above the value of x is 20, then

    3(20)+20+z=180 ....   Substitution
    60+20+z=180
    80+z=180 ....   Addition
    z=180-80=100 ....   Subtraction
    ∴ z=100°

Theorems 4-6:- Congruence of Segments
Segment congruence is reflexive, symmetric and transtive

Reflexive: For any segment PQ, $\overline{PQ} \cong \overline{PQ}$.

Symmetric   If $\overline{PQ} \cong \overline{RS}$, then $\overline{RS} \cong \overline{PQ}$.

Transtive   If $\overline{PQ} \cong \overline{RS}$ and $\overline{RS} \cong \overline{TU}$, then $\overline{PQ} \cong \overline{TU}$

Theorem 4-7   Congruence of Angles

Angle congruence is reflexive, symmetric, and transitive.

Reflexive: For any angle P, angle P is congruent to angle P.

Symmetric   If angle P is congruent angle Q, then angle Q is congruent to angle P.

Transitive   If angle P is congruent to angle Q and angle Q is congruent to angle R, then angle P is congruent to angle R.

Theorem 4-8 Right Angle Congruence Theorem

All right angles are congruent.

Example 6)   Find the value of x in each of the following.

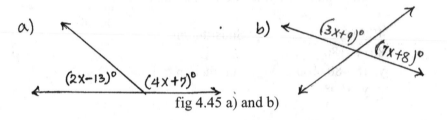

a)

$(2x-13)°$ $(4x+7)°$

b) $(3x+9)°$ $(7x+8)°$

fig 4.45 a) and b)

Solution

a) $(2x-13)°+(4x+7)°=180°$ ..... Straight angle.

2x-13+4x+7=180

6x-6=180 ....   Collecting like terms.

6x=186 ....   Division

x=31

b) $(3x+9)°+(7x+8)°=180°$ .... Definition of supplementary angles

3x+7x+9+8=180

10x+17=180 ....   Collecting like terms

10x=180-17=163 ....   Subtraction

10x=163 ....   Division

x=16.3

Example 7)   Find the values of the variables.

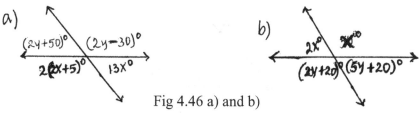

Fig 4.46 a) and b)

Solution

a) $(2y+50)°+(2y-30)°=180°$ .... Definition of supplementary angles
      $2y+50+2y-30=180$ ..
      $4y=180-20=160$ ....   Division
      $y=40$

      $(2y-30)°+13x°=180$ ....   Supplementary angles
      $2(40)-30+13x=180$ ...   Substitution
      $50+13x=180$
      $13x=130$ ....   Division
      $x=10$

b) $(2y+20)°+(5y+20)°=180°$ ..... Definition of supplementary.
      $2y+5y+20+20=180$ ....   Collecting like terms.
      $7y+40=180$ .....   Addition
      $7y=140$ ....   Division.
      $y=20$

      $2x°+(2y+20)°=180°$ .... Supplementary angles
      $2x+2(20)+20=180$ ....   Substitution
      $2x+60=180$ ... Addition
      $2x=180-60°$ .... Subtraction
      $x=60°$

Exercise 4-7

1)   If M is between A and B, the length of the segment $\overline{AB}$ is 48
      inches and $\overline{AM}$ =24 inches. What is the length of $\overline{MB}$?

2) In the figure below M(∠ABC)=100° and M(∠ABM)=80°, what is M(∠MBC)?

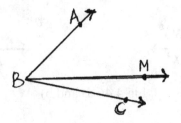

3) Suppose ∠a is complementary to ∠b, ∠c is complementary to ∠b, and m(∠b)=51°.

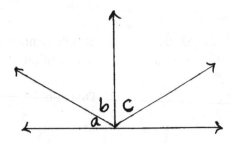

a) Find m(∠a)
b) Find m(∠c)
c) What can we say about m(∠a) and m(∠c)?

4) In the figures below, m(∠P)≅m(∠Q), then

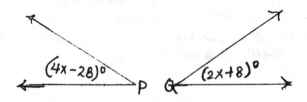

a) Solve for x
b) Find m(∠P) or m(∠Q)

5) If l_1 and l_2 are parallel lines crossed by a transversal l, find the values of x and y.

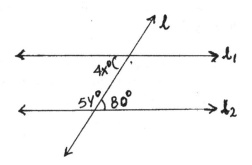

6) Find the values of x and y in each of the following

a)

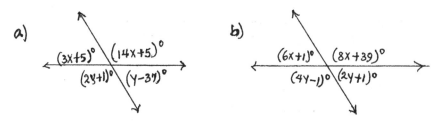

b)

7) If m($\angle$AOB)=3x°, m($\angle$BOC)=(x+6)° and m($\angle$AOC) is a right angle, then

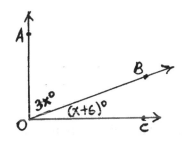

a) Find the value of x
b) Find m($\angle$AOB)
c) Find m($\angle$BOC)

4.29 Distance From a Point to a Line

The distance from a point to a line is the length of the perpendicular segment from the point to the line. It is calculated by the following formula.

If we have a line whose equation has the form ax+by+c=0 and a point (p, q), then the formula which gives the distance between a point (p, q) and a line ax+by+c=0 is:-

$$d = \frac{|ap+bq+c|}{\sqrt{a^2+b^2}}$$

Example 1) Find the distance from a point (2, 6) to a line 3x+4y+5=0.

Solution:- From the equation of a line 3x+4y+5=0
a=3, b=4 and c=5, and from a point (2, 6), p=2 and q=6

$$d = \frac{|ap+bq+c|}{\sqrt{a^2+b^2}}$$

$$d = \frac{|3(2)+4(6)+5|}{\sqrt{3^2+4^2}}$$

$$d = \frac{|6+24+5|}{\sqrt{25}}$$

$$d = \frac{35}{5}$$

$$d = 7$$

Example 2) Find the distance from a point (-2, 6) to a line x-y+3=0

Solution:- From the equation of a line x-y+3=0
a=1, b=-1 and c=3, and from a point (-2, 6), p=-2 and q=6.

$$d = \frac{|ap + bq + c|}{\sqrt{a^2 + b^2}}$$

$$d = \frac{|1(-2) + (-1)(6) + 3|}{\sqrt{1^2 + (-1)^2}}$$

$$d = \frac{|-6|}{\sqrt{2}}$$

$$d = \frac{6}{\sqrt{2}} = 3\sqrt{2}$$

(Rationalizing the denominator)

Example 3)   Find the value of k such that the line containing point (2, k) is perpendicular to the line y=2x-3 at point (4, 5)

Solution:-

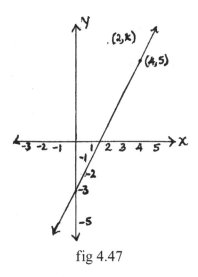

fig 4.47

The slope of a line y=2x-3 is 2 and the slope of a perpendicular segment from point (2, k) to line y=2x-3 is the negative reciprocal of 2, or $\frac{-1}{2}$. The segment from (4, 5) to (2, k) has slope $\frac{-1}{2}$. So, the segment is perpendicular to the line y=2x-3. To find the value of k, we use slope formula.

$$M = \frac{y_2 - y_1}{x_2 - x_1} ,$$

$$\frac{-1}{2} = \frac{5 - k}{4 - 2} \qquad \text{(Substitution)}$$

$$\frac{-1}{2} = \frac{5 - k}{2}$$

-2=10-2k

-2-10=-2k ...   Collecting like terms.

-12=-2k

$$k = \frac{-12}{-2} \ .... \qquad \text{Division}$$

k=6

Example 4)   If the shortest distance from a line containing a point (3, R) which is perpendicular to a line 4x+8y+5=0 is 6 units. What is the value of R?

Solution

$$d = \frac{|ap + bq + c|}{\sqrt{a^2 + b^2}}$$

$$6 = \frac{|4(3) + 8(R) + 5|}{\sqrt{4^2 + 8^2}}$$

$$6 = \frac{|12 + 8R + 5|}{\sqrt{80}}$$

$$6 = \frac{|8R + 17|}{\sqrt{80}}$$

$$|8R + 17| = 6\sqrt{80}$$

$$6\sqrt{80} = |8R + 17|$$

Case 1	Case 2

$$6\sqrt{80} = 8R + 17 \qquad\qquad 6\sqrt{80} = -(8R + 17)$$

$$8R = 6\sqrt{80} - 17 \qquad\qquad 6\sqrt{80} = -8R - 17$$

$$8R = 24\sqrt{5} - 17 \qquad\qquad 6\sqrt{80} + 17 = -8R$$

$$R = \frac{24\sqrt{5} - 17}{8} \qquad\qquad R = \frac{24\sqrt{5} + 17}{-8}$$

The values of R is $\dfrac{24\sqrt{5} - 17}{8}$ or $\dfrac{24\sqrt{5} + 17}{-8}$ units

Exercise 4-8

1) Find the distance from a point (6, 4) to a line y=3x-6.

2) Find the value of k such that the line containing point (5, k) is perpendicular to the line $y = \dfrac{1}{2}x - \dfrac{1}{2}$ at a point (6, 2•5)

3) If the shortest distance from a line containing a point (5, k), which is perpendicular to a line 2x+3y+4=0 is 7 units. What is the value of k?

4.30 Congruencies of Triangles

In geometry, two figures are congruent if they have the same shape and size. More formally, two sets of points are called congruent if, and only if, one can be transformed into the other by an isometry, i.e., a combination of translations, rotations and reflections.

The related concept of similarity permits a change in size.

4.30.1 Congruence and One to One Correspondence

Notation:- "A↔B" is read as "A corresponds to B."

Definition: Two triangles are congruent if their corresponding sides are equal in length and their corresponding angles are equal in size.

If triangle ABC is congruent to triangle DEF, the relationship can be written mathematically as:

△ABC≅△DEF (read as triangle ABC is congruent to triangle DEF) shows:

	Corresponding angles		Corresponding sides
1	∠A≅∠D	4	$\overline{AB} \cong \overline{DE}$
2	∠B≅∠E	5	$\overline{BC} \cong \overline{EF}$
3	∠C≅∠F	6	$\overline{AC} \cong \overline{DF}$

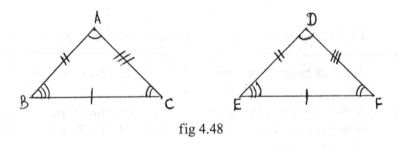

fig 4.48

The similar marks that we put on the figures show that the corresponding congruent parts.

Example) $\overline{AB} \cong \overline{DE}$ is indicated by inserting the mark "//" on both segments. Remember that two line segments are congruent if they have the same length. Corresponding vertices are often used to name the segments. The correspondence between the triangles can easily be remembered by the diagram.

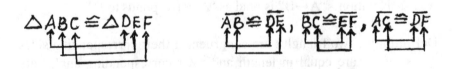

Note:- If the order of the vertices is ΔABC is changed, then the order of vertices in ΔDEF must be changed so that the correspondence is preserved. Therefore, ΔBCA≅ΔEFD, ΔACB≅ΔDFE. And so on. Since the order of the vertices indicates the correspondence. The corresponding parts can be determined from the written congruence without an illustration.

In addition to the congruency of the two triangles the statement ΔABC≅ΔDEF also tells us how to match the triangles, so that they fit on each other. In detail, it stays that the correspondence (abbreviated ABC↔DEF).

1) ∠A↔∠D
2) ∠B↔∠E
3) ∠C↔∠F
4) $\overline{AB} \leftrightarrow \overline{DE}$
5) $\overline{BC} \leftrightarrow \overline{EF}$
6) $\overline{AC} \leftrightarrow \overline{DF}$

From the above, if ΔABC≅ΔDEF, then

1)	ΔACB≅ΔDFE	∠A≅∠D, ∠C≅∠F, ∠B≅∠E
2)	ΔBAC≅ΔEDF	∠B≅∠E, ∠A≅∠D, ∠C≅∠F
3)	ΔBCA≅ΔEFD	∠B≅∠E, ∠C≅∠F, ∠A≅∠D
4)	ΔCAB≅ΔFDE	∠C≅∠F, ∠A≅∠D, ∠B≅∠E
5)	ΔCBA≅ΔFED	∠C≅∠F, ∠B≅∠E, ∠A≅∠D

Exercise 4-9

. 1)   If ΔPQR≅ΔSTU, then
      a) ∠P ≅ _____
      b) ∠Q ≅ _____
      c) ∠R ≅ _____
      d) $\overline{PR}$ ≅ _____
      e) $\overline{QR}$ ≅ _____

f) $\overline{PQ} \cong$ _____
g) $\triangle QPR \cong$ _____
h) $\triangle QRP \cong$ _____

2) If $\triangle XYZ \cong \triangle FGH$, then
   a) $\overline{XZ} \cong$ _____
   b) $\overline{XY} \cong$ _____
   c) $\triangle YXZ \cong$ _____
   d) $\triangle ZXY \cong$ _____

4.31 Congruence Postulates and Theorems
The Side-Angle-Side (SAS) Postulate

- SAS (Side-Angle-Side): If two pairs of sides of two triangles are equal in length, and the included angles are equal in measurement, then the triangles are congruent.

In this postulate we are concerned only on the three parts of each triangle, here our idea is if we know this much. It is enough to show that the two triangles are congruent and thereby the six facts given in the definition of congruence will be true.

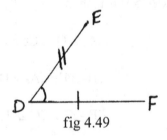

fig 4.49

1) If $\angle A \cong \angle D$

2) $\overline{BA} \cong \overline{ED}$

3) $\overline{AC} \cong \overline{DF}$ there is no way that $\overline{BC}$ is not congruent to $\overline{EF}$ and we can easily see that $\overline{BC} \cong \overline{EF}$ and $\triangle ABC \cong \triangle DEF$ by SAS postulate from these it follows that.

4) $\angle B \cong \angle E$

5) $\angle C \cong \angle F$ and

6) $\overline{BC} \cong \overline{EF}$ and the six conditions given in the definition will be fulfilled.

Example 1) Using the SAS congruence postulate show that
ΔXYZ≅ΔPQR

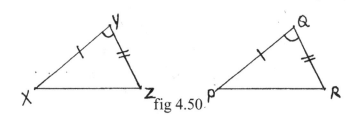

fig 4.50

Solution

	Statements	Reason
1)	$\overline{XY} \cong \overline{PQ}$ ...	Given
2)	$\overline{YZ} \cong \overline{QR}$	Given
3)	∠Y≅∠Q ...	Given
4)	ΔXYZ≅ΔPQR ...	by SAS postulate (Steps 1, 2 and 3)

Example 2)   Show that ΔPQS≅ΔRQS by SAS postulate.

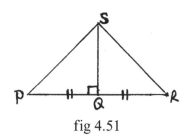

fig 4.51

	Statements	Reason
1)	$\overline{PQ} \cong \overline{RQ}$	Given
2)	$\overline{QS} \cong \overline{QS}$	Reflexive
3)	∠PQS≅∠RQS ...	Given
4)	ΔPQS≅ΔRQS ...	by SAS postulate.

Example 3) If $\overline{LJ} \cong \overline{IK}$; $\angle a \cong \angle b$

I is the mid point of $\overline{HJ}$, then show that $\Delta HLI \cong \Delta JKI$

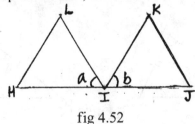

fig 4.52

	Statements	Reason
1)	$\overline{HI} \cong \overline{JI}$	I is the mid point of $\overline{HJ}$
2)	$\overline{IL} \cong \overline{IK}$	Given
3)	$\angle HIL \cong \angle JIK$ ...	Given
4)	$\Delta HIL \cong \Delta JIK$ ...	By SAS postulate.
5)	$\Delta HLI \cong \Delta JKI$ ...	By Step 4 and correspondence.

Example 4) Show that $\Delta ACB \cong \Delta ECD$ by SAS postulate.

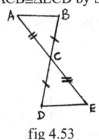

fig 4.53

	Statements	Reason
1)	$\overline{AC} \cong \overline{EC}$	Given
2)	$\overline{CB} \cong \overline{CD}$	Given
3)	$\angle ACB \cong \angle ECD$	Vertically opposite angles.
4)	$\Delta ACB \cong \Delta ECD$ ...	By SAS postulate.

Theorem 4-9: The Angle-Side-Angle Theorem (ASA).

(Two angles and included side theorem):- If two triangles have two angles and the included side of one congruent to two angles and the included side of the other, then the triangles are congruent.

Example 1)   Show that △ACB≅△EFD

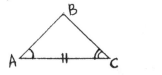 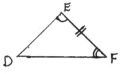

fig 4.54

	Statements	Reason
1)	∠A≅∠E	Given
2)	∠C≅∠F	Given
3)	$\overline{AC} \cong \overline{EF}$ ...	Given
4)	△ACB≅△EFD ...	by angle side angle (ASA)

Theorem 4-10 Side-Side-Side (SSS):- If two triangles have three sides of one congruent respectively to three sides of the other. Then the triangles are congruent.

Example 1)   Given: $\overline{DA} \cong \overline{BC}$,
$AB \cong CD$
Prove that △DAB≅△BCD

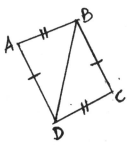

fig 4.55

	Statements	Reason
1)	$\overline{DA} \cong \overline{BC}$	Given
2)	$\overline{AB} \cong \overline{CD}$	Given
3)	$\overline{BD} \cong \overline{BD}$	Reflexive property of congruence
4)	△DAB≅△BCD ...	By SSS

Example 2)   Show that ΔFAB≅ΔDFE

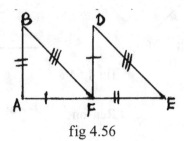

fig 4.56

	Statements	Reason
1)	$\overline{FA} \cong \overline{DF}$	Given
2)	$\overline{AB} \cong \overline{FE}$	Given
3)	$\overline{BF} \cong \overline{ED}$	Given
4)	ΔFAB≅ΔDFE	by SSS

Example 3)   Given: $\overline{DA} \cong \overline{BA}$
            C is the mid point of $\overline{DB}$.

Prove that ΔDAC≅ΔBAC

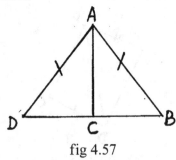

fig 4.57

	Statements	Reason
1)	$\overline{DA} \cong \overline{BA}$	Given
2)	C is midpoint of $\overline{DB}$	Given
3)	$\overline{DC} \cong \overline{BC}$	Definition of midpoint
4)	$\overline{AC} \cong \overline{AC}$	Reflexive property of congruence
5)	ΔDAC≅ΔBAC	by SSS

Two Angles and non-included side theorem

(In ΔPQR and ΔSTU, if ∠P≅∠S, ∠Q≅∠T, and $\overline{QR} \cong \overline{TU}$, then ΔPQR≅ΔSTU)

Two angles and a corresponding side theorem (AAS theorem):- if two angles and a side of one triangle are congruent to two angles and the corresponding sides of another, then the triangles are congruent.

Example 1) Given ∠P≅∠T
                    ∠a≅∠b,
                    $\overline{PR}$ bisects $\overline{QS}$
Prove that ΔPQR≅ΔTSR

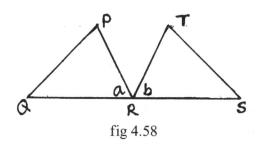

fig 4.58

	Statements	Reason
1)	∠P≅∠T	Given
2)	∠a≅∠b	Given
3)	$\overline{PR}$ bisects $\overline{QS}$	Given
4)	R is the midpoint of $\overline{QS}$	Definition of segment bisector
5)	$\overline{QR} \cong \overline{RS}$	Midpoint Theorem
6)	ΔPQR≅ΔTSR	By AAS Theorem

Example 2)   Prove that ΔABC≅ΔEFD

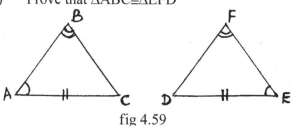

fig 4.59

Statements	Reason
1) $\angle A \cong \angle E$	Given
2) $\angle B \cong \angle F$	Given
3) $\overline{AC} \cong \overline{ED}$	Given
4) $\triangle ABC \cong \triangle EFD$	By AAS

Example 3) In the figure below, $\angle D \cong \angle B$ and $\overline{CB} // \overline{DA}$. prove that $\triangle DCA \cong \triangle BAC$

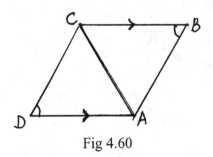

Fig 4.60

Statements	Reason
1) $\angle D \cong \angle B$	Given
2) $\overline{CB} // \overline{DA}$	Given
3) $\angle BCA \cong \angle DAC$	Alternate interior angles.
4) $\overline{AC} \cong \overline{AC}$	Reflexive property
5) $\triangle DCA \cong \triangle BAC$ ...	By AAS Congruence Theorem

Theorem 4-11: Hypotenuse-Acute-Angle (HA)
If the hypotenuse and acute angle of a right angled triangle are congruent to the corresponding parts of another right angled triangle, then the triangles are congruent.

Restatement

If $\triangle PQR$ is right angled at R and $\triangle STU$ is a right angled at T and the hypotenuse $\overline{PQ} \cong \overline{SU}$ and the acute angles P and S are congruent, then $\triangle PQR \cong \triangle SUT$

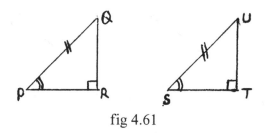

fig 4.61

Proof

	Statements	Reason
1)	∠P≅∠S	Given
2)	$\overline{PQ} \cong \overline{SU}$	Given
3)	∠R≅∠T	Right angles
4)	ΔPQR≅ΔSUT	AAS theorem

Theorem 4-12: The Leg and Acute Angle (LA)
If a leg and acute angle of one-right angled triangle are congruent to the corresponding parts of another right angled triangle, then the triangles are congruent.

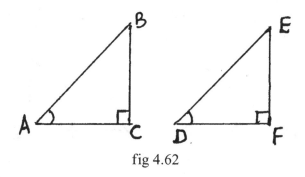

fig 4.62

Restatement

case -1) ΔABC is right angled at C.

ΔDEF is right angled at F.

∠A≅∠D and $\overline{AC} \cong \overline{DF}$

ΔABC≅ΔDEF by LA theorem.

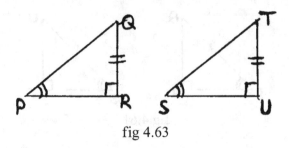

fig 4.63

case -2) ΔPQR is right angled at R.

ΔSTU is right angled at U.

∠P≅∠S

ΔPQR≅ΔSTU

Theorem 4-13: The Leg-Leg (LL)

If the two legs of one right angled triangle are congruent to the corresponding legs of the second, then the two triangles are congruent.

Restatement

In right angled triangles PQR and XYZ which are right angles at R and Z, respectively. If $\overline{PR} \cong \overline{XZ}$ and $\overline{RQ} \cong \overline{YZ}$, then ΔPQR≅ΔXYZ.

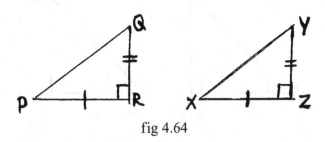

fig 4.64

	Statements	Reason
1)	$\overline{PR} \cong \overline{XZ}$	Given
2)	$\overline{QR} \cong \overline{YZ}$	Given
3)	∠R≅∠Z	Right angles
4)	ΔPQR≅ΔXYZ	By SAS theorem

Theorem 4-14: Right-Angled-Hypotenuse-Side Theorem (RHS) or HL:-
If in two right-angled triangles the hypotenuse and one side of one triangle are congruent respectively to the hypotenuse and one side of the other, then the two triangles are congruent.

Example 1)   Given $\overline{FD} \cong \overline{GI}$, $\overline{FE} \cong \overline{GH}$ and $\angle E \cong \angle H$.
               Prove that $\triangle FED \cong \triangle GHI$

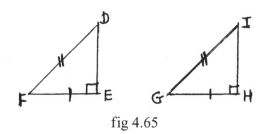

fig 4.65

	Statements	Reason
1)	$\overline{FD} \cong \overline{GI}$	Given
2)	$\overline{FE} \cong \overline{GH}$	Given
3)	$\angle E \cong \angle H$	Right angles
4)	$\triangle FED \cong \triangle GHI$	By RHS or HL theorem.

Example 2)   In the figure below prove that $\triangle ABD \cong \triangle CDB$

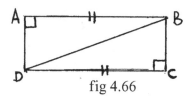

fig 4.66

	Statements	Reason
1)	$\overline{AB} \cong \overline{CD}$	Given
2)	$\angle A \cong \angle C$	Right angles
3)	$\overline{BD} \cong \overline{DB}$	Reflexive
4)	$\triangle ABD \cong \triangle CDB$	By RHS or HL theorem.

Example 3) If $\overline{RS} \cong \overline{PT}$, prove that ΔQPR and ΔQRP are congruent.

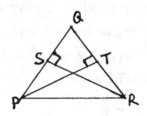

	Statements	Reason
1)	$\overline{PR} \cong \overline{RP}$	Reflexive (Hypotenuse)
2)	$\overline{RS} \cong \overline{PT}$	Given
3)	∠PSR≅∠RTP	Given right angles
4)	ΔPRS≅ΔRPT	By RHS or HL
5)	∠RPS≅∠PRT or ΔQPR≅ΔQRP...	Corresponding parts of congruent triangle.

Exercise 4-10

1)   Complete each of the following congruence statement.

   a) figures

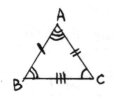

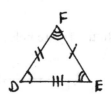

ΔBCA ≅ Δ_____

b) figures

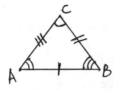

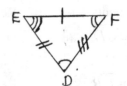

ΔBCA ≅ Δ_____

2)   Complete each of the following congruence statement.
    a) figure

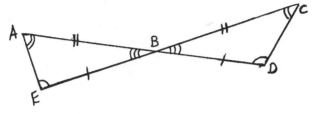

ΔCDB ≅ Δ_____

    b) figure

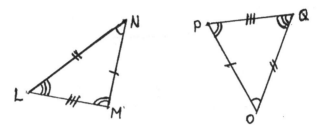

ΔNLM ≅ Δ_____

3)   Complete each congruence statement.
    a) figure

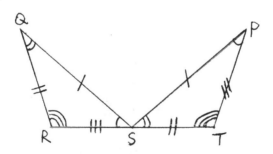

ΔPTS ≅ Δ_____

b) figure

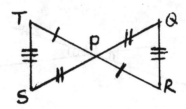

ΔTSP ≅ Δ_____

4)   Complete each congruence statement.

a) figure

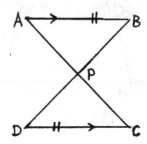

ΔABP ≅ Δ_____

b) figrue

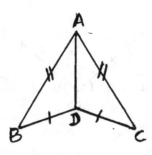

ΔBAD ≅ Δ_____

5) Complete each congruence statement.
    a) figure

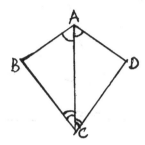

$\Delta BAC \cong \Delta$____

b) figure

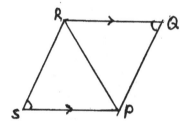

$\Delta SRP \cong \Delta$____

6) Complete each congruence statement.
    a) figure

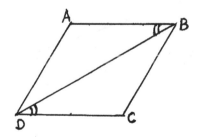

$\Delta ABD \cong \Delta$____

b) figure

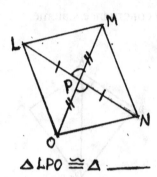

$\triangle LPO \cong \triangle$ _____

$\triangle LPO \cong \triangle$_____

7)   Complete each congruence statement.
   a) figure

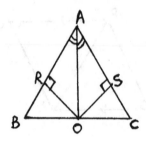

$\triangle OSA \cong \triangle$_____

b) figure

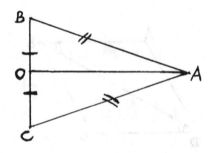

$\triangle AOB \cong \triangle$_____

8) Complete each congruence statements in the diagram shown below, Z is midpoint of XY, and $\overline{XB}$ and $\overline{YA}$ are both perpendicular to $\overline{CA}$.

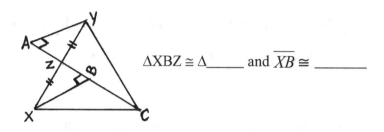

$\triangle XBZ \cong \triangle$_____ and $\overline{XB} \cong$ _____

9) Find the value of x for each pair of congruent triangles.

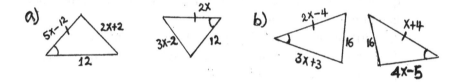

CHAPTER FIVE

5. RELATIONSHIPS IN TRIANGLES AND QUADRILATERALS

5.1 A quadrilateral is a polygon with four sides (or 'edges') and four vertices or corners. Sometimes, the term quadrangle is used, by analogy with triangle, and sometimes tetragon for consistency with pentagon (5-sided), hexagon (6-sided) and so on. The word quadrilateral is made of the word quad (meaning "four") and lateral meaning "of sides").

Quadrilaterals are simple (not self-intersecting) or complex (self-intersecting). Also called crossed. Simple quadrilaterals are either convex or concave.

The interior angles of a simple quadrilateral add up to 360 degrees of arc. This is a special case of the n-gon interior angle sum formula (n-2)×180°. In a crossed quadrilateral, the interior angles on either side of the crossing add up to 720°.

5.1.1 Convex quadrilaterals:- Parallelograms.

5.1.1.1 A parallelogram is a quadrilateral with two pairs of parallel sides. Equivalent conditions are that opposite sides are of equal length, that opposite angles are equal; or that the diagonals bisect each other. Parallelograms also include the square, rectangle, rhombus, and rhomboid.

5.1.1.2 Rhombus: or rhomb: all four sides are of equal length. Equivalent conditions are that opposite sides are parallel and opposite angles are equal, or that the diagonals perpendicularly bisect each other.

5.1.1.3 Rhomboid: A parallelogram in which adjacent sides are of unequal lengths and angles are oblique (not right angles).

5.1.1.4 Rectangle: all four angles are right angles. An equivalent condition is that the diagonals bisect each other and are equal in length.

5.1.1.5 Square (regular quadrilateral): All four sides are of equal length (equilateral), and all four angles are right angles. An equivalent condition is that opposite sides are parallel (a square is a parallelogram), that the diagonals perpendicularly bisect each other; and are of equal length. A quadrilateral is a square if and only if it is both a rhombus and a rectangle (four equal sides and four equal angles).

5.1.2 Convex quadrilaterals - Other

5.1.2.1 Kite:- two pairs of adjacent sides are of equal length. This implies that one diagonal divides the kite into congruent triangles, and so the angles between the two pairs of equal sides are equal in measure. It also implies that the diagonals are perpendicular. (It is common, especially in the discussions on planetessellations, to refer to the concave quadrilateral with these properties as a dart or arrowhead, with term kite being restricted to the convex shape.)

5.1.2.2 Orthodiagonal quadrilateral: the diagonals cross at right angles.

5.1.2.3 Trapezoid (American English) or Trapezium (British English):- One pair of opposite sides are parallel. It is a type of quadrilateral that is not a parallelogram.

5.1.2.4 Isosceles trapezium or Isosceles trapezoid: one pair of opposite sides are parallel and the base angles are equal in measure. This implies that the other two sides are of equal length, and that the diagonals are of equal length.

5.1.2.5 Cyclic quadrilateral:- the four vertices lie on a circumscribed circle. A quadrilateral is cyclic if and only if opposite angles sum to 180°.

5.1.2.6 Tangential quadrilateral: the four edges are tangential to an inscribed circle. Another term for a tangential polygon is inscriptible.

5.1.2.7 Biscentric quadrilateral: both cyclic and tangential.

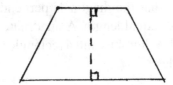

Trapezoid (American English)
Trapezium (British Eng.)

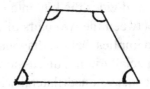

Isosceles trapezoid (Am. English)
Isosceles trapezium (British English)

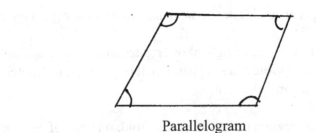

Parallelogram

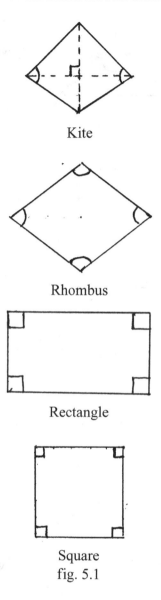

Kite

Rhombus

Rectangle

Square
fig. 5.1

5.2. A midsegment of a triangle is a segment joining the midpoints of two sides of a triangle.

Midsegment theorem: The segment connecting the midpoints of two sides of a triangle is parallel to the third side and is half as long as that side.

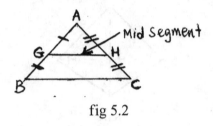

fig 5.2

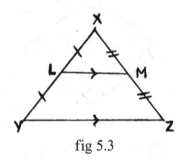

fig 5.3

L is midpoint of $\overline{XY}$, M is midpoint of $\overline{XZ}$, mid segment $\overline{LM}$; $\overline{LM}$ //

$\overline{YZ}$; $\overline{LM} = \dfrac{1}{2}\overline{YZ}$

Example 1) Given $\overline{OP}$ is the length of the midsegment. Find $\overline{MN}$.

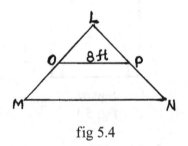

fig 5.4

Solution:

The midpoint segment, $\overline{OP} = 8\,ft$ is half of the third side, $\overline{MN}$, thus,

$$\overline{OP} = \frac{1}{2}\overline{MN}$$

$$8 = \frac{1}{2}\overline{MN}$$

$$\overline{MN} = 8 \times 2 = 16\,ft$$

Example 2)   In the diagram below, if $\overline{DE}$ and $\overline{EF}$ are midsegments of

ΔABC. Find the length of $\overline{AC}$ and $\overline{EF}$

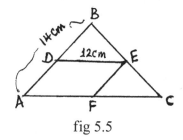

fig 5.5

Solution:

$$\overline{DE} = \frac{1}{2}\overline{AC}$$

$$\overline{AC} = 2\overline{DE}$$

$$= 2\left(12cm\right)$$

$$\overline{AC} = 24cm$$

$$\overline{EF} = \frac{1}{2}\overline{AB}$$

$$= \frac{1}{2}\left(14cm\right)$$

$$\overline{EF} = 7cm$$

Example 3)   If $\overline{PQ}$ is the midpoint of ΔXYZ. Find the value of y.

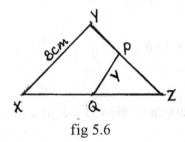

fig 5.6

Solution:

The midpoint segment, $\overline{PQ} = Y$, is half of the third side $\overline{XY} = 8cm$, thus,

$$\overline{PQ} = \frac{1}{2}\overline{XY}$$

$$y = \frac{1}{2}(8cm)$$

$$y = 4cm$$

y=½ (8 cm)
y=4cm

Example 4)   If ΔABC, below, if $\overline{DE}$ is the midsegment of ΔABC. If $\overline{DE} = 2x+12$ and $\overline{BC} = 2x+34$, find the value of x.

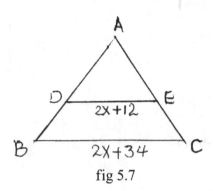

fig 5.7

Solution:

The midpoint segment $\overline{DE} = 2x+12$, is half of the third side, $\overline{BC} = 2x+34$, thus,

$$\overline{DE} = \frac{1}{2}\overline{BC}$$

$$2x+12 = \frac{1}{2}(2x+34)$$

2(2x+12)=2x+34
4x+24=2x+34
2x=10
x=5

NOTE:- If ΔDEF is the midsegment triangle of ΔABC, then the perimeter of ΔDEF is half of the perimeter of ΔABC.
   i.e. Perimeter of ΔDEF= $\frac{1}{2}$ (perimeter of ΔABC)

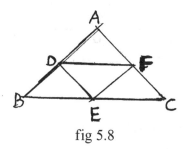

fig 5.8

Example 5)   In the figure below, if the perimeter of ΔABC is 40 cm. and ΔDEF is the midsegment triangle of ΔABC. Find the perimeter of ΔDEF.

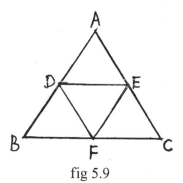

fig 5.9

Solution:

P(ΔDEF)=½P(ΔABC)

= ½× 40cm

P(ΔDEF)=20cm

Example 6)   Given: $\overline{PQ} = 54cm$, $\overline{PR} = 42cm$, $\overline{QR} = 36cm$, S, T and U are midpoints of ΔPQR. Find the perimeter of ΔSTU.

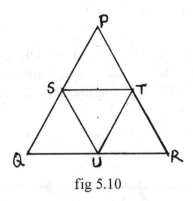

fig 5.10

Solution:

$$P(\Delta PQR) = \overline{PQ} + \overline{PR} + \overline{QR}$$

= 54cm+42cm+36cm

P(ΔPQR)=132cm

P(ΔSTU)=½P(ΔPQR)

P(ΔSTU)=½(132cm)

P(ΔSTU)=66cm

Example 7)   Given: P(3, 2), Q(1, 6) and R(7, 4) as the midpoints of the sides of triangle ABC. Find the coordinates of the vertices of triangle ABC.

Solution:-

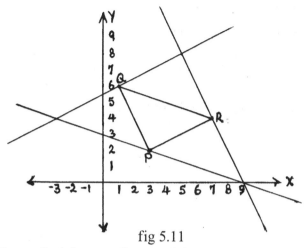

fig 5.11

We can find the coordinates of the vertices of triangle ABC by different methods. Here, let us use the midpoint formulas and solving using simultaneous equations.

Thus, let the coordinates of the vertices of triangle ABC be (x_1, y_1) (x_2, y_2) and (x_3, y_3)

$$\frac{x_1 + x_2}{2} = 3 \Rightarrow x_1 + x_2 = 6....(*)$$

$$\frac{x_1 + x_3}{2} = 7 \Rightarrow x_1 + x_3 = 14....(**)$$

$$\frac{x_2 + x_3}{2} = 1 \Rightarrow x_2 + x_3 = 2....(***)$$

Combine equation (*) and equation (**)

$$(-)\frac{\begin{cases} x_1 + x_2 = 6 \\ x_1 + x_3 = 14 \end{cases}}{x_2 - x_3 = -8}$$

Subtract eqn. (**) from eqn. (*)

Combine x_2-x_3=-8 with equation (***)

$$+\frac{\begin{cases} x_2 - x_3 = -8 \\ x_2 + x_3 = 2 \end{cases}}{2x_2 = -6} \qquad \text{Addition}$$

$x_2 = -3$ Division

Substitute $x_2 = -3$ in either eqn (*) or eqn. (***)

$x_1 + x_2 = 6$
$x_1 + (-3) = 6$ Substitution
$x_1 = 6 + 3 = 9$
$x_1 = 9$

$x_1 + x_3 = 14$
$9 + x_3 = 14$ Substitution
$x_3 = 14 - 9 = 5$
$x_3 = 5$

$$\frac{y_1 + y_2}{2} = 2$$

$$\frac{y_1 + y_2}{2} = 2 \Rightarrow y_1 + y_2 = 4 - - - - - (*)$$

$$\frac{y_1 + y_3}{2} = 4 \Rightarrow y_1 + y_3 = 8 - - - - - (**)$$

$$\frac{y_2 + y_3}{2} = 6 \Rightarrow y_2 + y_3 = 12 - - - - - (***)$$

Combine equation (*) and equation (**)

$$(-)\frac{\begin{cases} y_1 + y_2 = 4 \\ y_1 + y_3 = 8 \end{cases}}{y_2 - y_3 = 4 - 8 = -4} \qquad \text{Subtract eqn. (**) from eqn. (*)}$$

$$y_2 - y_3 = -4$$

Combine $y_2 - y_3 = -4$ with equation (***)

$$(+)\frac{\begin{cases} y_2 - y_3 = -4 \\ y_2 + y_3 = 12 \end{cases}}{2y_2 = 8} \qquad \text{Addition}$$

$y_2 = 4$ Division

Substitute $y_2 = 4$ in either eqn. (*) or eqn. (***)

$y_1 + y_2 = 4$
$y_1 + 4 = 4$ Substitution
$y_1 = 0$

$y_1 + y_3 = 8$
$0 + y_3 = 8$ Substitution
$y_3 = 8$

Thus, the coordinates of the vertices of triangle ABC are $(x_1, y_1) = (9, 0)$, $(x_2, y_2) = (-3, 4)$, and $(x_3, y_3) = (5, 8)$.

Example 8)   Given: D, E, and F are midpoints of the sides of $\triangle ABC$, if $\overline{DE} = 5x + 6$, $\overline{BA} = 6x + 4$, and $\overline{DF} = 4x + 7$, find

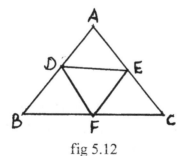

fig 5.12

a) $\overline{BC}$
b) $\overline{EF}$
c) $\overline{AC}$
d) The perimeter of $\triangle ABC$
e) The perimeter of $\triangle DEF$

Solution:

a) Given: $\overline{DE} = 5x + 6$, $\overline{BC} = ?$

$\overline{DE} = \dfrac{1}{2}\overline{BC}$

$\overline{BC} = 2\overline{DE}$

$\qquad = 2(5x + 6)$

$\overline{BC} = 10x + 12$

b) Given: $\overline{BA} = 6x + 4$, $\overline{EF} = ?$

$\overline{EF} = \dfrac{1}{2}\overline{AB}$

$\overline{EF} = \dfrac{1}{2}(6x + 4)$

$\overline{EF} = 3x + 2$

c) Given: $\overline{DF} = 4x + 7$, $\overline{AC} = ?$

$\overline{DF} = \dfrac{1}{2}\overline{AC}$

$\overline{AC} = 2\overline{DF}$

$\qquad = 2(4x + 7)$

$\overline{AC} = 8x + 14$

d) Perimeter of $\triangle ABC = \overline{AB} + \overline{BC} + \overline{AC}$
$\qquad\qquad = (6x+4)+(10x+12)+(8x+14)$
$\qquad\qquad = 24x+30$
$\qquad P(\triangle ABC) = 6(4x+5)$

e) Perimeter of $\triangle DEF = \frac{1}{2}\, P(\triangle ABC)$
$\qquad\qquad = \frac{1}{2}[6(4x+5)]$
$\qquad\qquad P(\triangle DEF) = 3(4x+5)$

5.3   Perpendicular Bisectors
        In Geometry, bisection is the division of something into two
        equal or congruent parts, usually by a line, which is then called
        a bisector. The most often considered types of bisectors are the
        segment bisector (a line that passes through the midpoint of a
        given segment) and the angle bisector (a line that passes through
        the apex of an angle, that divides it into two equal angles)

5.4   Line segment bisector
        A line segment bisector passes through the midpoint of the
        segment, particularly important is the perpendicular bisector
        of a segment, which, according to its name, meets the segment
        at right angles. The perpendicular bisector of a segment also
        has the property that each of its points is equidistant from the
        segment's endpoints.

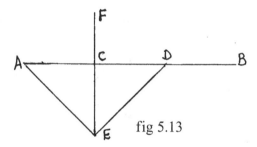

fig 5.13

Line DE bisects line AB at D, line EF is a perpendicular
bisector of segment AD at C and the interior bisector of right
angle AED.

5.5   Angle Bisector
        An angle bisector divides the angle into two angles with equal
        measures. An angle only has one bisector. Each point of an
        angle bisector is equidistant from the sides of the angle.

        The interior bisector of an angle is the half-line or line segment
        that divides an angle of less than 180° into two equal angles.
        The exterior bisector is the half-line that divides the opposite
        angle (of greater than 180°) into two equal angles.

5.5.1 Angle bisectors of a triangle

The angle bisectors of the angles of a triangle are concurrent in a point called the incenter of the triangle.

Isosceles Triangle Theorems

Isosceles Triangle is a triangle where two of its sides are congruent. In this section we supposed to study some of the properties of these triangles.

Theorem 5-1 (Isosceles Triangle Theorem) If two sides of a triangle are congruent, then the angles opposite those sides are congruent.
(The base angles of an isosceles triangle are congruent)

Given   In $\triangle ABC$, $\overline{AC} \cong \overline{BC}$

Required: To show $\angle A \cong \angle B$

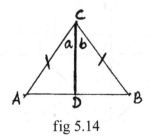

fig 5.14

__Construction

Let $\overline{CD}$ bisect angle C

	Statements	Reason
1)	$\angle a \cong \angle b$	$\overline{CD}$ bisects $\angle C$
2)	$\overline{AC} \cong \overline{BC}$	Given
3)	$\overline{CD} \cong \overline{CD}$	Reflexive property
4)	$\triangle ACD \cong \triangle BCD$	by SAS Steps (1), (2) and (3)
5)	$\angle A \cong \angle B$	by step (4) Corresponding angles.

The converse of an isosceles triangle theorem is also true. Theorem (converse of isosceles triangle theorem): if two angles of a triangle are congruent, then the sides opposite those angles are congruent.

If ∠P≅∠R, then $\overline{PQ} \cong \overline{RQ}$ or if ∠P≅∠R, the ΔPQR is isosceles.

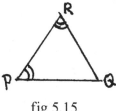

fig 5.15

The following are immediate consequences (corollaries) of the above two theorems.

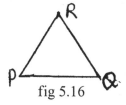

fig 5.16

Corollary: An equilateral triangle is also equiangular.
   i.e,   : In ΔPQR if $\overline{PQ} = \overline{QR} = \overline{PR}$, then ∠P=∠Q=∠R

Corollary: An equiangular triangle is equilateral.
   i.e,   : In ΔPQR if ∠P≅∠Q≅∠R, then $\overline{PQ} \cong \overline{QR} \cong \overline{PR}$

Theorem 5-2 A ray that bisects the vertex angle of an isosceles triangle
          is the perpendicular bisector of the base.

Restatement: If ΔABC is an isosceles triangle and $\overline{BD}$ bisects angle B,
          then $\overline{BD}$ is the perpendicular bisector of $\overline{AC}$.

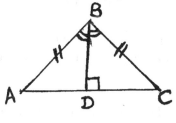

fig 5.17

Theorem 5-3 The bisector of one angle of a triangle is perpendicular to the opposite side if and only if it bisects the opposite side and the triangle is Isosceles.

Restatement: $\overline{BQ}$ is bisector of $\angle Q$ and $\overline{BQ}$ is perpendicular to $\overline{PR}$ if and only if it bisects $\overline{PR}$ and $\triangle PQR$ is isosceles.

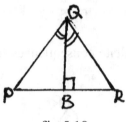

fig 5.18

Theorem 5-4 Any point on the perpendicular bisector of a segment is equidistant from the end points of the segment.

Restatement: If $\overline{NL}$ is perpendicular bisector of $\overline{KM}$ and O any point on $\overline{NL}$, then O is equidistant from the end points K and M or $\overline{KO} = \overline{MO}$.

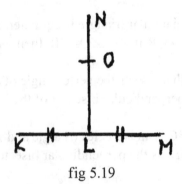

fig 5.19

Theorem 5-5:The perpendicular bisectors of a triangle intersect at a point that is equidistant from the the vertices of a triangle.

The perpendicular bisectors of the sides of a triangle intersect at a point called circumcenter of the triangle. This point is the center of the circumcircle, the circle passing through all three vertices.

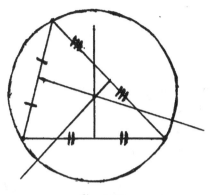

fig 5.20

The circumcenter is the centre of a circle passing through the three vertices of the triangle.

Example 1) In the figure shown below $\overline{SP}$ is the perpendicular bisector of $\overline{QR}$.

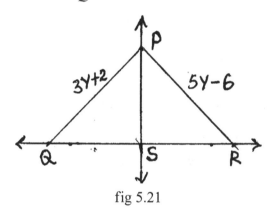

fig 5.21

a) Find the value of y
b) Find the length of $\overline{PR}$

Solution:

a) $\overline{PQ} \cong \overline{PR}$ ... perpendicular bisector theorem
$3y+2=5y-6$ ... substitute
$5y-3y=6+2$ ... collecting like terms.
$y=4$ .... solve for y.

b) $\overline{PR} = 5y - 6$

$\overline{PR} = 5(4) - 6$

$\overline{PR} = 14$

Example 2)   In the diagram shown below, $\overline{ST}$ is the perpendicular bisector of $\overline{PR}$. Find

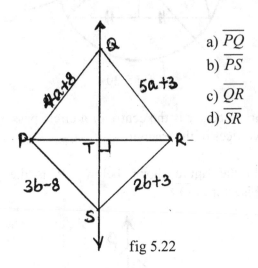

a) $\overline{PQ}$

b) $\overline{PS}$

c) $\overline{QR}$

d) $\overline{SR}$

fig 5.22

Solution

   a) Since $\overline{ST}$ is the perpendicular bisector of $\overline{PR}$, $\overline{PQ} \cong \overline{QR}$,

$\overline{PQ} = \overline{QR}$

4a+8=5a+3 .... Substitution

5a-4a=8-3 ....   Collecting like terms.

a=5

$\overline{PQ} = 4a + 8$

$\quad = 4(5) + 8$

$\overline{PQ} = 28$

b) Since $\overline{ST}$ is the perpendicular bisector of $\overline{PR}$, $\overline{PS} \cong \overline{SR}$

$\overline{PS} \cong \overline{SR}$

3b-8=2b+3....   Substitution

3b-2b=8+3 ....   Collecting like terms.

b=11

$\overline{PS}$ =3b-8

342

=3(11)8

=33-8

=25

c) $\overline{QR}$ =5a+3

=5a+3

=5(5)+3

=28

d) $\overline{SR}$ =2b+3

=2(11)+3

=22+3

=25

Example 3) In the diagram shown below, if $\overline{SQ}$ bisects ∠PSR, then find

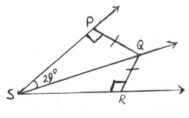

fig 5.23

a) m(∠QSR)

b) m(∠PSR)

Solution:

a) Since $\overline{SQ}$ is a bisector of ∠PSR, m(∠PSQ)≅m(∠QSR)

i.e. m(∠PSQ)≅(∠QSR)=29°

b) m(∠PSR)=m(∠PSQ)+m(∠QSR)

=°29+29°

=58°

Example 4) In the diagram shown below, if $\overline{XY}$ is a bisector of $\angle PXQ$, $\overline{PY} = 6$ and $\overline{XP} = 8$, find

a) $\overline{XY}$

b) $\overline{YQ}$

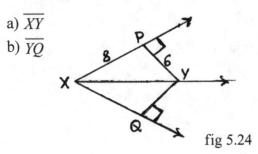

fig 5.24

Solution:

a) $\triangle XPY$ is a right angled triangle and hence,

$(XP)^2+(PY)^2=(XY)^2$ .....   Pythagoras theorem

$8^2+6^2=(XY)^2$ ....   Substitution

$64+36=(XY)^2$

$100=(XY)^2$ .....   Addition

$10=XY$

$XY=10$

b) Since $\overline{XY}$ is a bisector of $\angle PXQ$, $\triangle XPY \cong \triangle XQY$ by SAS. Thus

$\overline{PY} \cong \overline{QY} = 6$ $\cdots$ Corresponding sides

Example 5) Find the value of Y such that Q lie on the bisector of $\angle PSR$.

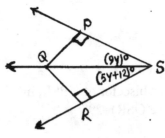

fig 5.25

Solution:

Since Q lie on the bisector of $\angle PSR$;

$M(\angle PSQ) \cong M(\angle RSQ)$

i.e $(9Y)° = (5Y+12)°$ ....   Substiuttion

$9Y-5Y=12$ ....   Collecting like terms.

$4Y=12$

$Y = \dfrac{12}{4}$ .....   Division

$Y=3$

Example 6)   In the figure shown below, if M is the incenter of $\triangle ABC$. What is the length of $\overline{ME}$?

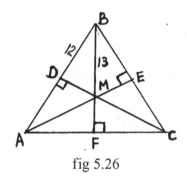

fig 5.26

Solution: By the concurrency of angle bisectors of triangle theorem, M, the incenter of $\triangle ABC$ is equidistant from the sides of $\triangle ABC$, thus to find $\overline{ME}$ we can find first $\overline{DM}$ in $\triangle BDM$ by using the Pythagorean theorem.

$$\left(\overline{MB}\right)^2 = \left(\overline{MD}\right)^2 + \left(\overline{BD}\right)^2$$

$c^2=a^2+b^2$ ....   Pythagorean Theorem.

$(13)^2=a^2+(12)^2$ ....   Substitution

$169=a^2+144$ ...   Multiply

$169-144=a^2$ ....   Subtract 144 from each side.

$a^2=25$

$a=5$ .....   Taking the positive square root of each side.

$\overline{MD} = 5$

Since $\overline{MD} = \overline{ME}$, $\overline{ME} = 5$

5.6   Altitude of Triangle

An altitude of a triangle is a straight line through a vertex and perpendicular to a line containing the base - (the opposite sides of a triangle). This line containing the opposite side is called the extended base of the altitude. The intersection between the extended base and the altitude is called the foot of the altitude. The length of the altitude, often simply called the altitude, is the distance between the base and the vertex.

Altitudes can be used to compute the area of a triangle: one half of the product of an altitude's length and its base's length equals the triangle's area.

In an isosceles triangle (a triangle with two congruent sides), the altitude having the incongruent side as its base will have the midpoint of that side as its foot. Also the altitude having the incongruent side as its base will form the angle bisector of the vertex.

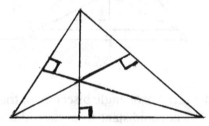

Three altitudes intersecting at the orthocenter.
fig 5.27

In a right triangle, the altitude with the hypotenuse as base divides the hypotenuse into two lengths p and q. If we denote the length of the altitude by h, we then have the relation.

$h^2 = pq$

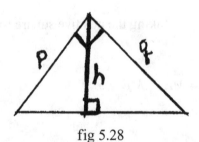

fig 5.28

5.6.1 The Orthocenter

The three altitudes intersect in a single point, called the orthocenter of the triangle. The orthocenter lies inside the triangle (and consequently the feet of the altitudes all fall on the triangle) if and only if the triangle is not obtuse (i.e. doesn't have an angle greater than a right angle).

The orthocenter, along with the centroid, circumcenter and center of the nine-point circle all lie on a single line, known as the Euler line. The center of the nine-point circle lies at the midpoint between the orthocenter and the circumcenter, and the distance between the centroid and the circumcenter is half that between the centroid and the orthocenter.

Unlike the centroid and circumcenter of a triangle, the orthocenter has no special characteristics (such as being equidistant from all sides or vertices).

More on triangle center

Incenter

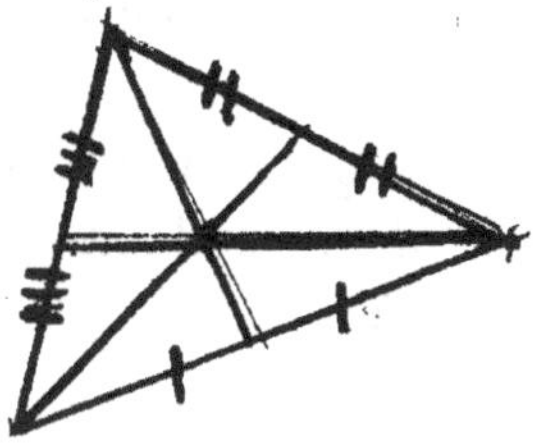

Centroid

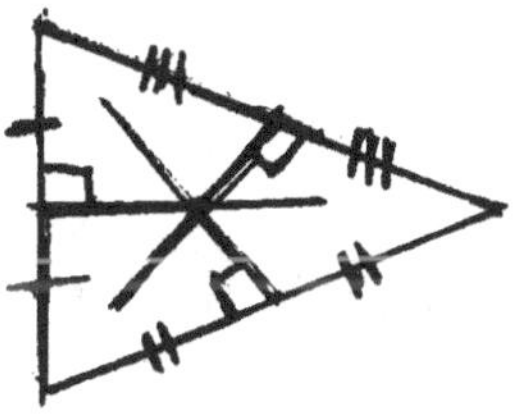

Circumcenter

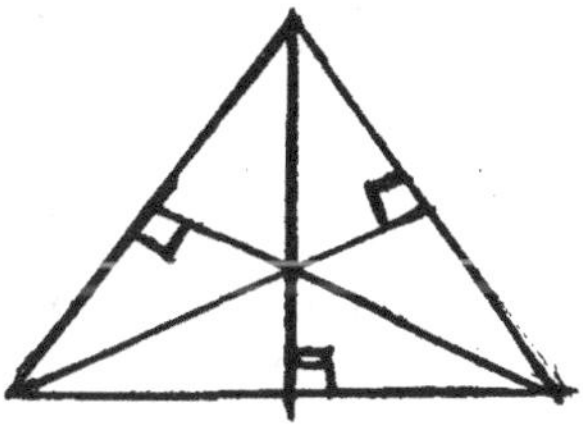

Orthocenter

fig 5.29

5.6.2
- Incenter:- The center of an inscribed circle; that point where the bisector of the angles of a triangle or a regular polygon intersect.

5.6.3
- Centroid:- The point where the median of a triangle intersect.

5.6.4
- Circumcenter: The center of a circumscribed circle, that point where any two perpendicular bisector of the sides of a polygon inscribed in the circle intersect.

- Orthocenter: The point of intersection of the three altitudes of a triangle.

Example 1)   The vertices of $\triangle PQR$ are P(0, 0), Q(4, 9), and R(10, 2). Find the coordinates of the centroid S of $\triangle PQR$.

Solution:- First draw $\triangle PQR$. Then use the midpoint formula to find the midpoint M of $\overline{PR}$ and draw the median $\overline{QM}$.

$$M\left(\frac{0+10}{2}, \frac{0+2}{2}\right)$$
$$= M(5, 1)$$

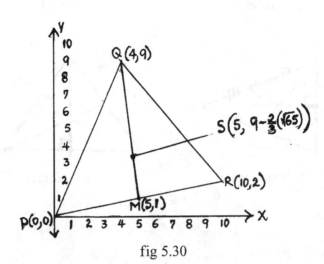

fig 5.30

Note:- The centroid is two-thirds of the distance from each vertex to the midpoint of the opposite side. The distance from the vertex Q(4, 9) to the midpoint M(5, 1) is $\sqrt{65}$ units.

Thus, the centroid is $\frac{2}{3}\left(\sqrt{65}\right)$ units down from Q on $\overline{QM}$.

The coordinates of the centroid S are $\left(5, 9 - \frac{2}{3}\left(\sqrt{65}\right)\right)$, which is approximately (5, 3•4).

Example 2)   The vertices of $\triangle ABC$ are A(3, 2), B(5, 10) and C(9, 6). Find the coordinates of the centroid P of $\triangle ABC$.

Solution:- First draw $\triangle ABC$. Then use the midpoint formula to find the midpoint M of $\overline{AC}$ and draw the median $\overline{BM}$.

$$M\left(\frac{3+9}{2}, \frac{2+6}{2}\right)$$

$$M(6, 4)$$

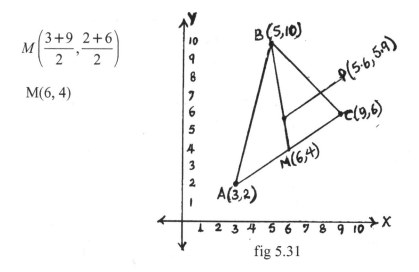

fig 5.31

The centroid is two-thirds of the distance from each vertex to the midpoint of the opposite side. The distance from the vertex

B(5, 10) to the midpoint M(6, 4) is $\sqrt{37}$ units. Thus, the centroid

is $\frac{2}{3}\left(\sqrt{37}\right)$ units down from B on $\overline{BM}$.

Therefore, the coordinates of the centroid P are

$\left(5\sqcap6, 10 - \frac{2}{3}\left(\sqrt{37}\right)\right)$, which is approximately (5•6, 5•9).

5.7   Triangle Inequality

The triangle inequality states that for any triangle, the sum of the lengths of any two sides must be greater than the third side.

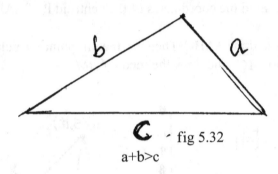

fig 5.32

a+b>c

Example 1)   In the figures below, if M(∠A)<M(∠B)<M(∠C), then find all possible values of X.

a)

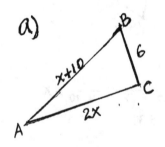

b)

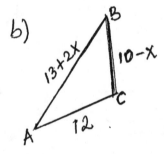

fig 5.33 a and b

Solution:

a) Since m(∠A)<m(∠B)<m∠C); $\overline{BC} < \overline{AC} < \overline{AB}$, thus,

$\overline{AC} + \overline{BC} > \overline{AB}$.

i.e, $\overline{AC} + \overline{BC} > \overline{AB}$ ... Triangle inequality.

2x+6>x+10 ....   Substitution

2x-x>10-6 ....   Collecting like terms

x>4

b) Since $m(\angle A) < m(\angle B) < m(\angle C)$; $\overline{BC} < \overline{AC} < \overline{AB}$, thus $\overline{AC} + \overline{BC} > \overline{AB}$.

i.e, $\overline{AC} + \overline{BC} > \overline{AB}$ .... Triangle inequality

12+10-x>13+2x ...   Substitution

-x-2x>13-12-10 ...   Collecting like terms.

-3x>-9

$x < \dfrac{-9}{-3}$ ....   Division

x<3   (Make sure the sides should be positive number)

Example 2)   Determine whether the following set of numbers can be the sides of a triangle. If not, explain why not.

a) 6, 5, 8
b) 3, 4, 9
c) 12, 15, 27
d) 12, 13, 19
e) 8, 8, 8

Solution:-

a) 6, 5, 8

6+5>8

Can be the sides of the triangle, because they satisfy the definition of triangle inequality.

b) 3, 4, 9

3+4<9

Cannot be the sides of the triangle, because they don't satisfy the definition of triangle inequality.

c) 12, 15, 27

12+15=27

cannot be the sides of the triangle, because they don't satisfy the definition of triangle inequality.

d) 12, 13, 19

12+13>19

can be the sides of the triangle, because they satisfy the definition of triangle inequality.

e) 8, 8, 8

8+8>8

can be the sides of the triangle, because they satisfy the definition of triangle inequality.

Example 3)   Determine the shortest and the longest sides of the triangle given below:-

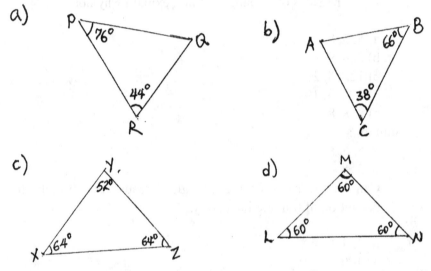

fig 5.34 a, b, c and d

Solution

a) In ΔPQR

m(∠Q)=180°-(76°+44°)

m(∠Q)=60°

Thus, the side opposite to angle P, $\overline{QR}$ is the longest side and the side opposite to angle R, $\overline{PQ}$ is the shortest side of ΔPQR.

b) In ΔABC

m(∠A)=180°-(38°+66°)

m(∠A)=76°

Thus, the side opposite to angle A, $\overline{BC}$ is the longest side and the side opposite to angle c, $\overline{AB}$ is the shortest side of ΔABC.

c) In ΔXYZ

m(∠X)≅m(∠Z)

Since m(∠X)≅m(∠Z), the side opposite to angle X, $\overline{YZ}$ and the side opposite to angle Z, $\overline{XY}$ are equal in length. The side opposite to angle Y $\overline{XZ}$ is the shortest side of ΔXYZ.

d) In ΔLMN

m(∠L)≅m(∠M)≅m(∠N)=60°

Since each angle of ΔLMN is 60°, all its sides are equal.

Remark:- ΔLMN is an equilateral triangle.

Example 4) Determine the smallest and the largest angles of each of the following triangles.

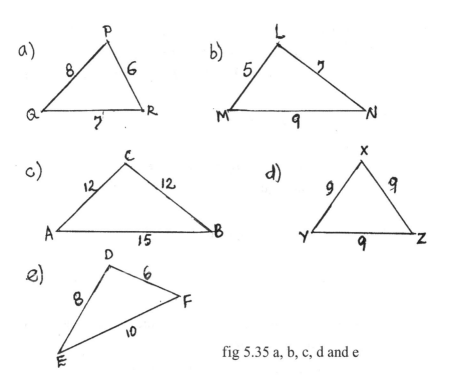

fig 5.35 a, b, c, d and e

Solution

a) In ΔPQR

$\overline{PQ} > \overline{QR} > \overline{PR}$

8>7>6

Thus, the angle opposite to side $\overline{QP}$ is the largest and the angle opposite to side $\overline{PR}$ is the smallest.

i.e., ∠R is the largest and ∠Q is the smallest.

b) In ΔLMN

$\overline{MN} > \overline{LN} > \overline{LM}$

Thus, the angle opposite to side $\overline{MN}$ is the largest and the angle opposite to side $\overline{LM}$ is the smallest.

i.e., ∠L is the largest and ∠N is the smallest.

c) In ΔABC

$\overline{AB} > \overline{AC}$, $\overline{AB} > \overline{BC}$, and $\overline{AC} \cong \overline{BC}$

Thus, the angle opposite to side $\overline{AB}$ is the largest and ∠A and ∠B are equal,

i.e, ∠C is the largest and ∠A≅∠B.

d) In ΔXYZ

$\overline{XY} \cong \overline{YZ} \cong \overline{XZ}$, showing $< X \cong < Y \cong < Z = 60°$.

i.e, all angles are equal.

Remark!- ΔXYZ is an equilateral triangle.

e) In ΔEDF

$\overline{EF} > \overline{ED} > \overline{DF}$

Thus, the angle opposite to side $\overline{EF}$ is the largest angle and the angle opposite to side $\overline{DF}$ is the smallest.

i.e, ∠D is the largest and ∠E is the smallest.

Exercise 5-1

1) Given $\overline{OP}$ is the length of the midsegment. Find $\overline{MN}$

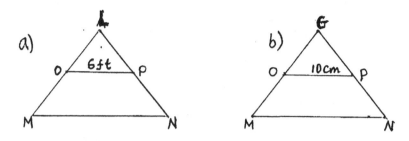

a)

b)

2) If $\overline{PQ}$ is the midsegment of $\triangle ABC$. Find the value of X.

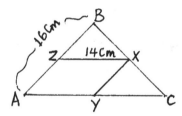

3) In the diagram below, if $\overline{XY}$ and $\overline{ZX}$ are midsegments of $\triangle ABC$. Find the length of $\overline{AC}$ and $\overline{XY}$

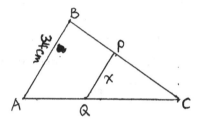

4) In $\triangle LMN$, below, if $\overline{DE}$ is the midsegment of $\triangle LMN$. If $DE = 3x + 2$ and $\overline{MN} = 5x + 8$, find the value of X.

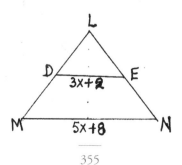

5) In the figure below, if the perimeter of △ABC is 34 cm. And △DEF is the midsegment triangle of △ABC. Find the perimeter of △DEF.

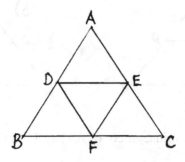

6) Given: $\overline{PQ} = 56\,ft$, $\overline{PR} = 36\,ft$, $\overline{QR} = 46\,ft$. X, Y and Z are midpoints of △PQR. Find the perimeter of △XYZ.

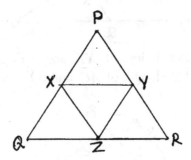

7) In the figure shown below $\overline{SP}$ is the perpendicular bisector of $\overline{QR}$.

a) Find the value of x.
b) Find the length of $\overline{PR}$

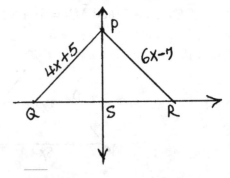

8)   In the diagram shown below, $\overline{SQ}$ is the perpendicular bisector of $\overline{PR}$. Find

a) $\overline{PQ}$
b) $\overline{PS}$
c) $\overline{QR}$
d) $\overline{SR}$

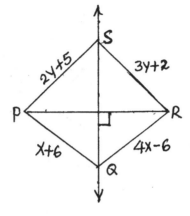

9)   In the figure shown below $\overline{XY}$ is the mid-segment of $\triangle ABC$, and $AB = 24cm$, $AC = 28cm$ and $XY = 18cm$. Find the perimeter of $\triangle ABC$

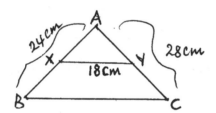

10)   In the diagram shown below, if $\overline{AD}$ is a bisector of $\angle BAC$, $BD = 3$ and $AB = 4$, find

a) $\overline{AD}$
b) $\overline{DC}$

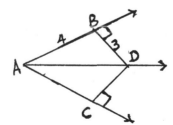

11) Find the value of X such that Q lie on the bisector of ∠PSR.

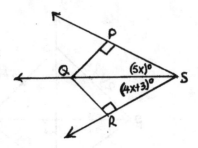

12) In the figure shown below. If M is the incenter of △ABC. What is the length of ME?

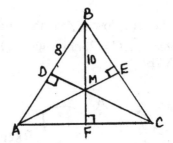

13) In each of the following right angled triangle, find the value of h.

a)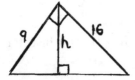

b)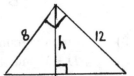

14) In each of the following right angled triangle find the value of p.

a)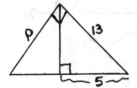

b)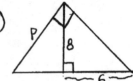

15) The vertices of ΔPQR are A(2, 2), B(7, 8), and C(10, 4). Find the coordinates of the centroid D of ΔABC.

16) In the figures shown below, if m(∠L)<m(∠M)<m∠N), then find all possible values of x.

a)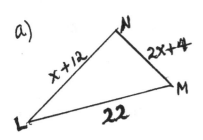

b)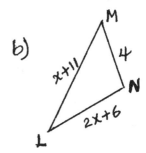

17) Determine whether the following set of numbers can be sides of a triangle. If not, explain why not.
a) 6, 4, 10
b) 8, 12, 17
c) 4, 4, 4
d) 8, 6, 6
e) 13, 15, 34
f) ½, ⅓, ¼

18) Determine the shortest and the longest sides of the triangle given below.

a)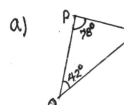

b)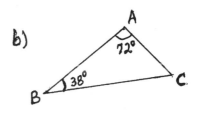

19) Determine the smallest and the largest angles of each of the following triangles.

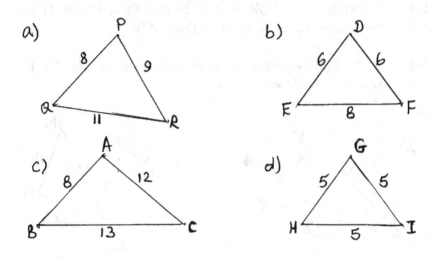

5.8 Polygon

A polygon is a geometric figure formed by three or more coplanar segments called sides. Each side intersects exactly two other sides, but only at their endpoints, and the intersecting sides must be non-collinear.

The vertices of a polygon are the intersecting points of the sides. The two endpoints of any side are called consecutive vertices. Sides that share a vertex are called consecutive sides. Segments with endpoints that are non-consecutive vertices are called - Diagonals.

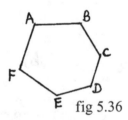

fig 5.36

A and B are consecutive vertices.
CD and DE are consecutive sides.
ABCDEF is a convex polygon

5.9   Convexity and types of non-convexity
        Polygons divide a plane into three parts. The polygon itself, its
        interior, and its exterior. A polygon can be classified as concave
        or convex.

5.9.1   Concave:- A line may be found which meets its boundary more
        than twice. In other words, it contains atleast one interior angle
        with a measure larger than 180°.

5.9.2   Simple:- The boundary of the polygon does not cross itself. All
        convex polygons are simple.

5.9.3   Star-shaped:- The whole interior is visible from a single point
        without crossing any edge. The polygon must be simple, and
        may be convex or concave.

5.9.4   Star-polygon:- A polygon which self-intersects in a regular
        way.

Symmetry

- Equiangular:- All its corner angles are equal.
- Cyclic:- All corners lie on a single circle
- Isogonal or vertex-transitive:- All corners lie with in the same
  symmetry orbit. The polygon is also cyclic and equiangular.
- Equilateral:- All edges are of the same length. (A polygon with
  five or more sides can be equilateral without being convex).
- Isotoxal or edge-transitive:- All sides lie with in the same
  symmetry orbit. The polygon is also equilateral.

5.10   REGULAR POLYGON

A regular polygon is a polygon that is equiangular (all angles are equal
in measure) and equilateral (all sides have the same length). Regular
polygons may be convex or star.

5.10.1 General properties

These properties apply to all regular polygons (both convex and star).

A regular n-sided polygon has rotational symmetry of order n.

All vertices of a regular polygon lie on a common circle (the circumscribed circle). i.e, they are concyclic points.

Together with the property of equal-length sides, this implies that every regular polygon also has an inscribed circle or incircle which is tangent to every side at the mid-point.

5.10.2 Regular convex polygons

All regular simple polygons (a simple polygon is one which does not intersect itself anywhere) are convex. Those having the same number of sides are also similar.

The table below shows names of Polygons

Prefix	Number of Sides	Name of the Polygon
	n	regular n-gon
tri-	3	equilateral triangle
quadri-	4	square (quadrilateral)
penta-	5	regular pentagon
hexa-	6	regular hexagon
hepta-	7	regular heptagon
octa-	8	regular octagon
nona-	9	regular nonagon
deca-	10	regular decagon
hedeca-	11	regular hendecagon
dodeca-	12	regular dodecagon
triskaideca-	13	regular triskaidecagon
tetradeca-	14	regular tetradecagon
pentadeca-	15	regular pentadecagon
hexadeca-	16	regular hexadecagon
heptadeca-	17	regular heptadecagon
octadeca-	18	regular octadecagon
enneadeca-	19	regular enneadecagon
icosa-	20	regular icosagon
hecta	100	regular hectagon
chilia	1000	regular chiliagon
myria	10000	regular myriagon
mega	1,000,000	regular megagon

5.11 Diagonals and Angle Measures of Polygons

In a polygon, a segment that joins nonconsecutive vertices is called a Diagonal.

In hexagon ABCDEF, $\overline{AE}$ and $\overline{AD}$ are diagonals drawn from vertex A. Two diagonals can be drawn from each vertex of this polygon.

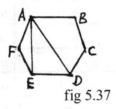

fig 5.37

Theorem 5-6:- If a convex polygon has n-sides, then the sum of the degree measures of its angle is (n-2)×180°.

Note:- An exterior angle forms a linear pair with an angle of the triangle. In △EFG, ∠EGH is an exterior angle. Likewise, an exterior angle of any convex polygon forms a linear pair with an angle of the polygon.

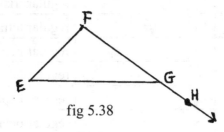

fig 5.38

Example 1) Find the sum of the degree measure of a regular hexagon.

Solution:
The sum of all interior angle is given by (n-2)×180°
n=6
(n-2)×180
(6-2)×180
4×180
720°

Example 2)   What is the number of sides of a regular polygon, if the sum of all its interior angle is 1080°?

Solution:- The sum of all interior angle is given by (n-2)×180°

(n-2)180°=1080°

$$n-2=\frac{1080°}{180°}$$

n-2=6

n=8

∴ A polygon has 8 sides.

Example 3)   Find the measure of each interior angle of a regular pentagon.

Solution:-

The sum of the measure of all interior angles of a regular polygon is (n-2)×180°.

The degree measure of each interior angle of a regular polygon

is given by $\dfrac{(n-2)\times180°}{n}$.

n=5

$$\frac{(n-2)\times180°}{n}$$

$$\frac{(5-2)\times180}{5}$$

$$=\frac{3\times180}{5}$$

$$=\frac{540}{5}$$ Divide the sum of the measures of the interior angles by the number of sides

$$=108°$$

Theorem 5-7:- In any convex polygon, the sum of the degree measures of the exterior angles, one angle at each vertex, is 360°.

Example 3) What is the sum of the measures of the exterior angles of a regular pentagon?

Solution:-

The sum of the measure of the exterior angles of a regular pentagon is 360°.

Example 4) What is the measure of each exterior angles of a regular hexagon?

Solution:-

The measure of each exterior angle of a regular hexagon is

$$\frac{360°}{6.} = 60°$$

Example 5) The measure of the five angles of a convex hexagons are 132°, 119°, 130°, 120°, and 105°. Find the measure of the sixth angle.

Solution:-

Let y=measure of the sixth angle.
132°+119°+130°+120°+105°+y=(n-2)×180°... Theorem 5-6, n=6
606°+y=(6-2)×180°
606°+y=720°
y=114°
∴ The measure of the sixth angle is 114°.

Example 6) Find the perimeter of a regular nonagon whose sides have a length of 14cm.

Solution:-

The perimeter of a polygon is the sum of the lengths of its sides.
P=(9)(14cm)
=126 cm (In aregular nonagon, all nine sides have the same measure.
∴ The perimeter of the regular nonagon is 126 cm.

Example 7) What is the measure of each interior angle of a regular dodecagon?

Solution:- The measure of each interior angle of a regular polygon

is given by:- $\dfrac{(n-2)\times180°}{n}$

$\dfrac{(n-2)\times180°}{n}$, n=12

$= \dfrac{(12-2)x180°}{12}$

$= \dfrac{1800°}{12}$

$=150°$

∴ The measure of each interior angle of a regular polygon is 150°.

Example 8) Find the value of x°, such that the measure of the exterior angles of a regular hexagon are x°, x°, 2x°, 2x°, 3x° and 3x°.

Solution:-

The measure of the exterior angle of a regular hexagon is 360°.

x°+x°+2x°+2x°+3x°+3x°=360°

12x°=360°

$x = \dfrac{360}{12}$

x=30°

Example 9) The sum of the measures of eleven exterior angles of
a regular dodecagon is 330°. Find the measure of the
twelveth angle.

Solution
The sum of the measure of the exterior angle of a regular
dodecagon is 360°.
Let $x°$ be the measure of the twelveth angle.
$x°+330°=360°$
$x=360°-330°$
$x=30°$
∴ The measure of the twelveth angle is 30°.

Example 10) Find the sum of the measures of the interior angles of the
regular polygon shown in the diagram.

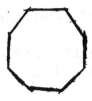

Solution:-
The polygon shown above has eight sides and the sum of its
interior angles is:-
$(n-2)×180°$; $n=8$
$(8-2)×180°$
$6×180°$
$1,080°$

Example 11) If the sum of the interior and exterior angles of a regular
polygon is 900°, then what is the number of sides of the
polygon?
Solution:- Let n be the number of sides of a polygon.
$(n-2)×180°+360°=900$
$(n-2)×180=900-360$
$(n-2)×180=540$

$$n - 2 = \frac{540}{180}$$

n-2=3

n=5

The polygon has 5 sides. (It is pentagon)

Example 12) If the sum of the angles of a regular polygon is four times the sum of its exterior angles, then find the number of sides of the polygon.

Solution:- Let n be the number of sides.

(n-2)×180°=4×360°

$$(n - 2) = \frac{4 \times 360}{180}$$

(n-2)=8

n=10

The polygon is decagon (10 sides polygon)

Example 13) If the sum of the interior angles minus the sum of the exterior angles of a polygon is negative, then what is the name of the polygon?

Solution:- Let the number of sides is n, then

(n-2)×180°-360°<0

(n-2)×180°<360°

$$n - 2 < \frac{360°}{180°}$$

n-2<2

n<4

∴ The polygon is triangle.

Example 14) Find the value of y in the diagram shown below.

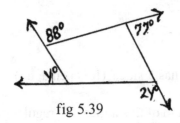

fig 5.39

Solution

By using the polygon exterior angles theorem. We can write and solve an equation involving y.

$y°+88°+77°+2y°=360°$...   Polygon exterior angle theorem.

$3y°=360°-165°$ ..   Combining like terms

$3y°=195°$

$y=\dfrac{195°}{3}$ ..   Division

$y=65°$ ..   Solve for y.

Example 15) Find the value of x in the diagram shown below.

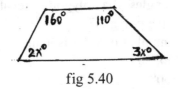

fig 5.40

Solution:- The polygon is quadrilateral, the sum of all interior angle is 360°, thus

$2x°+3x°+160°+110°=360°$ ...   Interior angles theorem

$5x°+270°=360°$ ...   Combine like terms

$5x°=90°$

$x°=18°$ Solve for x

Example 16) If the sum of the interior angles of a regular convex polygon is 900°. What is the number of sides of the polygon?

Solution:-

The sum of the interior angles of a regular polygon is given by
(n-2)180°

(n-2)×180°=900°

(n-2)=5

n=7

∴ A polygon has seven sides.

Example 17) Find the value of n, such that each exterior angle of the regular n-gon has a measure of 15°.

Solution:-

The sum of the exterior angles of a regular polygon is 360°. Thus,

$$\frac{360°}{n} = 15°$$

15n=360

$$n = \frac{360}{15}$$

n=24

∴ The polygon has 24 sides, i.e, n=24

Exercise 5-2

1) Find the sum of the degree measure of a regular octagon.

2) What is the number of sides of a regular polygon, if the sum of all its interior angle is 900°?

3) Find the measure of each interior angle of a regular dodecagon.

4) What is the sum of the exterior angle of a regular convex polygon having 20 sides?

5) What is the measure of each exterior angle of a regular nonagon?

6)   The measure of six angles of a regular heptagon are 122°, 128°, 148°, 140°, 143° and 126°. What is the measure of the seventh angle?

7)   Find the perimeter of a regular octagon whose sides have a length of 22 cm.

8)   What is the measure of each interior angle of a regular pentadecagon?

9)   Find the value of y°, such that the measure of each exterior angles of a regular nonagon is (y+10)°

10)   The sum of five exterior angles of a hexagon is 300. What is the measure of the sixth angle?

11)   Find the value of x°, such that the measures of the exterior angles of a quadrilateral are x°, 2x°, 3x° and 4x°.

12)   The sum of the measures of nine exterior angles of a regular decagon is 324°. Find the measure of the tenth angle.

13)   Find the sum of the measures of the interior angles of the regular polygon shown in the diagram below.

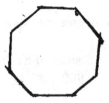

14)   Find the value of n, such that each exterior angle of the regular n-gon has a measure of 20°.

15)   Find the value of x° in the figures below.

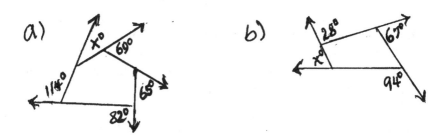

5.12   Properties of Parallelograms
A parallelogram is a quadrilateral with two pairs of parallel sides.

Example 1)

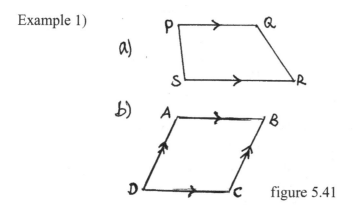

figure 5.41

a) $\overline{PQ}//\overline{SR}$ and $\overline{PS} \not\!/\!/ \overline{QR}$.
Quadrilateral PQRS has only one pair of parallel sides. Hence quadrilateral PQRS is a trapezium.

b) $\overline{AD}//\overline{BC}$ and $\overline{AB}//\overline{DC}$
Quadrilateral ABCD has two pairs of parallel sides. So that quadrilateral ABCD is a parallelogram.

Theorem 5-8:- opposite sides of a parallelogram are equal in length (congruent).

Theorem 5-9:- opposite angles of a parallelogram are equal in measure.

Theorem 5-10:- A diagonal of a parallelogram separates it into two congruent triangles.

Theorem 5-11:- Consecutive angles of a parallelogram are supplementary.

Theorem 5-12:- The diagonals of a parallelogram bisect each other.

Theorem 5-13. If two sides of a quadrilateral are parallel and congruent, then the quadrilateral is a parallelogram.

Example 2)   Find the value of x in the figure shown below.

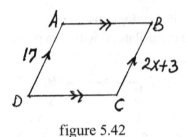

figure 5.42

Solution:-
   ABCD is a parallelogram, by definition of a parallelogram. To find the value of x, use Theorem 5-8.
   $\overline{AD} = \overline{BC}$ ....   opposite sides of a parallelogram ▢ are congruent
   17=2x+3 ....   Substitute 17 for AD and 2x+3 for $\overline{BC}$.
   x=7 ...   Solve for x.

Example 3)   Find the value of P in the figure below.

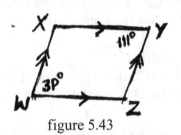

figure 5.43

Solution:-

XYZW is a parallelogram by definition of a parallelogram. To find the value of P, use Theorem 5-9.

$\angle W \cong \angle Y$ .... opposite angles of a parallelogram ▱ are congruent.

3P=111 .... Substitute 3P for $\angle W$ and 111 for $\angle Y$.

$P = \dfrac{111}{3}$ .... Division.

P=37° ... Solve for P.

Example 4)   Find the value of x and y in the figure below.

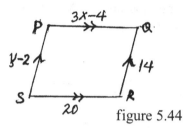

figure 5.44

Solution:-

PQRS is a parallelogram by definition of a parallelogram. To find the value of x and y. use Theorem 5-8.

• $\overline{PQ} \cong \overline{SR}$ .... opposite sides of a parallelogram ▱ are congruent.
   3x-4=20 .... Substitute 3x-4 for $\overline{PQ}$ and 20 for $\overline{SR}$.
   x=8 .... Solve for x.

• $\overline{PS} \cong \overline{QR}$ .... opposite sides of a parallelogram ▱ are congruent.
   y-2=14 .... Substitute y-2 for $\overline{PS}$ and 14 for $\overline{QR}$.
   y=16 .... Solve for y.

Example 5)   In parallelogram ▱ PQRS shown below, suppose $\overline{PT} = 5x-12$ and $\overline{RT} = 3x-2$, find $\overline{RP}$.

Solution:-

$\overline{PT} = \overline{TR}$ ··· Theorem 5-12

5x-12=3x-2 ... Substitution

5x-3x=12-2 .... Collecting like terms.

2x=10

x=5 .... Solve for x.

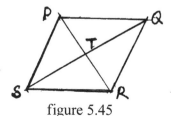

figure 5.45

Now substitute 5 for x to find either $\overline{PT}$ or $\overline{TR}$.

PT=5x-12

=5(5)-12

=25-12

=13

PT=RT=13

$\overline{RP} = 2(PT)$

=2(13)

$\overline{RP} = 26$

Example 6)   Given parallelogram ⧠ LMNO with diagonal $\overline{LN}$, show
that ΔLNO≅ΔNLM.

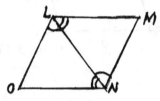

figure 5.46

Required to show: ΔLNO≅ΔNLM

Proof:-

Proof: $\overline{LM}//\overline{ON}$ and $\overline{LO}//\overline{MN}$ .... Defn. of parallelogram ⧠

∠LNO≅∠NLM .... Alternate interior angles are congruent.

∠OLN≅∠MNL ... Alternate interior angles are congruent.

LN≅LN .... Reflexive property of congruent segments

∠LNO≅∠NLM ... ASA

Example 7)   Find the value of a in the figure shown below.

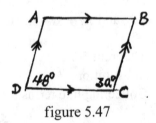

figure 5.47

Solution:- To find the value of a, use Theorem 5-11.

M(∠ADC)+M(∠BCD)=180°....   Consecutive angles in a parallelogram ▱ are supplementary

48°+3a°=180° ...   Substitute.

a=44° ...   Solve for a.

Example 8) The vertices of parallelogram ▱ PQRS are P(4, 5), Q(9, 5), R(7, 1) and S(2, 1). The diagonals of parallelogram ▱ PQRS intersect at point B. What are the coordinates of B?

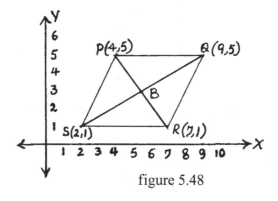

figure 5.48

Solution:-

Step 1) Sketch parallelogram ▱ PQRS in the coordinate plane.

Step 2) By Theorem 5-12, the diagonal of a parallelogram bisect each other. So, B is the midpoint of diagonals $\overline{PR}$ and $\overline{QS}$. Using the midpoint formula we can find the midpoint B of $\overline{PR}$.

Midpoint: $\left(\dfrac{4+7}{2}, \dfrac{5+1}{2}\right) = (5.5, 3)$

The coordinates of B are (5.5, 3)

Example 9) Using coordinate geometry show that quadrilateral EFGH is a parallelogram.

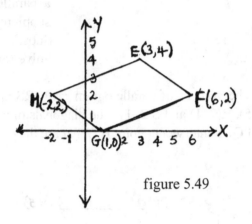

figure 5.49

Solution:

Step 1) Using the distance formula we show that $\overline{EF}$ and $\overline{HG}$ are congruent.

$$\overline{EF} = \sqrt{(6-3)^2 + (2-4)^2}$$

$$\overline{EF} = \sqrt{13}$$

$$\overline{HG} = \sqrt{(1-(2))^2 + (0-2)^2}$$

$$\overline{HG} = \sqrt{13}$$

Because $\overline{EF} \cong \overline{HG} = \sqrt{13}$, $\overline{EF} \cong \overline{HG}$

Step 2) Use the slope formula to show that $\overline{EF} // \overline{GH}$.

Slope of $\overline{EF} = \dfrac{2-4}{6-3} = \dfrac{-2}{3}$, slope of $\overline{GH} = \dfrac{0-2}{1-(-2)} = \dfrac{-2}{3}$

Because $\overline{EF}$ and $\overline{GH}$ have the same slope, they are parallel. $\overline{EF}$ and $\overline{GH}$ are congruent and parallel. So, EFGH is a parallelogram by Theorem 5-13

Example 10) In the figure shown below, what value of k makes the quadrilateral a parallelogram?

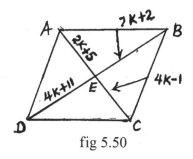

fig 5.50

Solution:-
In parallelogram □ ABCD, $\overline{AE} \cong \overline{CE}$ and $\overline{BE} \cong \overline{DE}$ by Theorem 5-12. i.e, the diagonals of a parallelogram bisect each other.

$\overline{AE} \cong \overline{CE}$ or $\overline{BE} \cong \overline{DE}$

2k+5=4k-1	7k+2≅4k+11
2k=6	7k-4k=11-2
k=3	3k=9
	k=3

Hence, k=3 makes the quadrilateral a parallelogram.

Exercise 5-3
1)   Find the value of y in the figures shown below

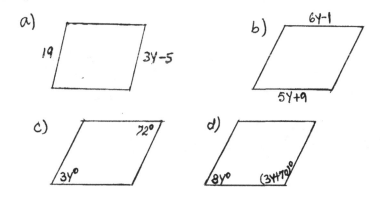

a)
19
3Y-5

b)
6Y-1
5Y+9

c)
72°
3Y°

d)
8Y°
(3Y+70)°

2)   Find the value of x in each of the following figures below.

a)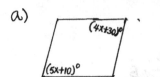

b)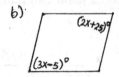

c)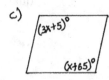

3)   Find the value of a in each of the following figures shown below.

a)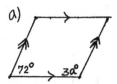

b)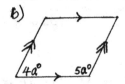

c)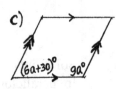

4)   The vertices of parallelogram ▱ ABCD are A(0, 7), B(5, 7), C(3, 3) and D(-2, 3). The diagonals of parallelogram ▱ ABCD intersect at a point R. What are the coordinates of R?

5)   In question number (4) above what are the lengths of sides?

6)   In each of the following find the value of x
a) perimeter=56cm
b) perimeter=45cm
c) perimeter=24cm

a) Perimeter = 56 cm

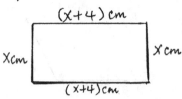

b) Perimeter = 45cm

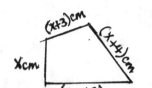

c) 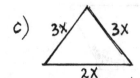 Perimeter = 24 cm

7)   Determine whether the following statements are true or false.
   a) opposite sides of a parallelogram are equal in length.
   b) opposite angles of a parallelogram are equal in measure.
   c) The diagonals of a parallelogram bisect each other.
   d) Consecutive angles of a parallelogram are complementary.
   e) The diagonal of a parallelogram separates it into two congruent triangles.
   f) If two sides of a quadrilateral are parallel and congruent, then the quadrilateral is a parallelogram.

5.13   Properties of Rhombuses, Rectangles, and Squares.
       Rhombuses, Rectangles and Squares are considered as special types parallelograms. Since they are parallelograms, they have al the characteristics of parallelograms. However, they have special properties of their own.

   •   A rhombus is a parallelogram with four congruent sides.
       Special properties of the diagonals of a rhombus.

Theorem 5-14: The diagonals of a rhombus are perpendicular.
Theorem 5-15: Each diagonal of a rhombus bisects a pair of opposite angles.
Theorem 5-16: Consecutive sides of a rhombus are congruent.

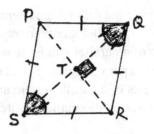

PQRS is a rhombus.
fig 5.51

- A rectangle is a parallelogram with four right angles.

Theorem 5-17:- The diagonals of a rectangle are congruent.

5.14   Some Properties of rectangle
- opposite sides of a rectangle are parallel.
- opposite angles of a rectangle are congruent.
- opposite sides of a rectangle are congruent.
- Diagonals of a rectangle bisect each other.
- Consecutive angles of a rectangle are supplementary.
- All four angles of a rectangle are right angles.

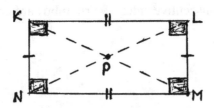

KLMN is a rectangle.
fig 5.52

- A square is a parallelogram with four congruent angles and four congruent sides.

Note:- A square is not only a parallelogram but also a rectangle and a rhombus. Therefore, all properties of parallelograms, rectangles, and rhombuses are properties of squares.

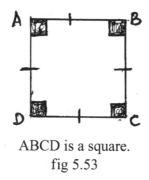

ABCD is a square.
fig 5.53

Theorem 5-18: A parallelogram is a rectangle if and only if its diagonals are congruent.

Theorem 5-19: A parallelogram is a rhombus if and only if its diagonals are perpendicular.

Theorem 5-20: A parallelogram is a rhombus if and only if each diagonal bisects a pair of opposite angles.

Example 1) The figure ABCD shown below is a rectangle. What is the length of the diagonal?

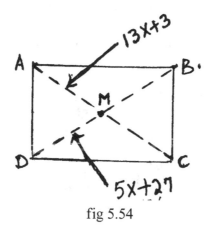

fig 5.54

Solution:-

The diagonals of a rectangle are congruent (Theorem 5-17), so $\overline{AC} \cong \overline{DB}$

13x+3=5x+27 ...   Substitution

x=3 ..   Solve for x.

$\overline{AC} = 13x + 3$

=13(3)+3

$\overline{AC} = 42$

Therefore, the length of the diagonal is 42.

Example 2)   The diagonals of rhombus ABCD intersects at E. If m(∠CBD)=40°, what is m(∠ABD)?

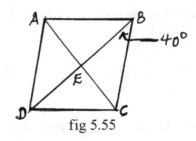

fig 5.55

Solution:

Each diagonal of a rhombus bisects a pair of opposite angles (Theorem: 5-15)

m(∠CBD)=40°

m(∠CBD)=m(∠ABD)=40°   (Theorem 5-15)

Example 3)   Find the value of x and y in the figure shown below

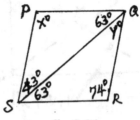

fig 5.56

Solution:-

• PQRS is a parallelogram and m(∠SPQ) and m(∠SRQ) are opposite angles, since opposite angles of a parallelogram are congruent. (Theorem 5-8).

m(∠SPQ)≅m(∠SRQ)
x°=74°

- m(∠RQS)≅m(∠PSQ) ... Alternate interior angles are congruent.
  y°=43°

Example 4)   If PQRS is a rhombus, find the value of x.

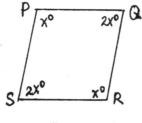

fig 5.57

Solution:-
    m(∠P)+m(∠Q)+m(∠R)+m(∠S)=360°
    x°+2x°+x°+2x°=360°
    6x°=360°
    x°=60°

Example 5)   If ABCD is a square, find the measure of the diagonal $\overline{AC}$.

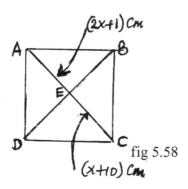

fig 5.58

Solution:-
    The diagonals of a square bisect each other. Thus, $\overline{AE} \cong \overline{CE}$,
    $\overline{AE} \cong \overline{CE}$
    (2x+1)=(x+10) ....   Substitution.

x=9 ...   Solve for x.

Thus, $\overline{AC} = AE + \overline{CE}$
=(2x+1)+(x+10)
=3x+11
=3(9)+11 ....   Substitute 9 for x.
$\overline{AC} = 38cm$

Example 6)   Find the value of a in $\square PQRS$

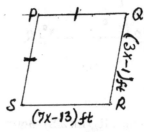

fig 5.59

Solution:-

As we see, all the sides of a $\square PQRS$ are congruent, thus, $\overline{SR} \cong \overline{QR}$.

$\overline{SR} \cong \overline{QR}$
(7x-13)=3x-1
7x-3x=13-1
4x=12
x=3

Example 7)   If LMNO is a parallelogram, such that $\overline{LN} = 3x+8$, and
$\overline{PM} = 2x-5$, find
a) The value of x.
b) The length of the diagonals.

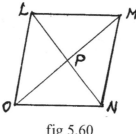

fig 5.60

Solution:-

a) Since LMNO is a parallelogram, its diagonals bisect each other. Thus, $\overline{PM} = \dfrac{1}{2}\left(\overline{LN}\right)$

$\overline{PM} = \dfrac{1}{2}\left(\overline{LN}\right)$

$2x-5=\dfrac{1}{2}(3x+8)$ ....   Substitute for $\overline{PM}$ and $\overline{LN}$

$4x-10=3x+8$

$x=18$ ...   Solving for x

b) $\overline{LN} = 3x+8$

$=3(18)+8$ ...   Substitute the value of x.

$\overline{LN} = 62$

∴ The length of the diagonal is 62.

Example 8)   If ABCD is a square whose side has a length of a units. What is the diagonal interms of a?

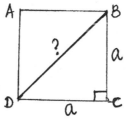

fig 5.61

Solution:-

In the square ABCD, the diagonal $\overline{BD}$ is the hypotenuse of right angled triangle BCD. To find the length of $\overline{BD}$, we have to apply Pythagoras formula. Thus,

$(BD)^2 = (BC)^2 + (DC)^2$

$= a^2 + a^2$

$(BD)^2 = 2a^2$

$\overline{BD} = \sqrt{2a^2}$

$\overline{BD} = a\sqrt{2}$ units

Exercise 5-4

1) Find the length of the diagonal in each of the following diagrams in terms of a.

a)
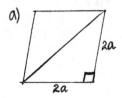

b)
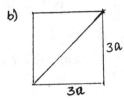

c)
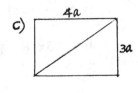

2) The figure shown below is a rectangle. What is the length of the diagonal?

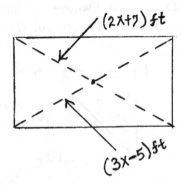

3)   The diagonal of rhombus PQRS intersects at E. If m(∠PQS)=46°.
        Find
        a) m(∠SQR)
        b) m(∠PQR)
        c) m(∠QSR)
        d) m(∠PSR)

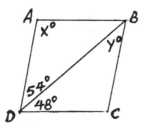

4)   Find the value of x and y in the ▱ ABCD   shown below

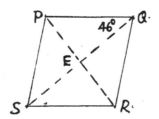

5)   If LMNO is a parallelogram, find the value of a

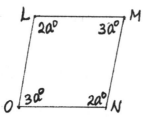

6)   ABCD is a parallelogram $\overline{UR}$ and $\overline{PT}$ are perpendicular to
        QS.

$\overline{PT} \cong$ _____

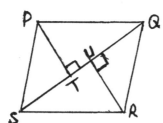

7)   If ABCD is a parallelogram. Find

a) The value of x and y.
b) m(∠ADC)
c) m(∠BCD).

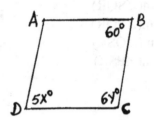

8)   Find the value of x and the length of the side of a rhombus shown below.

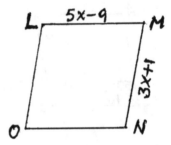

9)   Determine whether the following statement is true or false.
a) Every rhombus is a parallelogram.
b) Every square is a rhombus.
c) Every rectangle is a square.
d) Every square is a rectangle.
e) Rhombuses, rectangles and squares are all parallelograms.

5.15   Properties of Trapezoids and Kites

5.15.1 A trapezoid is a quadrilateral with exactly one pair of parallel sides. The parallel sides are known as the bases. The nonparallel sides of a trapezoid are called the legs of the trapezoid.

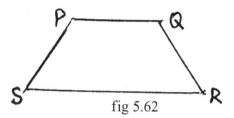

fig 5.62

- $\overline{PQ} // \overline{SR}$
- $\overline{PS} \not\parallel \overline{QR}$
- $\overline{PQ}$ and $\overline{SR}$ are the bases of trapezoid PQRS.
- $\overline{PS}$ and $\overline{QR}$ are the legs of trapezoid PQRS.
- For each of the bases of a trapezoid, there is a pair of base angles, which are the two angles that have that base as a side. The angle pairs ∠P and ∠Q, and ∠R and ∠S are base angles. If the legs of a trapezoid are congruent. The trapezoid is an isosceles trapezoid. In the above figure, if $\overline{PS} \cong \overline{QR}$, then trapezoid PQRS is an isosceles trapezoid.
  The midsegment of a trapezoid is the segment that connects the midpoints of its legs.

5.16 A Kite is a quadrilateral with two distinct pairs of congruent consecutive sides. But opposite sides are not congruent. The angles between two congruent sides are called vertex angles and the other two angles are called nonvertex angles.

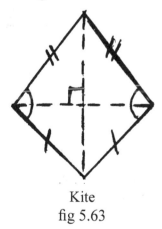

Kite
fig 5.63

5.16.1 Properties of Kite
  (1) The diagonals of a kite meet at a right angle.
  (2) Kites have exactly one pair of opposite angles that are congruent.

Theorem 5-21: A trapezoid is isosceles if and only if its diagonals are congruent.

Theorem 5-22: If a trapezoid is isosceles, then each pair of base angle is congruent.

Theorem 5-23: If a trapezoid has a pair of congruent base angles, then it is an isosceles trapezoid.

Midsegment Theorems for trapezoid.

Theorem 5-24: The midsegment of a trapezoid is the segment that connects the midpoints of its legs.

Theorem 5-25: The midsegment of a trapezoid is parallel to each base and its length is one half the sums of the lengths of the base.

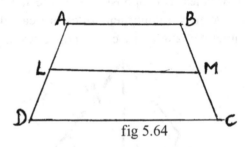

fig 5.64

ABCD is a trapezoid, $\overline{AB} // \overline{DC}$ and $\overline{LM}$ is the midpoint segment, then $\overline{LM} = \dfrac{1}{2}\left(\overline{AB} + \overline{DC}\right)$.

Theorem 5-26: If a quadrilateral is a kite, then exactly one pair of opposite angles are congruent.

Theorem 5-27: If a quadrilateral is a kite, then its diagonals are perpendicular.

Example: 1) Find the length of median PQ in trapezoid ABCD, if AB=16cm and CD=24cm.

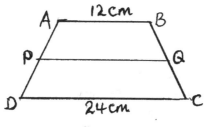

fig 5.65

Solution:

$$\overline{PQ} = \frac{1}{2}\left(\overline{AB} + \overline{DC}\right) \cdots \qquad \text{Theorem 5-25}$$

$$= \frac{1}{2}\ (12\text{cm}+24\text{cm}) \qquad \text{Substitution}$$

$$= \frac{1}{2}\ (36\text{cm})$$

$$\overline{PQ} = 18cm$$

Therefore, the length of the median of trapezoid ABCD is 18cm.

Example 2) If trapezoid PQRS is an isosceles. Find

a) m(∠S)
b) m(∠P)
c) m(∠Q)

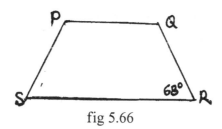

fig 5.66

Solution

a) Since trapezoid PQRS is an isosceles, its base angles are
   congruent. (Theorem: 5-22). Thus,

   m(∠S)≅m(∠R) Theorem:5-22

   m(∠S)=68° ...   Substitution

b) The sum of the degree measure of the angles of trapezoid PQRS
   is 360°.

   m(∠P)+m(∠Q)+m(∠R)+m(∠S)=360°

   m(∠P)+m(∠Q)+68°+68°=360° ...   Substitution.

   m(∠P)+m(∠Q)=360°-136°

   m(∠P)+m(∠Q)=224°

   2m(∠P) or 2m(∠Q)=224°, m(∠P)≅m(∠Q)..   Theorem. 5-22

   m(∠Q)=112°

   Therefore, m(∠P)=112°

   c) 112° ...   from the above (b)

Example 3)   If ABCD is a trapezoid with midsegment $\overline{PQ}$. Find the
             value of x.

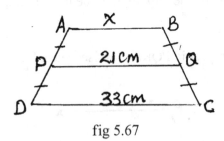

fig 5.67

Solution:

$$\overline{PQ} = \frac{1}{2}\left(\overline{AB} + \overline{DC}\right) \text{ ....}$$ Theorem: 5-25

$21cm = \frac{1}{2}\,(x+33)....$ Substitution

x+33=42

x=42-33=9cm....   Solve for x.

Example 4)   In the figure shown below, PQRS is an isosceles trapezoid and $\overline{TU}$ is the midsegment. Find m($\angle$P).

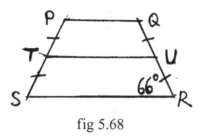

fig 5.68

Solution:-

Since $\angle$R and $\angle$Q are consecutive interior angles formed by $\overline{QR}$ intersecting two parallel lines, they are supplementary. Thus, m($\angle$P)=180°-66°=114°.

Example 5)   LMNO is a kite. Find m($\angle$O)

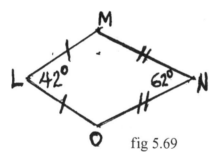

fig 5.69

Solution:-

By Theorem 5-26. LMNO has exactly one pair of congruent opposite angles. Since m($\angle$L)≠m($\angle$N), m($\angle$M) and m($\angle$O) must be congruent. Thus, m($\angle$M)≅m($\angle$O).

m($\angle$M)+m($\angle$O)+42°+62°=360° ....   Sum of angles of quadrilateral

m($\angle$O)+m($\angle$O)+104°=360° ....   Substitution

2m($\angle$O)=256° ...   Combine like terms.

m($\angle$O)=128° ...   Solve for $\angle$O.

Example 6) If ABCD is a kite, find the length of its side.

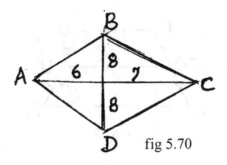

fig 5.70

Solution:-
   To find the sides of a kite, we use the Pythagorean Theorem.
   Thus

- $\left(\overline{AB}\right)^2 = 6^2 + 8^2$ ...   Pythagorean Theorem
   =36+64

   =100
   $\overline{AB} = \sqrt{100}$
   $\overline{AB} = 10$
   $\overline{AB} \cong \overline{AD} = 10$

- $(BC)^2 = 8^2 + 7^2$ ...   Pythagorean Theorem.
   =64+49
   $(BC)^2 = 113$
   $BC = \sqrt{113}$
   $\overline{BC} \cong \overline{DC} = \sqrt{113}$

Example 7)   In the figure below, find

    a) The value of y___
    b) The length of $\overline{EF}$

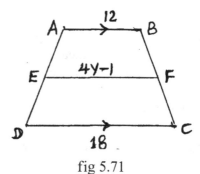

fig 5.71

Solution

a) $\dfrac{1}{2}$ (12+18)=4y-1 ...   Theorem 5-25

15=4y-1
4y=16 ....   Combining like terms
y=4 ...   Solve for y.

b) The length of $\overline{EF}$ is equal to (4y-1)

$\overline{EF} = 4y - 1$
    $= 4(4) - 1$
$\overline{EF} = 15$

Example 8)   Find the value of x in the figure shown below.

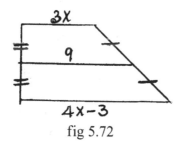

fig 5.72

Solution:
    9=½(3x+4x-3) ...   Theorem 5-26
    7x-3=18

7x=21 ...   Collecting like terms.

x=3 ...   Solve for x

Example 9) In an isoseles trapezoid, if one pair of base angles is four times the measure of the second pair of base angles. What are the measures of the angles?

Solution:-

Let x be the measure of the angle, then

$(x°+x°)+(4x°+4x°)=360°$ ....   Sum of all angles of trapezoid.

$10x°=360°$ ...   Adding like terms.

$x=36°$ ...   Solve for x.

Therefore, the measures of the angles are 36°, 36°, 144° and 144°.

Example 10) In the figure shown below: find

      a) Y

      b) X

      c) Z

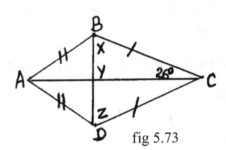

fig 5.73

Solution:-

    The figure shown above ABCD is a kite, thus;

a) The diagonals of a kite are perpendicular, so, m(∠Y)=90°

b) $x°+y°+26°=180°$...   Sum of all angles of triangle.

$x°+90°+26°=180°$ ...   Substitution

$x°=180°-116°$ ...   Collecting like terms.

$x°=64°$ ...   Solve for x.

c) m(∠X)≅m(∠Z) ...   Base angles of isosceles ΔCBD.

5.17 Relationships among special quadrilaterals

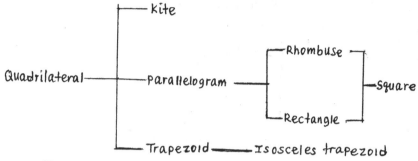

Exercise 5-5

1) Find the length of median RS in trapezoid ABCD, if AB=16cm and DC=26cm

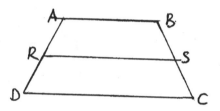

2) If trapezoid ABCD is an isosceles. Find

a) m(∠A)
b) m(∠B)
c) m(∠D)

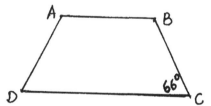

3) In the figure shown below find

a) ∠X
b) ∠Y
c) ∠Z

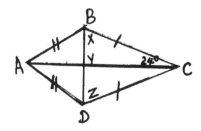

4) Find the value of x in the figure below

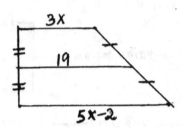

5) In the figure below find

a) The value of Y___
b) The length of $\overline{MN}$

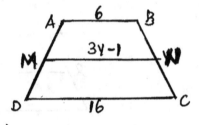

CHAPTER SIX

6. Statistics

6.1   Statistics: is the study of the collection, organization, and interpretation of data. It deals with all aspects of this, including the planning of data collection in terms of the design of Surveys and Experiments.

       A statistician: is someone who is particularly well versed in the ways of thinking necessary for the successful application of statistical analysis. Such people have often gained this experience through working in any of a wide number of fields.

       -To present data first measurements will be done.

6.2   Raw Data:- is a term for data collected on source which has not been subjected to processing or any other manipulation, it is also known as primary data. It is a relative term.

6.3   Frequency Distributions:- is an arrangement of the values that one or more variables take in a sample. Each entry in the table contains the frequency or count of the occurrences of values with in a particular group or interval, and in this way, the table summarizes the distribution of values in the sample.

Example 1)   The following raw data shows the number of books sold by a small shop for 30 days.

6, 3, 5, 4, 6, 2, 5, 3, 4, 2, 6, 5, 3, 6, 4, 3, 5, 4, 6, 3, 3, 5, 4, 5, 3, 6, 5, 5, 6, 7

We can give the frequency distribution table as follows

No of pencils sold	Tally	Frequency (number of days)
2	II	2
3	IHI II	7
4	IHI	5
5	IHI III	8
6	IHI II	7
7	I	1

From the frequency distribution table we can read for two days the shop sales 2 books, for 7 days the shop sales 3 books every day...

Example 2) Here is a raw data showing scores obtained by 40 students in physics examination taken out of 10, prepare a frequency distribution table.

5, 9, 10, 5, 4, 3, 9, 7, 5, 2, 1, 7, 6, 5, 4, 3, 1, 1, 8, 5, 6, 7, 7, 6, 4, 3, 2, 7, 9, 8, 7, 7, 6, 5, 6, 5, 6, 7, 7, 10

Solution:- The frequency distribution is shown below.

Score	1	2	3	4	5	6	7	8	9	10
Frequency	3	2	3	3	7	6	9	2	3	2

6.4   Fundamental Principle of Counting or Rule of Product

The rule of product or multiplication principle is a basic counting principle. (The fundamental principle of counting). Stated simply, it is the idea that if we have a ways of doing something and b ways of doing another thing, then there are a•b ways of performing both actions.

$$\{A, B, C\} \qquad \{X, Y\}$$

To choose one of these and one of these

Is to choose one of   these
$$\{AX, AY, BX, BY, CX, CY\}$$

In this example the rule says: Multiply 3 by 2, getting 6. The sets {A, B, C} and {X, Y} in these examples are disjoint, but that is not necessary. The number of ways to choose a member of {A, B, C}, and then to do so again. In effect choosing an ordered pair each of whose components in {A, B, C}, is 3x3=9.

In set theory, this multiplication principle is often taken to be the definition of the product of cardinal numbers. We have $|S_1| \cdot |S_2|...|S_n| = |S_1 x S_2 x ... x S_n|$

where x is the Cartesian product operator. These sets need not be finite, nor it is necessary to have only finitely many factors in the product.

Example 1)   When you decide to order pizza, you must first choose the type of crust: thin or deep dish (2 choices). Next, you choose the topping: cheese, pepperoni, or sausage (3 choices).

Using the rule of product, you know that there are 2x3=6 possible combinations of ordering a pizza.

Example 2)   If Alex wants to travel from city A to city B then to city C, then in how many different ways can he travel?, if the means of transportation from city A to city B is bus or plane and from city B to city C the means of transportation are ship or plane or taxi.

Solution:- Let us use letters Bus (B), Train (T), Ship (S) plane (P) and Taxi (X). Here we have two acts to be performed one after the other they are:-

Act-1: To travel from city A to city B and it can be done in two different ways.

Act-2: To travel from city B to city C and if can be done in three different ways.

The total act can be performed in 2x3=6 different ways .... $(A_1 x A_2)$.

If we are interested to see the details we can use tree diagram.

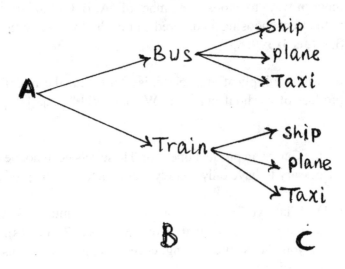

Our possibility set is
{(B, S), (B, P), (B, X), (T, S), (T, P), (T, X)}

We have 6 possible ways, in our possibility sets (B, S) stands for taking bus from city A to city B and ship from city B to city C.

Here we may be asked what is the probability to use train in the first trip and taxi in the second trip.

We can easily see $P(T, X) = \frac{1}{6}$

Example 3)   If event A can occur in 4 ways and event B can occur in 5 ways, then A followed by B can occur in 4x5=20 ways.

Example 4)   In a decimal system of notation how many different three digit numerals can we write, if we are allowed to repeat a digit?

Solution:- In decimal system of notation we are using the numbers 0, 1, 2, 3, 4, 5, 6, 7, 8, 9, our hundreds place can be selected in 9 ways (we cannot start with 0).

Our tens place can be written in ten different ways and our unit can be written in 10 different ways.

There are 9x10x10=900 different three digits numerals.

Example 5)   College faculty are to be issued special coded identification cards that consist of four letters of the alphabet. How many different ID cards can be issued if the letters can be used more than once?

Solution:- Because four letters are to be used, there are four positions to fill. Any of the 26 letters of the alphabet can be used in each of the four code positions. Therefore the total number of identification cards that can be made will be:-

(26)(26)(26)(26)=456, 976 can be made.

Example 6)   If the letters in college faculty identification codes not allowed to repeat, how many different identification cards can be formed?

Solution:- Since the letters are not allowed to repeat, once a letter is used, it cannot be used again. Hence, there will be 26 options for the first position, 25 options for the second position, 24 for the third, and 23 for the fourth.

(26)(25)(24)(23)=358,800 different identification cards can be performed.

Example 7)   A man has three daughters to visit. In how many different ways can he makes his rounds. If he visits each daughter once?

Solution:- For his first visit, he could visit any of the 3 daughters.

For his second visit, he can visit either of the 2 remaining daughters.

On his last visit, there will be only one daughter left to visit.

The number of different ways that he can make his round will be:-
(1st visit)(2nd visit)(3rd visit)=3x2x1=6

6.5   The Addition Counting Principle

- If one group contains m objects and a second group contains n objects, and the groups have no objects in common, then there are m+n total objects to choose from. This is known as the addition counting principle.

Example 1)   A high school library has 48 text books on Maths and 60 text books dealing with Chemistry. By the rule of addition, a student at this school can select among 48+60=108 text books in order to learn more about one or the other of these two subjects.

Example 2)   Suppose there are 6 different kinds of soda and 8 difference flavored water. How many selections does a customer have?

Solution: In this case, an event is "selecting a soda and a flavored water." There are 6 outcomes for the soda event and 8 outcomes for flavored water event. According to the addition principle there are 6+8=14 possible selections.

Example 3)   A university gives 5 courses in the morning and 4 in the evening. Find the number of ways a student can select exactly one course, either in the morning or in the evening.

Solution:- The student has five choices from the morning courses out of which he/she can select one course in 5 ways.

For the evening course, he/she has 4 choices out of which he/she can select one course 4 ways.

Therefore, he/she has total number of 5+4=9 choises.

6.6 Permutation

The notion of permutation is used with several slightly different meanings. All related to the act of permuting (rearranging in an ordered fashion) objects or values. A permutation is a rearrangement of a set of objects. All permutations on a given set from a group under composition. For example, there are six permutations of the set {1, 2, 3}, namely (1, 2, 3), (1, 3, 2), (2, 1, 3), (2, 3, 1), (3, 1, 2), and (3, 2, 1).

The number of permutations of n distinct objects is $n \times (n-1) \times (n-2) \times ... \times 2 \times 1$, which number is called "n factorial" and written "n!".

n! read as n-factorial or factorial n.
$n! = n(n-1)(n-1)(n-3) .... x3x2x1$ and
$o! = 1$

Example 1) 4! (four factorial) is given as $4! = 4x3x2x1 = 24$

Example -2) $6! = 6x5x4x3x2x1 = 720$

Example 3) $2! = 2x+1 = 2$

For simplification of calculations, we can use the following formulae.
$n! = n \cdot (n-1)!$
$n! = n(n-1)(n-2)!$
$n! = n(n-1)(n-2)(n-3)! ... etc$

Example 4) Simplify each of the following

a) $\dfrac{8!}{8} = \dfrac{8 \times 7!}{8} = 7! = 7 \times 6 \times 5 \times 4 \times 3 \times 2 \times 1 = 5,040$

b) $\dfrac{7!4!}{4!6!} = \dfrac{7 \times 6! \times 4!}{4!6!} = 7$

c) $\dfrac{9!8!}{6!5!} = \dfrac{9\times8\times7\times6!\times8\times7\times6\times5!}{6!5!} = 9\times8\times7\times8\times7\times6 = 169,344$

d) $\dfrac{n!}{(n-2)!} = \dfrac{n(n-1)(n-2)!}{(n-2)!} = n(n-1) = n^2 - n$

e) $\dfrac{n!n!}{(n-2)!(n-1)!} = \dfrac{n(n-1)!n(n-1)(n-2)!}{(n-2)!(n-1)!} = n \bullet n(n-1) = n^3 - n^2$

Example 5) $\dfrac{n!}{(n-2)!} = 42$, then what is the value of n if n≥2?

Solution: $\dfrac{n!}{(n-2)!} = 42$

$= \dfrac{n \bullet (n-1)(n-2)!}{(n-2)!} = 42$

n²-n=42

n²-n-42=0

(n+6)(n-7)=0

n-7=0 $\Rightarrow$ n=7

Exercise 6-1

1) A man has four sons to visit. In how many different ways can he makes his rounds. If he visits each son once?

2) A college library has 52 text books on Physics and 42 text books dealing with Biology. How many books can select a student in this school?

3) Suppose there are 8 different kinds of soda and 9 different flavored water. How many selections does a customer have?

4) If event m can occur in 6 ways and event n can occur in 8 ways, then m followed by n can occur in how many ways?

5) How many five-digit numerals can be made where all digits are uniques. If the possible digits are 0 through 9?

6) If we toss three coins A, B, and C one after the other, then how many possible sequences of heads and tails can we find?

7) How many five digit numerals can be made where all digits are unique. If the possible digits are 0 through 9 and if the first digit cannot be a 0?

8) Compute:
a) 10!
b) 6!
c) 0!
d) 5!

9) Simplify

a) $\dfrac{13!}{9!3!}$

b) $\dfrac{9!}{7!6!}$

c) $\dfrac{(16-11)!}{8!}$

d) $\dfrac{n!}{(n-2)!}$

e) $\dfrac{n!}{(n-3)!}$

10) If $\dfrac{n!}{(n-2)!} = 20$, then what is the value of n, if n≥2?

11) If $\dfrac{n!}{(n-1)!} = 10$, then what is the value of n, if n≥1?

12) If $\dfrac{n!}{(n-2)!} = 30$, then what is the value of n, if n≥2?

6.7 Principles of Permutations

An arrangement of n-objects taken r at a time is given by nP$_r$ (read as n permutation r) where r<n, in such arrangements order is very important or order makes difference on the total number of arrangements.

Here $n P_r = \dfrac{n!}{(n-r)!}$

6.7.1 Notation: nP$_r$ can also be written as P(n, r).

Example 1) Calculate each of the following.

a) 6P$_4$
b) 8P$_3$
c) 7P$_5$
d) 5P$_3$

Solution:-

a) $6P_4 = \dfrac{6!}{(6-4)!} = \dfrac{6!}{2!} = \dfrac{6\times5\times4\times3\times2!}{2!} = 360$

b) $8P_3 = \dfrac{8!}{(8-3)!} = \dfrac{8!}{5!} = \dfrac{8\times7\times6\times5!}{5!} = 336$

c) $7P_5 = \dfrac{7!}{(7-5)!} = \dfrac{7!}{2!} = \dfrac{7\times6\times5\times4\times3\times2!}{2!} = 2{,}520$

d) $5P_3 = \dfrac{5!}{(5-3)!} = \dfrac{5!}{2!} = \dfrac{5\times4\times3\times2!}{2!} = 60$

Example 2) In how many different ways can five books be arranged on a shelf if they are selected from nine books?

Solution:- A permutation of 9 objects taken 5 at a time is equal to:-

$9P_5 = \dfrac{9!}{(9-5)!} = \dfrac{9!}{4!} = \dfrac{9\times8\times7\times6\times5\times4!}{4!} = 15{,}120$

Example 3) If we have four chairs in a row and four students in how many different ways can we give seats for the students?

Solution:- It is an arrangement of four objects taken all at a time.

There are $4P_4 = \dfrac{4!}{(4-4)!} = 4! = 24$ different ways of giving seats.

Example 4) In how many ways can a president and vice president be selected for a club that has 10 members?

Solution:- A permutation of 10 objects taken 2 at a time is equal to:-

$$10P_2 = \dfrac{10!}{(10-2)!} = \dfrac{10!}{8!} = \dfrac{10\times9\times8!}{8!} = 90$$

Example 5) In how many ways can 6 boys sit in a round table of 6 chairs?

Solution:- It is round table one student will take one of the chairs first there are (n-1)!

Arrangements=(6-1)!=5!=120

Example 6) Using the digits 1, 3, 5, 7, 8 and 9 how many four digit numerals can we write if we use a digit at most once.?

Solution:- From the six digits given we need four at a time, so it is an arrangement of six objects taken four at a time is given by $6P_4$.

$$6P_4 = \dfrac{6!}{(6-4)!} = \dfrac{6!}{2!} = \dfrac{6\times5\times4\times3\times2!}{2!} = 360 \quad \text{four digit numerals.}$$

Example 7) From a mathematics department of eight members in how many different ways can a chairman and a secretary be selected?

Solution:- It is an arrangement of eight taken two at a time. It is given by:-

$$8P_2 = \frac{8!}{(8-2)!} = \frac{8!}{6!} = \frac{8 \times 7 \times 6!}{6!} = 56 \text{ different ways of election.}$$

Example 8) How many natural numbers can be written using the digits 1, 5, 7 and 9. If a digit is used at most once in a numberal.

Solution:- Here the number of digits is not restricted, we can write one digit, 2 digits, 3 digits or 4 digits numerals.

There are $4P_1 = 4$ ...... 1 digit numerals.
There are $4P_2 = 6$ ...... 2 digit numerals.
There are $4P_3 = 24$ ...... 3 digit numerals.
There are $4P_4 = 24$ ...... 4 digit numerals.
Totally, there are 4+6+24+24=58 numerals.

6.7.2 When some of the items are identical, a different permutation formula must be used:-

We use the following permutation formula when some of the item possibilities are alike.

If there are n items with n_1 alike, n_2 alike, n_3 alike, ..., n_k alike, the number of permutations is calculated by dividing n factorial by the product of the factorials of the number of occurrences of each of the like items.

$$\frac{n!}{(n_1!)(n_2!)(n_3!)...(n_n!)}, \text{ when n=r } \quad nP_r = n!$$

Example 1) How many permutations are there for the letters in the word, STATISTICS?

Solution:- The letter S repeats three times (3!).
The letter T repeats three times (3!),
and the letter I repeats two times (2!)

$$\frac{10!}{3!3!2!} = \frac{10 \times 9 \times 8 \times 7 \times 6 \times 5 \times 4 \times 3!}{3!3 \times 2 \times 1 \times 2!} = 50,400 \text{ permutations}$$

Example 2) How many permutations are there for the letters in the word, ENGINEER?

Solution:- The letter E repeats three times (3!)
The letter N repeats two times (2!)

$$\frac{8!}{3!2!} = \frac{8\times7\times6\times5\times4\times3!}{3!2\times1} = 3,360 \text{ permutations.}$$

Example 3) Calculate nP_n

Solution: $nP_n = \frac{n!}{(n-n)!} = \frac{n!}{0!} = \frac{n!}{1} = n!$

Example 4) How many different 5-letter words can be formed from the word APPLE?

Solution: The letter P repeats twice.

$$\frac{5P_5}{2!} = \frac{5!}{2!} = \frac{5\times4\times3\times2!}{2!} = 60 \text{ words.}$$

Example 5) How many permutations are there for the letters in the word, MATHEMATICS?

Solution:- There are eleven total letters of which, M is repeated 2 times, A is repeated 2 times, T is repeated 2 times. Hence the number of permutation is equal to:-

$$\frac{11!}{2!2!2!} = \frac{11\times10\times9\times8\times7\times6\times5\times4\times3\times2\times1!}{2\times1\times2\times1\times2\times1} = 4,989,600$$

Exercise 6-2

1) Calculate each of the following
   a) $8P_4$
   b) $6P_3$
   c) $9P_5$
   d) $7P_2$

2) In how many different ways can seven books be arranged on a shelf. If they are selected from ten books?

3) In how many ways can a president and vice president be selected for a club that has 12 members?

4) In how many ways can seven girls sit in a round table of seven chairs?

5) Using the digits 1, 2, 3, 4, 5, 6 and 7 how many five digit numerals can we write if we use a digit at most once?

6) From a Physics department of ten members in how many different ways can a chairman and a secretary be selected?

7) How many permutations are there for the letters in the word MOROCCO?

8) How many permutations are there for the letters in the word TELECOMMUNICATION?

6.8   COMBINATION

A combination is a way of selecting several things out of a larger group, where (unlike permutations) order does not matter. In smaller cases it is possible to count the number of combinations. For example given three fruit, say an apple, orange and pear, there are three - combinations of two that can be drawn from this set: an apple and a pear; an apple and an orange; or a pear and an orange.

6.8.1   Combination Formula

A combination has the following formula:-

C(n, r) ..... (read as n combination r or combination of n, r)

$$C(n,r) = \frac{n!}{r!(n-r)!} \text{ , for } r \leq n$$

    n: number of available items or choices.

    r: the number of items to be selected.

Sometimes we can use the notation $\binom{n}{r}$ for C(n, r)

Example 1)   Compute each of the following
            a) C(5, 3)
            b) C(8, 2)
            c) C(9, 4)
            d) C(8, 5)

Solution:-

a) $C(5,3)\dfrac{5!}{3!(5-3)!}=\dfrac{5!}{3!2!}=\dfrac{5\times4\times3!}{3!2\times1}=10$

b) $C(8,2)\dfrac{8!}{2!(8-2)!}=\dfrac{8!}{2!6!}=\dfrac{8\times7\times6!}{2\times1\times6!}=28$

c) $C(9,4)\dfrac{9!}{4!(9-4)!}=\dfrac{9!}{4!5!}=\dfrac{9\times8\times7\times6\times5!}{4\times3\times2\times1\times5!}=126$

d) $C(8,5)\dfrac{8!}{5!(8-5)!}=\dfrac{8!}{5!3!}=\dfrac{8\times7\times6\times5!}{5!3\times2\times1}=56$

Example 2)   Compute each of the following
            a) C(6, 4)
            b) C(6, 2)
            c) C(10, 3)
            d) C(10, 7)

Solution:-

a) $C(6,4)\dfrac{6!}{4!(6-4)!}=\dfrac{6!}{4!2!}=\dfrac{6\times5\times4!}{4!2\times1}=15$

b) $C(6,2)\dfrac{6!}{2!(6-2)!}=\dfrac{6!}{2!4!}=\dfrac{6\times5\times4!}{2!4!}=15$

c) $C(10,3)\dfrac{10!}{3!(10-3)!}=\dfrac{10!}{3!7!}=\dfrac{10\times9\times8\times7!}{3\times2\times1\times7!}=120$

d) $C(10,7)\dfrac{10!}{7!(10-7)!}=\dfrac{10!}{7!3!}=\dfrac{10\times9\times8\times7!}{7!\times3\times2\times1}=120$

* What do you observe from the above solutions for questions a-d?

Thus, we observe that:-

C(6, 4)=C(6, 2)

C(10, 3)=C(10, 7),

Generally; C(n, r)=C(n, n-r)

We can also use: $C(n,r) = \dfrac{n!}{r!(n-r)!}$ to find number of

arrangements of n-objects where r of them are of one type and the rest (n-r) are of the same and of the second type.

Example 1) On a mathematics exam, students are required to answer five algebra questions from a list of eight questions. How many possible combinations of five questions will there be?

Solution:- The question represents a combination of 8 objects taken five at a time and it will be

$$8C_5 = (8,5) = \frac{8!}{5!(8-5)!} = \frac{8!}{5!3!} = \frac{8\times7\times6\times5!}{5!3\times2\times1} = 56$$

Example 2) From five girls in a class in how many different ways can we select two girls?

Solution:- This represents a combination of 5 girls taken 2 at a time disregarding order.

Thus, it is given by:-

$$C(5,2) = \frac{5!}{2!(5-2)!} = \frac{5!}{2!3!} = \frac{5\times4\times3!}{2\times1\times3!} = 10$$

Example 3) In how many ways can a group of 4 students be selected from a class of 18 students?

Solution:- It is a combination of 18 objects taken 4 at a time and it is given by:-

$$C(18,4) = \frac{18!}{4!(18-4)!} = \frac{18!}{4!14!} = \frac{18 \times 17 \times 16 \times 15 \times 14!}{4 \times 3 \times 2 \times 1 \times 14!} = 3,060$$

Example 4) If we have 8 balls where 3 are red and the remaining are blue. In how many different ways can we arrange the balls in a row based on their colors.?

Solutions:- This represents an arrangement of 8 objects where 3 are of the same type and 5 are of the same and of other type. The total arrangement is:-

$$C(8,3) = \frac{8!}{3!(8-3)!} = \frac{8!}{3!5!} = \frac{8 \times 7 \times 6 \times 5!}{3 \times 2 \times 1 \times 5!} = 56 \text{ arrangements}$$

Example 5) There are 6 students in class A and 8 students in class B if we want to select 3 students from each class. In how many different ways could this selection be made?

Solution:- Here we have two, acts to be performed, first act is selection of 3 from 6 which is C(6, 3), and the second act is selection of 3 from 8 which is C(8, 3). Thus, the total act can be done in:-

$$C(6,3) \times c(8,3) = \frac{6!}{3!(6,3)!} \times \frac{8!}{3!(8,3)!} = 1,120$$

Example 6) If 15 people come from different places and shake hands in how many different ways could this be done?

Solution:- This represents a selection of 2 from 15 which is:-

$$C(15,2) = \frac{15!}{2!(15-2)!} = \frac{15!}{2!13!} = \frac{15 \times 14 \times 13!}{2 \times 1 \times 13!} = 105$$

Example 7) In how many ways can a physics teacher select six students to work questions on the board from a class of twenty one students?

Solution:- This is a combination of 21 objects taken 6 at a time and is equal to:-

$$C(21,6) = \frac{21!}{6!(21-6)!} = \frac{21!}{6!15!} = \frac{21 \times 20 \times 19 \times 18 \times 17 \times 16 \times 15!}{6 \times 5 \times 4 \times 3 \times 2 \times 15!} = 54,264$$

Example 8)   There are 3 girls and 2 boys in a room if we want to select two individuals for a given work. How many of these events contain at least one girl?

Solution:- Atleast one girl means 1 girl or 2 girls, and this selection is selection of exactly one or 2 girls.

C(3; 1) x C(2, 1) + C(3, 2) x C(2, 0)
=3x2+3x1=9 different ways.

Exercise 6-3
1)   Compute each of the following:-
   a) C(6, 3)
   b) C(9, 6)
   c) C(10, 8)
   d) C(7, 3)

2)   Compute each of the following:-
   a) C(8, 3)
   b) C(8, 5)
   c) C(10, 2)
   d) C(10, 8)

3)   What do you observe in question number 2, between
   a) a and b
   b) c and d

4)   On a Chemistry exam., students are required to answer six questions from a list of twelve questions. How many possible combinations of six questions will there be?

5)   From six boys in a class in how many different ways can we select two boys?

6)   In how many ways can a group of five students be selected from a class of 20 students?

7)   If we have 10 balls where 4 are white and the remaining are blue. In how many different ways can we arrange the balls in a row based on their colors?

6.9   Probability

Probability is a way of expressing knowledge or belief that an event will occur or has occurred.

- An event is one or more outcomes of an experiment.
- An outcome is the result of a single trial of an experiment
- The set of all possible outcomes of a random experiment is called the sample space and it is represented by the symbol "S".
- Each outcome in a sample space is called an element or a member of the sample space.
- An experiment is any procedure that can be infinitely repeated and has a well-defined set of outcomes.
- An experiment is said to be random experiment if it has more than one possible outcomes.
- If an experiment has only one possible outcome then it is known as Deterministic Experiment.
- An experiment is composed of one or more trials. A trial with two mutually exclusive outcomes is known as Bernoulli trial.
- A population is the set representing all measurements of interest to the sample collector.
- A sample space for an experiment consists of all possible events.
- If P denotes a probability and A, B and C denote specific events. Then P(A) denotes the probability of event A occurring.
- Possibility set of an experiment is a set which contain all the possible outcomes of an experiment.

- Probability of an event A can be expressed by a real number P where $0 \leq P(A) \leq 1$, for any event A.
- The probability of an impossible event is 0.
- The probability of an event that is certain to occur is 1.
- The complement of an event A, denoted by $\overline{A}$, consists of all outcomes in which event A does not occur.

Let's see the following examples

Example -1) Consider an experiment of tossing a coin. If we toss a fair coin the possible outcomes are head or tail.
In this experiment the possibility set is {h, t} where h is occurrence of head and t is the occurrence of tail. Here {t} or {h} is an event. Occurrence of tail is an event of this experiment; also occurrence of head is an event of this experiment.

Example 2) Consider an experiment of tossing three coins and observing the sequence of heads and tails.
In this experiment the possibility set is:- {(h, h, h), (h, h, t), (h, t, h), (h, t, t), (t, h, t), (t, t, h), (t, t, t)}
Thus, there are 8 different possible outcomes in this experiment.

Example 3) Consider an experiment of tossing a die and observing the number appearing on the upper face of the die. In this experiment the possibility set is {1, 2, 3, 4, 5, 6}, there are six possible outcomes where 1 is the occurrence of number 1 on the upper face of the die, 2 is the occurrence of number 2 on the upper face of the die, 3 is the occurrence of number 3 on the upper face of the die, ... etc.
Thus, in this experiment we can consider different events such as:-

- Observing an odd number on the upper face of the die {1, 3, 5}.
- Observing a number less than 5, {1, 2, 3, 4}.
- Observing a number greater than 2, {3, 4, 5, 6}.
- Observing a number which are a multiple of 3, {3, 6}
- Observing an even number, {2, 4, 6}... etc.

Exercise 6-4
1) Define each of the following terms
    a) event
    b) outcome
    c) experiment

2) In an experiment of tossing two coins and observing the sequences of heads and tails, what is the possibility set?

6.10   Expressing Probability
If U is possibility set of an experiment and if each element of U is equally likely to occur, then the probability of the event E is given by P(E), where:-

$$P(E) = \frac{Number\ of\ Elements\ in\ E}{Number\ of\ Elements\ in\ U} \quad or \quad P(E) = \frac{n(E)}{n(U)}$$

Example -1) If you toss a fair dice, what is the probability to see
    a) The number 4
    b) An odd number
    c) An even number
    d) A number that is less than 7.
    e) A number that is a multiple of 2.
    f) A number that is greater than 6.

Solution:
    a) Here our possibility set is U={1, 2, 3, 4, 5, 6} and our event is {4}.

$$P(E) = \frac{n(E)}{n(U)} = \frac{1}{6} \quad \text{... (probability to see the number 4)}$$

    b) Here our possibility set is U={1, 3, 5}.

$$P(E) = \frac{n(E)}{n(U)} = \frac{3}{6} = \frac{1}{2} \quad \text{... (probability to see odd numbers)}$$

    c) Here our possibility set is U={2, 4, 6}.

$$P(E) = \frac{n(E)}{n(U)} = \frac{3}{6} = \frac{1}{2} \quad \text{... (probability to see even numbers)}$$

d) Here our possibility set is U={1, 2, 3, 4, 5, 6}

$$P(E) = \frac{n(E)}{n(U)} = \frac{6}{6} = 1 \quad \text{... (probability to see numbers less than 7)}$$

e) Here our possibility set is U={2, 4, 6}

$$P(E) = \frac{n(E)}{n(U)} = \frac{3}{6} = \frac{1}{2} \quad \text{... (probability to see numbers which are multiple of 2)}$$

f) Here it is an impossible event, there is no number greater than 6 in the possibility of an experiment. The probability is 0. Thus,

$$P(E) = \frac{n(E)}{n(U)} = \frac{0}{6} = 0 \quad \text{... (probability to see numbers greater than 6)}$$

Example -2) If you toss a coin, then what is the probability to see head?

Solution:- our possibility set U={h, t} and our event E={h}. Thus

$$P(E) = \frac{n(E)}{n(U)} = \frac{1}{2} \quad \text{... (probability to see head)}$$

Note:- It is equal to probability to see tail.

Example -3) If there are 5 oranges, 4 apples, and 2 bananas in the basket, what is the probability of choosing an orange, without looking the basket?

Solution:

$$P(\text{choosing orange}) = \frac{5}{5+4+2} = \frac{5}{11}$$

Example 4) From a jar which contains 12 similar toys where 4 are red and 8 are yellow, what is the probability to draw yellow toy. If we draw one ball at random?

Solution:- Here our possibility set containing 12 similar toys and our event contains 8 similar toys.

Thus,

$$P(E) = \frac{P(E)}{P(U)} = \frac{8}{12} = \frac{2}{3}$$

Example-5) From 13 similar cards in a box numbered (1-13). If we draw one card at random, then what is the probability to draw an even numbered card?

Solution:- our possibility set contains 13 elements i.e, {1, 2, 3, 4, 5, 6, 7, 8, 9, 10, 11, 12, 13}, where {2, 4, 6, 8, 10, 12} is our event. Thus,

$$P(E) = \frac{P(E)}{P(U)} = \frac{6}{13} \quad \text{... (probability to draw even numbered card)}$$

Example 6) In the above, example number 5, what is the probability to draw an odd numbered card?

Solution:

$$P(\text{to draw odd numbered card}) = 1 - \frac{6}{13} = \frac{13-6}{13} = \frac{7}{13}$$

Example 7) Two coins are tossed, what is the probability that two tails are obtained?

Solution:-
- Let U be our sample space and it is given by:-
  U={H, T), (H, H), (T, H), (T, T)}
- Let E be the event "Two tails are obtained."
  E={(T, T)}, Thus
  $$P(E) = \frac{P(E)}{P(U)} = \frac{1}{4} \quad \text{....(probability of obtaining two tails)}$$

Example 8) There are 6 white, 9 red and 5 blue balls in a basket. One ball is picked up randomly. What is the probability that it is neither white nor blue?

Solution:-

Our possibility set contains 20 elements, where the total number of balls=6+5+9=20 balls.

Let E be an event that the ball drawn is neither white nor blue. i.e, Event that the ball drawn is red. Thus, n(E)=9

Therefore, $P(E) = \dfrac{n(E)}{n(U)} = \dfrac{9}{6+5+9} = \dfrac{9}{20}$

Remark:- We can also use by the following method

$$P(E) = 1 - \dfrac{(6+5)}{20}$$

$$= \dfrac{9}{20}$$

Exercise 6-5

1) In a box there are 13 similar balls where 8 are blue and the remaining are red. If we picked up one ball randomly, what is the probability to draw a red ball?

2) If a coin is tossed two times, then what is the probability to obtain head in succession?

3) A certain factory can produce 100,000 canned food in five hours. 5% of the production is spoiled. If one canned food is selected at random. What is the probability that

a) It is spoiled?

b) It is non-spoiled?

4) If you toss a die and observe the number appearing on the upper face of a die, then what is the probability to see
a) An even number?
b) An odd number?
c) A number less than 6?
d) A number less than or equal to 5?
e) A number greater than or equal to 1?
f) A number greater than 6?

5)   If there are 5 red, 6 yellow and 7 blue balls in a box. One ball is drawn at random. What is the probability that it is neither red nor blue?

6.11   More on Probabilities and Rules of Probabilities.
A compound event is an event that consists of two or more events that are not mutually exclusive. It combines two or more events, using the word and or the word or. Tossing two dice is an example of compound event.

6.11.   Definition:- Two events A and B are said to be mutually exclusive if they cannot occur at the same time.

Example 1)   Tossing a coin once, which can result in either heads or tails, but not both.

Example 2)   From pack of 52 playing cards if we draw one card it cannot be red and black card, the two events, appearance of red card or appearance of black card are two mutually exclusive events.

Example 3)   From a box which contains 12 similar cards numbered (1-12), if we pick one card at random and observe an odd number or a number less than eight, the two events are not mutually exclusive, because they have common elements.

6.12   RULES OF ADDITION
The probability that one or the other of two mutually exclusive events will occur is the sum of the probabilities of the individual events. Or
If A and B are two mutually exclusive events, then the probability of A or the probability of B is given by P(A)+P(B).

P(A or B)=P(A)+P(B)

Note: If two events A and B are not mutually exclusive events (if they have some common events). Then the probability of A or the probability of B is given by:-

$$P(A \text{ or } B)=P(A)+P(B)-P(A \cap B)$$

Example 1) If we toss a die and observe the number appearing on the upper face of the die, then what is the probability to see 3 or 4?

Solution:- Let event A be observing a number 3 event. This is given by the set {3}.

Let event B be observing a number 4 event. This is given by the set {4}.

The two events A and B have no common elements or they are mutually exclusive events.

$$P(A \text{ or } B)=P(A)+P(B)=\frac{1}{6}+\frac{1}{6}=\frac{2}{6}=\frac{1}{3}$$

Example 2) If we toss a die and observe the number appearing on the upper face of the die, then what is the probability to see a number less than 5 or a number greater than 5?

Solution: Let event A be observing a number less than 5. This event is given by the set {1, 2, 3, 4}.

Let event B be observing a number greater than 5. This event is given by the set {6}.

The two events A and B have no common elements or they are mutually exclusive events.

$$P(A \text{ or } B)=P(A)+P(B)=\frac{4}{6}+\frac{1}{6}=\frac{5}{6}$$

Example 3) If we toss a die and observe the number appearing on the upper face of the die, then what is the probability to see an even number or a number greater than 3?

Solution:- Let event A be observing even number. This event is given by the set={2, 4, 6}

Let event B be observing a number greater than 3. This event is given by the set={4, 6}.

Here Event A and Event B are not mutually exclusive events.

$A \cap B = \{4,6\}$

Here P(A or B)=P(A)+P(B)-P(AnB)

$$P(A \, or \, B) = \frac{3}{6} + \frac{2}{6} - \frac{2}{6} = \frac{3}{6} = \frac{1}{2}$$

Example 4)   If we toss a coin what is the probability to see a head or a tail?

Solution: The two events are mutually exclusive events each with probability half.

$$P = \frac{1}{2} + \frac{1}{2} = 1$$

Such events with probability 1 are called sure events (events that will occur definitely.)

Example 5)   If we toss a die and observe the number appearing on the upper face of the die, then what is the probability to see a number 1 or 6?

Solution:- Let event A be observing a number 1. This event is given by the set={1}

Let event B be observing a number 6. This event is given by the set={6}.

Here Event A and Event B are mutually exclusive events.

$$P(A \, or \, B) = P(A) + P(B) = \frac{1}{6} + \frac{1}{6} = \frac{2}{6} = \frac{1}{3}$$

Exercise 6-6

1)   From a pack of 52 playing cards if we draw one card at random. What is the probability to draw number 6 card or a Jack?

2)   If we toss a dice and observe the number appearing on the upper face of the dice, then what is the probability to see an odd number or a number less than 4?

3)   If we toss a die and observe the number appearing on the upper face of the die, then what is the probability to see a number less than 3 and a number greater than 6?

6.13 Multiplication Rule

We need to have some concept about independent and dependent events to use the rule of multiplication to find probability.

6.13.1 Independent Event:- Two events are said to be independent if the occurrence or non occurrence of the first event has no effect on the probability of the occurrence of the second event.

Example 1) If we toss two fair coins one after the other and observe the sequences of heads and tails. We can see that the occurrence of tail on the first coin has no effect on the occurrence or non occurrence of tail on the second coin. The same thing will be true for head.

Example 2) The event of getting a 6 the first time a die is rolled and the event of getting a 6 the second time are independent.

Example 3) If two cards are drawn with replacement from a deck of cards, the event of drawing a red card on the first trial and that of drawing a red card on the second trial are independent.

6.13.2 Dependent Event: Two events are said to be dependent if the occurrence or non-occurrence of the first affect the probability of occurrence of the second event.

Example 4) Taking out a marble from a basket containing some marbles and not replacing it. And then taking out a second marble are dependent events.

Example 5) In a basket there are 12 balls, where 7 are red and 5 are yellow. If we draw two balls one after the other without replacement and if we want to find the probability for the second ball drawn to be red, first we should know the color of the first ball drawn. This shows the two events are dependent events, because the color of the ball drawn first will affect the probability of the second ball drawn to be red.

6.14   Rule of Multiplication to find Probability
If two events A and B are independent events the probability that both events will occur is the product of their probabilities.
or P(A and B)=P(A)•P(B)

Example 6)   If fair die is tossed twice. Find the probability of getting a 3 or 7 on the first toss and a number 4, 5, or 6 in the second toss.

Solution: $P(A) = P(3 \, or \, 7) = \frac{2}{6} = \frac{1}{3}$

$P(B) = P(4, 5 \, or \, 6) = \frac{3}{6} = \frac{1}{2}$

They are independent events, thus,

$$P(A \, and \, B) = P(A) \cdot P(B) = \frac{1}{3} \times \frac{1}{2} = \frac{1}{6}$$

Example 7)   A bag contains 6 red and 9 yellow balls, if 2 balls are drawn at random. What is

a)   The probability of obtaining both balls, red, if the first ball drawn is replaced?
b)   The probability of obtaining both the balls is red, if the first ball drawn is not replaced?

Solution:-
a)   The probability of drawing a red ball in the first draw is $\frac{6}{15}$. The probability in the second draw is also $\frac{6}{15}$, because the ball is replaced.

Therefore, probability of obtaining both balls

$red = \frac{6}{15} \times \frac{6}{15} = \frac{36}{225} = \frac{4}{25}$

b) The probability of drawing a red ball in the first draw is $\frac{6}{15}$. Since the first ball drawn is red and not replaced in the bag, we have 5 red balls and 9 yellow balls.

Thus, the probability of drawing a red ball in the second draw $= \dfrac{5}{14}$.

In this case, the probability of drawing a red ball in the second draw depends on the occurrence and non-occurrence of the event in the first draw.

Therefore, probability of both the balls

$$red = \dfrac{6}{15} \times \dfrac{5}{14} = \dfrac{30}{210} = \dfrac{1}{7}$$

Example 8) If we toss two coins one after the other and observe sequences of heads and tails, then what is the probability to see tail in succession?

Solution: Here the two events are independent, A occurrence of tail on the first coin and B occurrence of tail on the second coin. Thus,

$$P(A \, and \, B) = P(A) \bullet P(B) = \dfrac{1}{2} \times \dfrac{1}{2} = \dfrac{1}{4}$$

Example 9) A coin is tossed and a die is thrown, what is the probability to see a tail on the coin and a even number on the upper face of the die?

Solution:-
Here, the two events, A, tossing a coin and observing a tail on the coin={T} and B throwing a die and observing an even number ={2, 4, 6} are independent events. Thus,

$$P(A \, and \, B) = P(A) \cdot P(B) = \dfrac{1}{2} \times \dfrac{3}{6} = \dfrac{3}{12} = \dfrac{1}{4}$$

Example 10) If two dice are thrown, then what is the probability to see number 5 on the upper faces of both dice?

Solution:- The two events are independent events.

$$P \text{ (to see number 5 on both dice)} = \dfrac{1}{6} \times \dfrac{1}{6} = \dfrac{1}{36}.$$

Example 11) If two dice are thrown. What is the probability to see even number on the upper faces of both dice?

Solution:- Here the two events are independent events,

$$P \text{ (to see even numbers on both dice)} = \frac{3}{6} \times \frac{3}{6} = \frac{9}{36} = \frac{1}{4}$$

Example 12) If two dice are thrown. What is the probability to see a number greater than 2 on the upper faces of both dice?

Solution:- Here the two events are independent events,

$$P \text{ (to see a number greater than 2 on both dice)} = \frac{4}{6} \times \frac{4}{6} = \frac{4}{9}$$

Example 13) In a certain sardine factory 2 sardines are spoiled out of 24 sardines. If we choose two sardines one after the other with replacement. What is the probability that both are spoiled?

Solution:- Here the two events are independent events (spoiled) = $\frac{2}{24} = \frac{1}{12}$

$$P \text{ (two spoile in succession)} = \frac{1}{12} \times \frac{1}{12} = \frac{1}{144}$$

Example 14) If three coins are tossed, then what is the probability to see tail in succession?

Solution:- The three events are independent events.

P (to see tail in three succession)=½x½x½=⅛

Exercise: 6-7

1)   If fair die is tossed twice. Find the probability of obtaining a 5 or 6 on the first toss and a number 3, 5, or 6 in the second toss

2) A box contains 5 white and 8 black balls. If two balls are drawn at random what is
   a) The probability of obtaining both balls, white, if the first ball drawn is replaced?
   b) The probability of obtaining both the balls is white, if the first ball drawn is not replaced?

3) If we toss two coins one after the other and observe sequences of heads and tails, then what is the probability to see head in succession?

4) A coin is tossed and a die is thrown, what is the probability to see a head on the coin and a number less than 5 on the upper face of the die?

5) In a certain canned tomatoes company one canned tomato is spoiled out of 28 canned tomatoes. If we choose two canned tomatoes one after the other with replacement. What is the probability that both are spoiled?

6.15   Conditional Probability

When two events A and B are dependent events we will use the concept of conditional probability.

- P(B/A) indicates the probability of event B given that event A has occurred and $P(B/A) = \dfrac{P(A \, and \, B)}{P(A)}$

- P(A/B) indicates the probability of event A given that Event B has occurred and $P(A/B) = \dfrac{P(A \, and \, B)}{P(B)}$, P(A/B) is read as

  "the probability of A, given B"

Example -1) If we toss a die and observe the number on the upper face of the die what is the probability to see an odd number given that a number less than five appears on the upper face of the die?

Solution:- Let event A=observe an odd number={1, 3, 5} and let Event B=observe a number less than five={1, 2, 3, 4}, AnB={1, 3}.

$$P(A/B) = \frac{P(AnB)}{P(B)} = \frac{\left(\frac{2}{6}\right)}{\left(\frac{4}{6}\right)} = \frac{2}{6} \times \frac{6}{4} = \frac{2}{4} = \frac{1}{2}$$

Example 2) From a pack of playing cards if we draw one card at random what is the probability to see an ace card given that a red card is drawn?

Solution:- Event A=an ace card - there are 4 cards
Event B=26 red cards
AnB=2 red ace cards

$$P(A) = \frac{4}{52}, \quad P(B) = \frac{26}{52}, \quad P(AnB) = \frac{2}{52}$$

$$P\left(\frac{A}{B}\right) = \frac{P(AnB)}{P(B)} = \frac{\left(\frac{2}{52}\right)}{\left(\frac{26}{52}\right)} = \frac{2}{52} \times \frac{52}{26} = \frac{1}{13}$$

Example 3) A bag contains white and black balls. Two balls are drawn without replacement. The probability of selecting a white ball and then a black all is 0.34. The probability of selecting a white ball on the first draw is 0.8. What is the probability of selecting a black ball on the second draw, given that the first ball drawn was white?

Solution:-

$$P(Black / white) = \frac{P(Black\ and\ white)}{P(white)}$$

$$= \frac{0.34}{0.8}$$

$$= 0.425$$

6.16 Complement Event to Find Probability

The complement of an Event A is the set of all outcomes that is not A. The complement of event A is written as $\overline{A}$. (Read as: A bar).

The probability of an event and its complement always has the sum equal to 1. (An event either occurs or doesn't occur.)

$$P(A)+P(\overline{A})=1$$

$$P(\overline{A})=1-P(A)$$

Example 1) In an experiment of tossing a die and observing the number on the upper face, the probability to see an even number is equal to:-

Solution:-

P(to see even number)=1-P(to see odd number)

$$=1-\frac{3}{6}$$

$$=\frac{1}{2}$$

Example 2) In a class of 60 students where 35 are boys and 25 are girls the probability that a girl will be elected as a monitor is given by:-

P(a girl will be elected)=1-P(a boy will be elected)

$$=1-\frac{35}{60}$$

$$=\frac{5}{12}$$

Example 3) When tossing a fair die, the probability of not getting a number 4 is given by:-

$$P(\overline{4})=1-P(4)$$

$$=1-\frac{1}{6}$$

$$P(\overline{4})=\frac{5}{6}$$

Example 4) In an experiment of tossing three coins and observing the sequence of heads and tails the probability to see at least one head is equal to:-

P(to see at least one head)=1-P(to see three tails)

$$= 1 - \frac{1}{8}$$

$$= \frac{7}{8}$$

Exercise 6-8

1) When tossing a fair die, the probability of not getting a number 6 is _____

2) In a class of 48 student where 28 are girls and 20 are boys the probability that a girl will be elected as a monitor is _____

3) In an experiment of tossing a die and observing the number on the upper face, the probability to see an odd number is equal to

_____

Finding Probabilities Using Combinations

Example 1) If a committee of 3 persons is to be randomly chosen from a group of 4 men and 2 women, then what is the probability that exactly one member of the committee is a woman?

Solution:
Choosing 3 from 6 can be done in C(6, 3) different ways, if one of them is a woman it can be done in C(4, 2)•C(2, 1).

Hence, P (exactly one woman) $= \dfrac{C(4,2)\cdot C(2,1)}{C(6,3)} = \dfrac{6\times 2}{20} = \dfrac{3}{5}$

Example 2) A committee of 5 persons is to be chosen randomly from a group consisting of 6 men and 4 women. What is the probability that exactly two of the members of the committee are women?

Solution:- Two women from 4 women are chosen in C(4, 2) ways. Since 5 persons are chosen 3 of them are men. 3 men are chosen from 6 men in C(6, 3) ways, and 5 persons are chosen from 10 in C(10, 5) ways:-

$$P \text{ (choosing exactly 2 women)} = \frac{C(4,2) \cdot C(6,3)}{C(10,5)} = \frac{10}{21}$$

Example 3) A bag contains 3 white and 4 black balls. If two balls are drawn at random, what is the probability that one of them is white and the other is black?

Solution:- There are 7 balls, we can select 2 balls out of 7 balls in C(7, 2)=21 different ways.

$$P(\text{one white and one black}) = \frac{C(4,1) \cdot C(3,1)}{C(7,2)} = \frac{4 \times 3}{21} = \frac{4}{7}$$

Example 4) If a fair coin is tossed three times, then what is the probability of exactly one head?

Solution:-

$$P(\text{exactly one head}) = \frac{C(3,1)}{8} = \frac{3}{8}$$

Example 5) From a box contains five red and three white balls. What is the probability to draw at least two red balls if three balls are drawn at random?

Solution: At least two red balls means two or three red balls, from the three balls.

P(at least two red balls)=P(two red balls)+P(three red balls)

$$P(\text{at least two red balls}) = \frac{C(5,2) \cdot C(3,1)}{C(8,3)} + \frac{C(5,3) \cdot C(3,0)}{C(8,3)}$$

$$= \frac{30}{56} + \frac{10}{56}$$

$$= \frac{40}{56}$$

Exercise 6-9

1) There are 11 balls in a box where 2 are red, 5 are yellow, and 4 are green. If two balls are drawn at the same time, then what is the probability that they will be of the same colors?

2) A box contains 12 balls of the same size where five are red, three are black, and the rest are white colored balls. If we select 2 balls at random and observe their color, what is the probability to see balls of different colours?

6.17 Measures of Central Tendency and Dispersion

An average, or central tendency of a data set is a measure of the "middle" value of the data set. It is higher degree of summarization of data. Data tend to accumulate at a certain position between the extremes it is this point which is called the measure of central tendency. There are many different descriptive statistics that can be chosen as a measurement of the central tendency of the data items. These include mean (arithmetic mean), the median and the mode. Other statistical measures such as the standard deviation and the range are called measures of spread and describe how spread out the data is.

6.17.1 Mean

Arithmetic Mean (Mean): is the average value of the distribution which is obtained by adding the observations of a variable and then dividing the sum by the number of observations.

Notation:- We can use μ(mu) for mean

Arithmetic Mean $\mu = \dfrac{\sum x}{N} = \dfrac{x_1 + x_2 + x_3 + x_4 + ... x_n}{N}$

Where, $\sum_x$ =sum of individual variable

N=Total number of observation.

Example 1) In a certain high school the final physics test result of 12 students was 70, 46, 48, 75, 60, 36, 85, 56, 90, 58, 67 and 81. Find the mean

Solution:

$$M = \frac{\sum x}{N} = \frac{70 + 46 + 48 + 75 + 60 + 36 + 85 + 56 + 90 + 58 + 67 + 81}{12}$$

$$= \frac{772}{12}$$

$\therefore$ Mean=64.3

Example 2) The age of ten students are: 26, 38, 18, 34, 28, 27, 24, 27, 30 and 31. Find the arithmetic mean of the data.

Solution:-

$$M = \frac{\sum x}{N} = \frac{26 + 38 + 18 + 34 + 28 + 27 + 24 + 27 + 30 + 31}{10}$$

$$Mean = \frac{283}{10}$$

Mean = 28.3

Example 3) The number of cars sold by a dealer for eight days was: 11, 13, 17, 10, 5, 21, 3 and 16. What will be the mean?

Solution:-

$$M = \frac{\sum x}{N} = \frac{11 + 13 + 17 + 10 + 5 + 21 + 3 + 16}{8}$$

$$Mean = \frac{96}{8}$$

Mean=12

Example 4) The driving speed of 8 different cars on the same driving road was: 70mph, 55mph, 68mph, 64mph, 59mph, 56mph, 57mph and 66mph. Find the mean of the data.

Solution:-

$$M = \frac{\sum x}{N} = \frac{70+55+68+64+59+56+57+66}{8}$$

$$Mean = \frac{495}{8}$$

Mean≈61.9

Example 5) The following table is frequency distribution of a certain population. Find the mean

X	F		To find Mean	X	F	XF
4	3			4	3	12
10	5			10	5	50
8	6			8	6	48
7	10			7	10	70
5	13			5	13	65
9	4			9	4	36
				SUM	41	281

$$M = \frac{\sum x_i f_i}{\sum f_i}$$

$$= \frac{281}{41}$$

Mean=6.85

6.17.1.1 Properties of Mean

1)   If we add a constant (N) to each number of a population having mean μ; the new mean will be (μ+N).

2)   If we subtract a constant (N) from each member of the population having mean μ the new mean will be (μ-N).

3)   If each member of a population function with mean μ is multiplied by N the new mean will be μN.

4) If each member of a population function with mean μ is divided by N the new mean will be $\dfrac{\mu}{N}$ or $\dfrac{1}{N}\mu$.

5) The sum of the deviations of all values of the distribution from their arithmetic mean is zero.

$$\Sigma(x_i-\mu)=0$$

Example: Consider the number of pencils in the bag of six students: 8, 10, 13, 12, 11 and 6.

Here mean $\mu = \dfrac{8+10+13+12+11+6}{6}$

$$\mu = \dfrac{60}{6}$$

Mean=10

1) If we add 3 pencils for each student the new mean will be

$$\dfrac{(8+3)+(10+3)+(13+3)+(12+3)+(11+3)+(6+3)}{6}$$

$$=\dfrac{78}{6}$$

$=13$

i.e, it is simply the previous mean plus three (10+3)

2) If we take away 3 pencils from each student the new mean will be:-

$$\dfrac{(8-3)+(10-3)+(13-3)+(12-3)+(11-3)+(6-3)}{6}=\dfrac{42}{6}=7$$

i.e, it is simply the previous mean minus three. (10-3)

3) If we multiply the number of pencils in the bag of each students by 3, the new mean will be:-

$$\dfrac{(8\times3)+(10\times3)+(13\times3)+(12\times3)+(11\times3)+(6\times3)}{6}=\dfrac{180}{6}=30$$

i.e, it is simply the previous mean multiplied by 3.. (10x3)

4)   If the number of pencils in the bag of each students is divided by 3 the new mean will be:-

$$\frac{\frac{8}{3}+\frac{10}{3}+\frac{13}{3}+\frac{12}{3}+\frac{11}{3}+\frac{6}{3}}{6}=\frac{\frac{60}{3}}{6}=\frac{60}{18}=\frac{10}{3}$$

i.e, it is simply the previous mean divided by 3. $\left(\frac{10}{3}\right)$

5)   The sum of the deviations of all values of the distribution from their arithmetic mean is zero.

(8-10)+(10-10)+(13-10)+(12-10)+(11-10)+(6-10)
= -2+0+3+2+1+(-4)=0

6.17.2 Mode

Mode:- is one of the measures of central tendency, and it is the most frequent value. It is possible to have no mode, one mode, or more than one mode in a given variable.

- A set of values which has only one mode is called unimodal.
- A set of values which has two modes is called bimodal
- A set of values which has more than two modes is called multimodal.

Note:- To find the mode from a frequency distribution table we can take the value which corresponds to the highest frequency.

Example 1)   The age of 10 students in a certain class is given as: 25, 18, 23, 17, 18, 19, 30, 27, 18, 29. From the given data the mode is 18 which occurs 3 times and it is called multimodal.

Example 2)   The number of cars sold by a dealer for ten days is given as: 7, 8, 3, 2, 4, 10, 3, 9, 6, 11. Here, 3 is the most frequent value, it appears twice and it is modal value and it is called bimodal.

Example 3) The amount of money in the pocket of 11 students in dollar is given as:- 10, 20, 13, 22, 11, 7, 16, 30, 40, 18, 19. There is no mode in this data.

6.17.3 Median

Median:- is measure of central tendency, it is that value that separates the highest half of the sample from the lowest half. After the data is arranged in an increasing order (from lowest value to highest value). If the number of observations is odd the median is the middle value, and if the number of observations is even the median is the mean of the two middle variables.

Example 1) Find the median of the following list of values: 15, 11, 20, 9, 17, 10, 7, 24, 13,

Solution: First arrange in an increasing order.

7, 9, 10, 11, 13, 15, 17, 20, 24, the number of observations is odd, we have 9 observations the middle variable is

$$Median = \left(\frac{n+1}{2}\right)^{th} = \left(\frac{9+1}{2}\right)^{th} \text{ or } 5^{th} \text{ value} = 13$$

Example 2) Find the median of the following list of values.

6, 19, 12, 8, 27, 33, 16, 20, 10, 14, 36, 23

Solution: Here the number of observations is even, the median is the mean of the two middle observations.

First arranging in an increasing order gives:-

6, 8, 10, 12, 14, 16, 19, 20, 23, 27, 33, 36

$$Median = \left[\frac{\left(\frac{n}{2}\right)^{th} + \left(\frac{n}{2}+1\right)^{th}}{2}\right]$$

$$Median = \left[\frac{\left(\frac{12}{2}\right)^{th} + \left(\frac{12}{2}+1\right)^{th}}{2}\right]$$

$$Median = \frac{6^{th} + 7^{th}}{2} = \frac{16+19}{2} = \frac{35}{2} = 17.5$$

Note:- The median is the mean of 6th and 7th observations.

Example 3)   The following is a frequency distribution of a population function V. Find the median.

V	4	6	10	16	20	24
F	3	4	7	5	8	1

Solution:-

The total number of observation is: $(3+4+7+5+8+1)=28$ it is even the median is mean of the 14th and the 15th values when the data is arranged in an ascending or descending order.

The first 3 members have a value 4, the 4th, 5th, 6th and 7th members have a value 6, from 8th up to 14th the value is 10, from 15th up to 19th the value is 16, from 20th up to 27th the value is 20 and the 28th value is 24.

$$Median = \frac{14^{th} \ value + 15^{th} \ value}{2} = \frac{10+16}{2} = \frac{26}{2} = 13$$

Exercise 6-10

1)   The weight of 8 students in kilograms, was as follows 85, 67, 72, 85, 90, 62, 69, 85

a) What is the mean of the data?

b) What is the median?

c) What is the mode?

d) What will happen to the mean if each students lose 5 kilograms?

e) What will happen to the mean if each students gain 5 kilograms?

2)   A student scored 62, 95, 77, 85 and 91 in five subjects. If her mean score for six subjects is 77, then what is her score in the sixth subject?

3)   The mean score on a set of 9 scores is 73. What is the sum of the 9 test scores?

4)   The mean score on five of a set of six scores is 79. The sixth score is 84. What is the sum of the six scores? What is the mean of the six scores?

5)   Given below are the reading scores of the seventh grade students:
64, 78, 94, 67, 88, 81, 94, 79, 77, 67, 58
a) What is the mean of the reading scores?
b) What is the mode of the reading scores?
c) What is the median of the reading scores?

6)   Consider the following table which shows the frequency distribution of a population function V.

V	9	18	23	24	30	32
F	5	4	7	4	3	3

a) Find the mean
b) Find the mode
c) Find the median

6.18   Measures of Dispersion
The measures of dispersion or measures of scatter variability indicates how dispersed a set of observation is, and it tells how far the measures of central tendency are reliable.

6.18.1 Range
Range is a measure of dispersions which is the difference between the largest and the smallest observed values.

Example 1)   The height in centimeters of 8 students was measured as follows: 176, 154, 172, 182, 166, 172, 184, 164. What is the range?

Solution:- Range=Maximum value-Minimum value
Range=184-154
Range=30

6.18.2 Standard Deviation

The range only uses two extreme values, it is not considered to be the best measure of dispersion. Standard deviation is:- a widely used measurement of variability or diversity used in statistics and probability theory. It shows how much variation or "dispersion" there is from the average (mean, or expected value). A low standard deviation indicates that the data points tend to be very close to the mean, whereas high standard deviation indicates that the data are spread out over a large range of values.

6.18.2.1 To find standard deviation for a given set of data
- Find the mean for the given set of data.
- Find the deviation from the mean for each value in the set of data.
- Square each deviation.
- Find the mean of the squared deviation (which is called variance).
- Find the square root of the variance.

Notation: We used (σ or s) for standard deviation.

$$\text{Variance } S^2 = \frac{\sum (x - M)^2}{N}$$

Standard deviation=positive square root of variance
$$S.D. = \sqrt{S}$$

Example 1) Find the standard deviation of the following data which shows the test taken by 6 students.
5, 9, 10, 4, 8, 6

Solution: Mean $\mu = \dfrac{5+9+10+4+8+6}{6} = 7$

Value v	v-μ	(v-μ)²
4	4-7=-3	(-3)²=9
5	5-7=-2	(-2)²=4
6	6-7=-1	(-1)²=1

8	8-7=1	$(1)^2=1$
9	9-7=2	$(2)^2=4$
10	10-7=3	$(3)^2=9$
		$\Sigma(v-\mu)^2=28$

$$\text{Variance} = \frac{\Sigma(v-\mu)^2}{N} = \frac{28}{6} = 4.67$$

$$\therefore \text{ Standard deviation} = \sqrt{\frac{28}{6}} = 2.16$$

Example 2) The number of cars sold by a dealer for nine days was: 8, 6, 12, 4, 11, 9, 10, 3, 18.
Find the variance (σ^2) and the standard deviation (σ).

Solution:- $Mean = \dfrac{3+4+6+8+9+10+11+12+18}{9} = \dfrac{81}{9} = 9$

To find variance (σ^2)

Value v	v-μ	$(v-\mu)^2$
3	3-9=-6	$(-6)^2=36$
4	4-9=-5	$(-5)^2=25$
6	6-9=-3	$(-3)^2=9$
8	8-9=-1	$(-1)^2=1$
9	9-9=0	$(0)^2=0$
10	10-9=1	$(1)^2=1$
11	11-9=2	$(2)^2=4$
12	12-9=3	$(3)^2=9$
18	18-9=9	$(9)^2=81$
		$\Sigma(v-\mu)^2=166$

$$S^2 = \frac{\Sigma(v-M)^2}{N} = \frac{166}{9} = 18.44$$

The standard deviation is a positive square root of variance.

Standard deviation $S = \sqrt{18.44}$

To find standard deviation from single valued frequency distribution table, we use variance $= S = \dfrac{\sum (x_i - \mu)^2 f_i}{\sum f_i}$

And standard deviation is the positive square root of the variance.

Example -3) Given below is the frequency distribution of a population function V. What is the standard deviation.

V	1	2	3	4	13
F	4	3	2	1	1

Solution:- First find the mean

$$Mean = \frac{(1\times 4)+(2\times 3)+(3\times 2)+(4\times 1)+(13\times 1)}{4+3+2+1+1} = \frac{33}{11} = 3$$

Then find the variance

V	F	V-M	(V-M)²	F(V-M)²
1	4	1-3=-2	(-2)²=4	4x4=16
2	3	2-3=-1	(-1)²=1	3x1=3
3	2	3-3=0	(0)²=0	2x0=0
4	1	4-3=1	(1)²=1	1x1=1
13	1	13-3=10	(10)²=100	1x100=100
	$\sum F=11$			$\sum F(V\text{-}M)^2=120$

$$\text{Variance} = \frac{\sum F(v-M)^2}{\sum F} = \frac{120}{11} = 10.9 \; ,$$

and standard deviation $= \sqrt{10.9}$

6.19   Quartile

A quartile is one of three points that divide a data set into four equal groups, each representing a fourth of the distributed sampled population:

Definitions

6.19.1   First quartile (designated Q_1)=lower quartile=cuts off lowest 25% of data=25th percentile.

$$Q_1 = x\left[\frac{n}{4} + \frac{1}{2}\right] \quad ..... \text{ first quartile.}$$

6.19.2   Second quartile (designed Q_2)=median=cuts data set in half=50th percentile.

$$Q_2 = x\left[\frac{2n}{4} + \frac{1}{2}\right] \quad .... \text{ second quartile.}$$

6.19.3   Third quartile (designated Q_3)=upper quartile=cuts off highest 25% of data, or lowest 75%=75th percentile.

$$Q_3 = x\left[\frac{3n}{4} + \frac{1}{2}\right] \quad ...... \text{ Third quartile.}$$

The difference between the upper and lower quartile is called the interquartile range.

Example 1)   Consider the following set of data.

Data set: 6, 47, 49, 15, 42, 41, 7, 39, 43, 40, 36

Find Q_1, Q_2, Q_3

Solution:- First arrange the data set in an increasing order: 6, 7, 15, 36, 39, 40, 41, 42, 43, 47, 49

$$Q_1 = x\left[\frac{n}{4} + \frac{1}{2}\right] \quad ..... \text{ n = 1 (number of observation)}$$

$$Q_1 = x\left[\frac{11}{4} + \frac{1}{2}\right]$$

Q_1=x(3.25), $\frac{25}{100} = \frac{1}{4}^{th}$ of the way between the 3rd and 4th values.

3rd value is 15 and 4th value is 36, 36-15=21, half of 21=10.5

$$\therefore Q_1=15+10.5=25.5$$

$$Q_2 = x\left[\frac{2n}{4}+\frac{1}{2}\right]$$

$$Q_2 = x\left[\frac{2(11)}{4}+\frac{1}{2}\right]$$

$$Q_2 = x\left[\frac{22}{4}+\frac{1}{2}\right]$$

$$= x\left[\frac{24}{4}\right]$$

Q_2=6x This is half way between the 5th and 7th values.

5th value is 39 and 7th value is 41, 41-39=2
half of 2=1

Thus, Q_2=39+1=40

$$Q_3 = x\left[\frac{3n}{4}+\frac{1}{2}\right]$$

$$Q_3 = x\left[\frac{3(11)}{4}+\frac{1}{2}\right]$$

$$Q_3 = x\left[\frac{33}{4}+\frac{1}{2}\right]$$

$$Q_3 = x\left[\frac{35}{4}\right]$$

Q3=8.75x, $\dfrac{75}{100}=\dfrac{3^{rth}}{4}$ of the way between the 8th and 9th values.

8th value is 42 and 9th value is 43, 43-42=1
half of 1=0.5

$$\therefore Q_3=42+0.5=42.5$$

Example 2) Consider the following set of data.

Data set: 36, 7, 15, 41, 39, 40

Find Q_1, Q_2, Q_3.

Solution: First arrange the data set in an increasing order: 7, 15, 36, 39, 40, 41

$$Q_1 = x\left[\frac{n}{4}+\frac{1}{2}\right] \qquad \text{n=6 (number of observation).}$$

$$= x\left[\frac{6}{4}+\frac{1}{2}\right]$$

$$= x\left[\frac{8}{4}\right]$$

Q_1=2x, This is half way between the 1st and 3rd values. 1st value is 7 and 3rd value is 36, 36-7=29, half of 29=14.5

$\therefore$ Q_1=7+14.5=21.5

$$Q_2 = x\left[\frac{2n}{4}+\frac{1}{2}\right] \qquad \text{n=6 (number of observation)}$$

$$= x\left[\frac{2(6)}{4}+\frac{1}{2}\right]$$

$$= x\left[\frac{7}{2}\right]$$

Q_2=3.5x, This is half way between the 3rd and 4th values. 3rd value is 36 and 4th value is 39, 39-36=3, half of 3=1.5

$\therefore$ Q_2=36+1.5=37.5

$$Q_3 = x\left[\frac{3n}{4}+\frac{1}{2}\right]$$

$$= x\left[\frac{3(6)}{4}+\frac{1}{2}\right]$$

$=5x$, This is half way between the 4th and 6th values. 4th value is 39 and 6th value is 41, 41-39=2, half of 2=1

$\therefore Q_3=39+1=40$.

6.20   Decile

Decile:- is any of nine values that divide the sorted data into ten equal parts, so that each part represents $\frac{1}{10}$ of the sample or population.

i.e. Deciles divide a variable into ten equal parts.

$$D_1 = x\left(\frac{n}{10} + \frac{1}{2}\right) \quad ..... \text{ deciles-one}$$

$$D_2 = x\left(\frac{2n}{10} + \frac{1}{2}\right) \quad ..... \text{ deciles-two}$$

$$D_3 = x\left(\frac{3n}{10} + \frac{1}{2}\right) \quad ..... \text{ deciles-three}$$

$$\vdots$$

$$D_8 = x\left(\frac{8n}{10} + \frac{1}{2}\right) \quad ..... \text{ deciles-eight}$$

$$D_9 = x\left(\frac{9n}{10} + \frac{1}{2}\right) \quad ..... \text{ deciles nine}$$

Example:-   Consider the following set of data 13, 41, 20, 45, 18, 25, 35, 33. Find D_3 and D_9.

Solution:- First arrange in an increasing order:- 13, 18, 20, 25, 33, 35, 41, 45.

$$D_3 = x\left(\frac{3n}{10} + \frac{1}{2}\right)$$

$$= x\left(\frac{3\times 8}{10} + \frac{1}{2}\right), \text{ n=8 (number of observation)}$$

$$= 2.9x$$

$= \dfrac{9}{10}$ of the way between 2nd value and 3rd value, 2nd value=18 and 3rd value=20, 20-18=2 and $\left(\dfrac{9}{10}\right) \times 2 = 1.8$

$D_3 = 18 + 1.8 = 19.8$

$D_9 = x\left(\dfrac{9 \times 8}{10} + \dfrac{1}{2}\right) = 7.7x, \dfrac{7}{10}$ of the way between 7th and 8th values.

$= (45 - 41) \times \dfrac{7}{10} = 2.89$

$D_9 = 41 + 2.8 = 43.8$

6.21 Percentiles

Percentile or (centile):- is the value of a variable below which a certain percent of observations fall. For example, the 20th percentile is the value (or score) below which 20 percent of the observations may be found. The term percentile and the related term percentile rank are often used in the reporting of scores from norm referenced tests.

The 25th percentile is also known as the first quartile (Q_1), the 50th percentile as the median or second quartile (Q_2), and the 75th percentile as the third quartile (Q_3).

Thus, percentile divide observation into 100 equal parts.

$$P_1 = x\left[\dfrac{1n}{100} + \dfrac{1}{2}\right] \quad \ldots\ldots \text{ percentile-1 (1st percentile)}$$

$$P_2 = x\left[\dfrac{2n}{100} + \dfrac{1}{2}\right] \quad \ldots\ldots \text{ percentile-2 (2nd percentile)}$$

$$P_3 = x\left[\dfrac{3n}{100} + \dfrac{1}{2}\right] \quad \ldots\ldots \text{ percentile-3 (3rd percentile)}$$

$\vdots$

$$P_{99} = x\left[\dfrac{99n}{100} + \dfrac{1}{2}\right] \text{ percentile-99 (99th percentile)}$$

Example 1) For the following observation shown below 50, 30, 38, 90, 10, 45, 16, 25, 80.
Find P_3, P_{30}, P_{60}

Solution:- First arrange in an increasing order 10, 16, 25, 30, 38, 45, 50, 80, 90

$P_3 = x\left[\dfrac{3n}{100}+\dfrac{1}{2}\right] = x\left[\dfrac{3\times9}{100}+\dfrac{1}{2}\right]$, where, n=9, number of observation.

$= x\left[\dfrac{27}{100}+\dfrac{1}{2}\right] = 0.77.$

$\left(\dfrac{77}{100}\right)\times10 = 7.7$

$\therefore P_3 = 7.7$

• $P_{30} = x\left[\dfrac{30n}{100}+\dfrac{1}{2}\right] = x\left[\dfrac{30\times9}{100}+\dfrac{1}{2}\right] = ; \dfrac{2}{10}$ of the way

between 3rd and 4th values, 30-25=5 and $\left(\dfrac{2}{10}\right)\times5 = 1$
$\therefore P_{30} = 25+1 = 26$

• $P_{60} = x\left[\dfrac{60n}{100}+\dfrac{1}{2}\right] = x\left[\dfrac{60\times9}{100}+\dfrac{1}{2}\right] = 5.9x$, This is $\dfrac{9}{10}$

of the way between 5th and 7th values.
(7th value)-(5th value)=50-38=12 and

$\left(\dfrac{9}{10}\right)\times12 = 10.8$

$\therefore P_{60} = (5^{th}$ value)+10.8
$= 38+10.8$
$P_{60} = 48.8$

Example 2) For the following observation shown below: 32, 46, 13, 48, 24, 19, 56, 39, find P_{24}, P_{56}, P_{80}

Solution:- First arrange in an increasing order: 13, 19, 24, 32, 39, 46, 48, 56

- $P_{24} = x\left[\dfrac{24n}{100} + \dfrac{1}{2}\right]$

$= x\left[\dfrac{24\times 8}{100} + \dfrac{1}{2}\right]$, where n=8 (number of observation)

$P_{24} = x\left[\dfrac{24\times 8}{100} + \dfrac{1}{2}\right] = x[2.42] = 2.42x$, this is $\dfrac{42}{100}$ of the way between 2nd and 4th values.

(4th value)-(2nd value)=32-19=13 and $\left(\dfrac{42}{100}\right)(13) = 5.46$

$\therefore P_{24}$=19+5.46=24.46

- $P_{56} = x\left[\dfrac{56n}{100} + \dfrac{1}{2}\right]$

$= x\left[\dfrac{56\times 8}{100} + \dfrac{1}{2}\right]$

$= x[4.48+0.5]=4.98x$, this is $\dfrac{98}{100}$ of the way between 4th and 6th values.

(6th value)-(4th value)=46-32=14, and $\dfrac{98}{100}\times 14 = 13.72$
$\therefore P_{56}$=32+13.72=45.72

- $P_{80} = x\left[\dfrac{80n}{100} + \dfrac{1}{2}\right]$

$= x\left[\dfrac{80\times 8}{100} + \dfrac{1}{2}\right] = x[6.4+0.5] = 6.9x$, this is $\frac{9}{10}$ of the way between 6th and 8th values.

$(8^{th}$ value$)-(6^{th}$ value$)=56-46=10$ and $\left(\dfrac{9}{10}\right)(10)=9$

$\therefore P_{80}=46+9=55$

6.22 Absolute Deviation

The absolute deviation of an element of a data set is the absolute difference between that element and a given point. Typically the point from which the deviation is measured is a measure of central tendency, most often the median or sometimes the mean of the data set.

$Di=|x_i-\mu|$

where

Di is the absolute deviation,

x_i is the data element

and μ is the chosen measure of central tendency of the data set - sometimes the mean (μ), but most often the median.

6.23 Average Absolute Deviation

The average absolute deviation, or simply mean deviation or average deviation of a data set is the average of the absolute deviations and is a summary statistic of statistical dispersion or variability. It is also called the mean absolute deviation.

The average absolute deviation of a set $\{x_1, x_2, x_3, ..., x_n\}$ is given by:-

$$\text{Average Absolute Deviation} = \frac{\sum\limits_{i=1}^{n}|x_i - \mu|}{n}.$$

The choice of measure of central tendency, μ, has a marked effect on the value of the average deviation.

Example:- For the data set {3, 3, 4, 5, 15}, find the average absolute deviation.

Solution:-

Measure of central tendency μ	Average absolute deviation
Mean=6	$\dfrac{\|3-6\|+\|3-6\|+\|4-6\|+\|5-6\|+\|15-6\|}{5}=3.6$
Median=4	$\dfrac{\|3-4\|+\|3-4\|+\|4-4\|+\|5-4\|+\|15-4\|}{5}=2.8$
Mode=3	$\dfrac{\|3-3\|+\|3-3\|+\|4-3\|+\|5-3\|+\|15-3\|}{5}=3$

The average absolute deviation from the median is less than or equal to the average absolute deviation from the mean. In fact, the average absolute deviation from the median is always less than or equal to the average absolute deviation from any other fixed number.
- The average absolute deviation from the mean is less than or equal to the standard deviation.

6.23.1 Mean Absolute Deviation
The mean absolute deviation (MAD) is the mean absolute deviation from the mean.

6.23.2 Median Absolute Deviation (MAD)
The median absolute deviation is the median of the absolute deviation from the median. It is a robust estimator of dispersion.

For example: {3, 3, 4, 5, 15}: 4 is the median, so the absolute deviations from the median are {1, 1, 0, 1, 11}, (recorded as {0, 1, 1, 1, 11} with a median of 1, in this case unaffected by the value of the outlier 15, so the median absolute deviation (also called MAD) is 1.

6.24   Maximum Absolute Deviation

The maximum absolute deviation about a point is the maximum of the absolute deviations of a sample from that point. It is realized by the sample maximum or sample minimum and cannot be less than half the range.

Example 1)   Consider the following observation: 3, 6, 7, 8, 11 and find the average absolute deviation from the mean and median.

Solution:- $Mean = \dfrac{3+6+7+8+11}{5} = 7$

Average absolute deviation from the mean.

x	x-μ	|v-μ|
3	3-7=-4	4
6	6-7=-1	1
7	7-7=0	0
8	8-7=1	1
11	11-7=4	4
		Σ|x-μ|=10

Average absolute deviation from the mean = $\dfrac{\sum |x-\mu|}{N}$

$\dfrac{10}{5}$

$=2$

Average Mean deviation from the median
The median =7

x	x-m(x)	|x-m(x)|
3	2-7=-5	5
6	6-7=-1	1
7	7-7=0	0

8	8-7=1	1
11	11-7=4	4
		$\Sigma|x-m(x)|=10$

$$\text{Average deviation from median} = \frac{\sum |x-m(x)|}{N}$$

$$= \frac{10}{5}$$

$$= 2$$

Example 2) Consider the following observation: 3, 6, 9, 12, 15, 15 and find the average absolute deviation from the mean.

$$\text{Mean} = \frac{3+6+9+12+15+15}{6} = 10$$

x	x-μ	$|x-\mu|$
3	3-10=-7	7
6	6-10=-4	4
9	9-10=-1	1
12	12-10=2	2
15	15-10=5	5
15	10-10=5	5
		$\Sigma|x-\mu|=24$

$$\text{Average absolute deviation from mean} = \frac{\sum |x-\mu|}{N}$$

$$= \frac{24}{6}$$

$$= 4$$

EXERCISE 6-11

1) Find the variance and standard deviation for the given data: 6, 12, 8, 5, 4, 15, 20, 10, 7, 3.

2) For the observation:

6, 12, 24, 20, 16, 8, 10, 26, 35, 9, 28

Find)   a) Q_1

b) Q_3

c) D_4

d) D_6

e) P_{60}

f) P_{80}

3) For the frequency distribution of the population function V. Find

a) variance

b) standard deviation

c) Average absolute deviation

d) Range

V	0	3	5	8	10	15
F	4	10	7	5	3	1

4) Given below is the data which shows the test taken by seven students: 5, 2, 2, 8, 16, 11, 19

Find:-

a) Mean

b) Mode

c) Median

d) range

e) variance

f) standard deviation.

g) Average absolute deviation from mean, mode and median

Review Solved Problems on Chapter Six

1) Given below is the frequency distribution of population function V.

V	1	2	3	4	5
f	6	5	4	3	2

Find:- a) mode b) mean c) median d) range

Soln:- a) The highest frequency is 6, this shows the mode is 1.

b)

V	1	2	3	4	5	sum
F	6	5	4	3	2	20
V.F	6	10	12	12	10	50

$$\mu = \frac{\sum x_i f_i}{\sum f_i} = \frac{50}{20} = 2.5$$

c) The total number of observation is:- $6+5+4+3+2=20$, it is even the median is mean of the 10th and the 11th values. When the data is arranged in an increasing order.

$$Median = \frac{10^{th}\ value + 11^{th}\ value}{2}$$

$$Median = \frac{2+2}{2} = \frac{4}{2} = 2$$

d) Range=maximum value-minimum value
   $= 5-1$
Range=4

2) Given below is the frequency distribution of the population to the nearest year of the ages of a certain 10th grade class students.

v	15	16	17	18	19	20	22
f	4	2	14	16	10	3	1

a) What percent of the students in the class is younger than the modal age?

b) What percent of the students in the class is older than the mean?

c) What percent of the students in the class equal to the modal age?

Solution:-

a) $\Sigma f = 4+2+14+16+10+3+1=50$

There are 50 students in the class the mode is 18

There are $(4+2+14)=20$ students younger than 18 years. In

percent: $\dfrac{20}{50} \times 100\% = 40\%$

$\therefore$ 40% of students in the class are younger than the modal age.

b)

$$Mean = \frac{(15\times4)+(16\times2)+(17\times14)+(18\times16)+(19\times10)+(20\times3)+(22\times1)}{50}$$

$$= \frac{890}{50}$$

M=17.8

There are $(16+10+3+1)=30$ students in the class older than the mean age.

In percent: $\dfrac{30}{50} \times 100\% = 60\%$

$\therefore$ 60% of the students in the class are older than the mean age.

c) The mode is 18, there are 16 students whose age is 18, in

percent: $\dfrac{16}{50} \times 100\% = 32\%$

∴ 32% of the students in the class have the age equal to the modal age.

3)   In a final examination a student took six examinations and scored, 72, 67, 68, 90 and 80 in five of the subjects. If his average score is 79, what was his score in the sixth subject?

Solution:- Number of subjects is six.
           Average=79,
           Sum of five scores=72+67+68+90+80=377
           Total sum of six subject=79x6=474
           Thus, the score of the 6th subject=474-377=97

Check $\dfrac{72+67+68+80+90+97}{6} = 79$

           Exercise 6-12
Review Supplementary Problems on Chapter Six
I   True or False

1)   In a given frequency distribution it is possible to find two medians.

2)   The mean, the mode and the median of a given distribution are always different.

3)   Mean, mode and median are measure of central tendency.

4)   Range and standard deviations are measures of dispersion.

5)   Adding a constant number to each member of the population will not change the standard deviation

6)   If each member of a population function with mean μ is multiplied by k the new mean will be μk.

7)   The standard deviation is the positive square root of variance.

8)   If we odd a constant (k) to each member of a population having mean μ; the new mean will be (μ+k).

9)   If we subtract a constant (k) from each member of the population having mean μ the new mean will be (μ-k)

II-   Answer the following questions.
10   Given below is a frequency distribution of a population function V.

V	-5	-4	0	3	8	9
F	12	16	21	8	27	10

A) What is the mean?
B) What is the mode?
C) What is the median?
D) What is the range?
E) What percent of the population is non negative?
F) What percent of the population is negative valued?

11)   Given below is the frequency distribution of a population function V.

V	-2	-1	0	2	4	6
F	6	8	4	3	2	1

A) What is the mode?
B) What is the median?
C) What is the mean?
D) What is the standard deviation?
E) What is the range?
F) What percent of the population is negative valued?
G) What percent of the population is non zero valued?

ANSWERS

CHAPTER-1

EXERCISE 1.1

1. a) -2
   b) $\frac{7}{6}$
   c) $\frac{-5}{6}$
   d) $\frac{7}{4}$

   d) 1
   e) 1
   f) 0

2. 10

3. -10

4. a) $\frac{5}{6}$
   b) 0
   c) 2

5. a) $\frac{-3c}{d}$
   b) $\frac{d}{c}$
   c) $\frac{-d}{c}$
   d) $\frac{2-dc}{cd}$
   e) c

EXERCISE 1.2

1. $y=-5x-6$

2. -6

3. $\frac{4}{5}$

4. $y=5x+13$

5. 4

6. $\frac{1}{3}$

7 fig a and b

a)

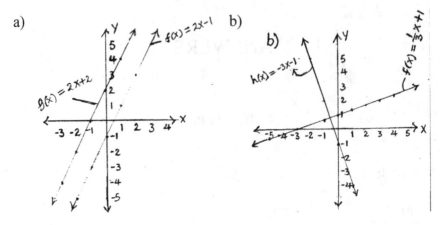

b)

EXERCISE 1.3

1 figure

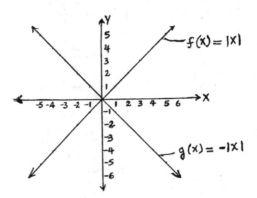

2 figure

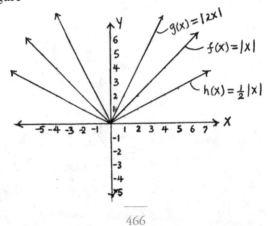

3. $(3, 2)$

4. $(1, 3)$

5. $\left(\dfrac{-1}{2}, \dfrac{-1}{3} \right)$

EXERCISE 1.4

1. a) 4

   b) $\dfrac{1}{4}$

   c) -7

   d) -13

   e) $\dfrac{5}{4}$

2. a) 6

   b) 23

   c) $\dfrac{1}{8}$

3. a) Slope $= \dfrac{10}{3}$
      y-intercept=-2

   b) slope=-2
      y-intercept $= \dfrac{1}{2}$

   c) slope $= \dfrac{1}{2}$
      y-intercept $= \dfrac{-7}{6}$

   d) slope=-1
      y-intercept=2

e) slope=-10
   y-intercept=-18

4. a) positive slope
   b) negative slope
   c) slope=0
   d) no slope

5. a) y=5x-13

   b) y=-3x+17

   c) $y = \dfrac{1}{4}x + 7$

   d) y=-x+6

   e) y=2x

   f) $y = \dfrac{1}{3}x + \dfrac{5}{4}$

6. a) $y = \dfrac{1}{2}x + 5$

   b) y=6x-3

   c) $y = \dfrac{1}{2}x$

   d) y=3x-6

   e) $y = \dfrac{-1}{4}$

7. a) $\dfrac{2}{5}$

   b) $\dfrac{-3}{2}$

   c) $\dfrac{5}{9}$

   d) $\dfrac{-3}{4}$

   e) $\dfrac{2}{3}$

c) Linear function
d) Not linear function
e) Linear function
f) Not linear function
g) Linear function
h) Linear function
i) Not linear function

9. a) y=3
   b) y=5
   c) y=2
   d) $y = \dfrac{1}{3}$
   e) y=6

8. a) Linear function
   b) Not linear function

10 fig a and b

a)

b)

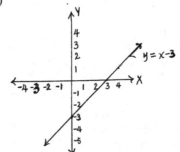

c)

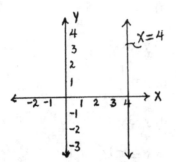

d)

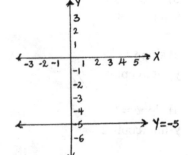

11 a) x=-3
   b) x=1
   c) x=2
   d) x=6
   e) x=0

12 a) $y = \dfrac{2}{5}x + \dfrac{23}{5}$

   b) $y = \dfrac{-5}{2}x - 7$

13. $\dfrac{1}{3}$

14. $\dfrac{5}{3}$

15. a)   figure

x	-3	-2	-1	0	1	2	3
y	-5	-3	-1	1	3	5	7

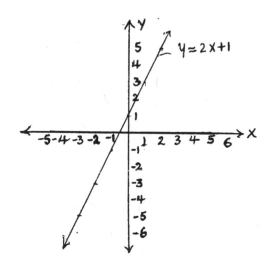

b   figure

x	-3	-2	-1	0	1	2	3	4	5
y	-6	-5	-4	-3	-2	-1	0	1	2

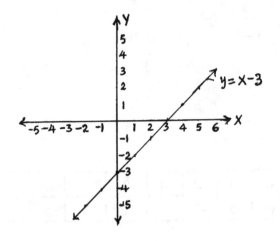

16   figure

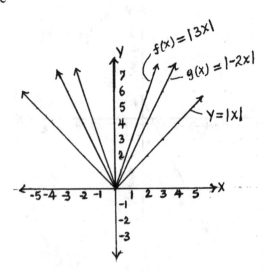

17   figure

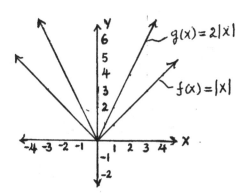

18   figure

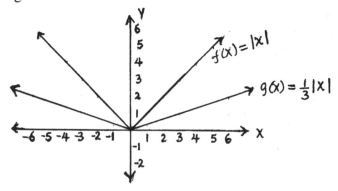

19   figure

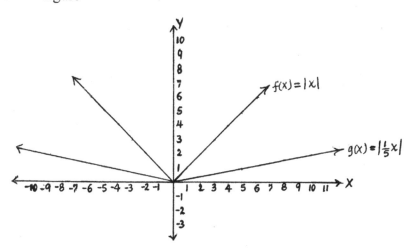

20   figure

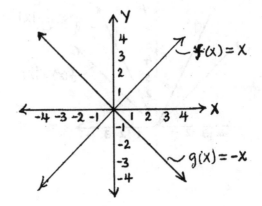

21.   a) (4, 2)
        b) (6, -5)
        c) (-4, -3)
        d) (1, 1)

22.   a) (5, 4)
        b) (-6, 0)
        c) (-3, -2)
        d) (1, -3)
        e) (7, 4)

23.   a) (2, -3)
        b) (3, -¼)
        c) (-¼, ½
        d) (-5, -2)
        e) (7, -8)

24.   a) true
        b) true
        c) true
        d) true
        e) true

CHAPTER-2

EXERCISE 2.1

A.1.   a) 3
        b) 5
        c) -1

2)   a) 1
      b) -4
      c) 1

3)   a) 5
      b) 4
      c) -9

4)   a) 2
      b) -5
      c) -5

5.   a) 0
      b) Has no leading
      coefficient
      c) 0

6.   a) 8
      b) -41
      c) 3

7.   a) 2
      b) 2
      c) 98

8.   a) 6
      b) 4
      c) -½

B.   1) Multinomial
      2) Monomial

3) Binomial
4) Trinomial
5) Trinomial
6) Multinomial

C.   1. Has no degree
      2. zero
      3. 1
      4. 1

D.   1) 7
      2) 5
      3) 11
      4) 3

E.   1) $-3x^2-2x+8$
      degree=2
      Leading coefficient=-3

      2) $-8a^7+3a^5-2a^3+4a^2+7a$
      Degree=7
      Leading coefficient=-8

      3) $-3b^4+2b^2-5b+10$
      Degree=4
      Leading coefficient=-3

      4) $6c^6+7c^5+18c^2-13c$
      Degree=6
      Leading coefficient=6

      5) $2y^7+3y^5+4y^3-4y+2$
      Degree=7
      Leading coefficient=2

F. a) $x^4+4x^3-2x^2+x+36$
    b) $6x^2-17x+7$

G. 1)   a) $(3x^2-2x+5)+(2x^2+4x+6)=5x^2+2x+11$

    b) $3x^2 - 2x + 5$
    $+2x^2 + 4x + 6$
    $\overline{5x^2 + 2x + 11}$

    2)   a) $(2x^4-3x^3+4x^2+2)+(6x^4+5x^2+6)=8x^4-3x^3+9x^2+8$

    b) $2x^4 - 3x^3 + 4x^2 + 2$
    $+6x^4 + 5x^2 + 6$
    $\overline{8x^4 - 3x^3 + 9x^2 + 8}$

    3)   a) $(2a^5+3a^2-6)+(4a^5+2a^4+4a^2+2a-1)=6a^5+2a^4+7a^2+2a-7$

    b) $2a^5 + 3a^2 + 6$
    $+4a^5 + 2a^4 + 4a^2 + 2a - 1$
    $\overline{6a^5 + 2a^4 + 7a^2 + 2a - 7}$

    4)   a) $(d^2-16d+10)+(12d^2+8d+3)=13d^2-8d+13$

    b) $d^2 - 16d + 10$
    $+12d^2 + 8d + 3$
    $\overline{13d^2 - 8d + 13}$

    5)   a) $(11y^2+6y-1)+(2y^2-7y+5)=13y^2-y+4$

    b) $11y^2 + 6y - 1$
    $+2y^2 - 7y + 5$
    $\overline{13y^2 - y + 4}$

H. 1) a) $(2x^2+7x+2)-(6x-2)=2x^2+x+4$

b) $2x^2 + 7x + 2$
$$\underline{-(6x-2)}$$
$2x^2 + x + 4$

2) a) $(5x^3+4x^2-6x+1)-(3x^2+3x+7)=5x^3+x^2-9x-6$

b) $5x^3 + 4x^2 - 6x + 1$
$$\underline{-(3x^2 + 3x + 7)}$$
$5x^3 + x^2 - 9x - 6$

3) a) $(3y^2-9y+2)-(y^2-2)=2y^2-9y+4$

b) $3y^2 - 9y + 2$
$$\underline{-\left(y^2 - 2\right)}$$
$2y^2 - 9y + 4$

4) a) $(-4c^5-2c^4+3c^2+2)-(2c^4+c^2+4)=-4c^5-4c^4+2c^2-2$

b) $-4c^5 - 2c^4 + 3c^2 + 2$
$$\underline{-\left(+2c^4 + c^2 + 4\right)}$$
$-4c^5 - 4c^4 + 2c^2 - 2$

5) a) $(11x^2+6x-1)-(3x-5)=11x^2+3x+4$

b) $11x^2 + 6x - 1$
$$\underline{-(3x-5)}$$
$11x^2 + 3x + 4$

EXERCISE 2.2

a) figure

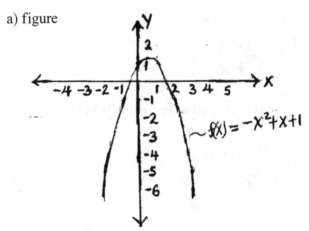

b) The graph open down.
c) 0
d) Doesn't have minimum value.
e) V(0, 0)
f) Y≤0

a) figure

$$\sim f(x) = -x^2 + x + 1$$

b) It opens down.
c) $x = \frac{1}{2}$

d) $\frac{5}{4}$

e) $v\left(\dfrac{1}{2}, \dfrac{5}{4}\right)$

f) $y \le \dfrac{5}{4}$

a) figure

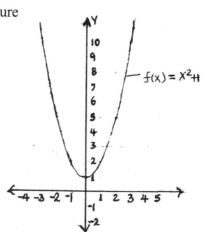

$f(x) = x^2 + 1$

b) opens up
c) 0
d) 1
e) (0, 1)
f) y≥1

4) a) $3(x+4)^2 - 38$

b) $(x-3)^2 - 5$

c) $3\left(x - \dfrac{10}{6}\right)^2 - \dfrac{49}{3}$

d) $-3\left(x - \dfrac{3}{2}\right)^2 + \dfrac{35}{4}$

e) $2\left(x + \dfrac{3}{2}\right)^2 - \dfrac{11}{2}$

5) v(-12, 44) and maximum value=44

6) a) $v\left(-\frac{1}{6}, \frac{73}{12}\right)$

    b) $-\frac{1}{6}$

    c) $\frac{73}{12}$

    d) $y \leq \frac{73}{12}$

7) x=-6

8) $2(x+3)^2-13$

EXERCISE 2.3

1) $\left\{\frac{1}{2}, -4\right\}$

2) $\left\{\frac{1}{2}, -\frac{2}{3}\right\}$

3) $\left\{7, \frac{4}{3}\right\}$

4) $\left\{0, -\frac{9}{2}\right\}$

5) $\left\{\frac{1}{2}, -\frac{4}{7}\right\}$

EXERCISE 2.4

1) a) {3, 0}
    b) {28, 0}
    c) {3, 0}
    d) {0, -5}

    e) $\left\{0, -\frac{1}{2}\right\}$

2) a) x(3x-5)

    b) $x\left(\frac{1}{4}x - 9\right)$

    c) x(x-25)

    d) $x\left(2x - \frac{1}{4}\right)$

    e) $x\left(2 - \frac{3}{2}x\right)$

EXERCISE 2.5

1)   a) $\left(x+\dfrac{1}{9}\right)\left(x-\dfrac{1}{9}\right)$

   b) $\left(x+\sqrt{2}\right)\left(x-\sqrt{2}\right)$

   c) (6+ab)(6-ab)
   d) (9+5x)(9-5x)
   e) (0.5x+0.8)(0.5x-0.8)

   f) $\left(\sqrt{2}x+12\right)\left(\sqrt{2}x-12\right)$

2)   a) $\left\{\sqrt{3},-\sqrt{3}\right\}$

   b) $\left\{\dfrac{9}{2},-\dfrac{9}{2}\right\}$

   c) $\left\{\sqrt{24},-\sqrt{24}\right\}$

   d) {4, -4}
   e) {0.6, -0.6}
   f) {5, -5}

EXERCISE 2.6

1)   a) -3 and 7
   b) 4 and -5
   c) 6 and 3
   d) $\dfrac{1}{4}$ and $\dfrac{2}{4}$
   e) -3 and -5
   f) 7 and -1
   g) 14 and -3

2)   a) {-3, -4}
   b) {3, 5}
   c) {1, -⅓}
   d) {-¼, -½}
   e) {5, -½}

3) a) (x+5)(x-5)
    b) $(x-6)^2$
    c) $(y-1)^2$
    d) $(x-4)^2$
    e) (x-5)(x-4)
    f) (x-3)(x-4)

4) a) $\left\{ \frac{1}{2}, -\frac{1}{2} \right\}$

    b) $\left\{ \frac{1}{2}, -2 \right\}$

    c) {5, 3}
    d) {5, 4}
    e) {4, -3}

5) a) $\left\{ \frac{2}{3}, -\frac{1}{2} \right\}$

    b) $\left\{ \frac{2}{5}, -2 \right\}$

    c) $\left\{ 1, -\frac{1}{7} \right\}$

    d) {-3, -4}
    e) {7, -4}

EXERCISE 2.7

1) a) $(4y^2-9)$
    b) $(16x^2-1)$

    c) $\left(a^2 - \frac{1}{16} \right)$

    d) $(49x^2-9)$

2) a) $(4x^2-4y^2)$
    b) $(4-36x^2)$
    c) $(25-16x^2)$

3)   a) $(x-3)^2$
         b) $(x-\frac{1}{2})(x-\frac{1}{4})$
         c) $(y-1)^2$
         d) $(x+4)^2$

4)   a) Perfect square
         b) Perfect square
         c) Not perfect square
         d) Not perfect square
         e) Not perfect square
         f) Not perfect square

EXERCISE 2.8

1)   a) $\{7, -7\}$

         b) $\left\{2\sqrt{3}, -2\sqrt{3}\right\}$

         c) $\left\{\dfrac{1}{2}\sqrt{5}, \dfrac{-1}{2}\sqrt{5}\right\}$

         d) $\{8, -8\}$

         e) $\left\{\dfrac{1}{11}, \dfrac{-1}{11}\right\}$

2   a) $\left\{2+\sqrt{17}, 2-\sqrt{17}\right\}$

         b) $\{\ \}$

         c) $\left\{\frac{5}{4}, \frac{-3}{4}\right\}$

         d) $\{\ \}$

         e) $\left\{-1, \frac{-5}{3}\right\}$

3) a) $\left\{-3+\sqrt{6},-3-\sqrt{6}\right\}$

 b) { }

 c) $\left\{\dfrac{11+\sqrt{109}}{2},\dfrac{11-\sqrt{109}}{2}\right\}$

 d) $\left\{-4+\sqrt{11},-4-\sqrt{11}\right\}$

 e) $\left\{\dfrac{3+\sqrt{17}}{4},\dfrac{3-\sqrt{17}}{4}\right\}$

 f) {-1, -2}

4) a) $\left\{-2+\sqrt{6},-2-\sqrt{6}\right\}$

 b) { }

 c) $\left\{\dfrac{-\left(1+2\sqrt{7}\right)}{9},\dfrac{-\left(1-2\sqrt{7}\right)}{9}\right\}$

 d) { }

 e) $\left\{\dfrac{-27+3\sqrt{73}}{4},\dfrac{-27-3\sqrt{73}}{4}\right\}$

EXERCISE 2.9

1) a) {3}

 b) { }

 c) $\left\{\dfrac{5+\sqrt{13}}{6},\dfrac{5-\sqrt{13}}{6}\right\}$

 d) $\left\{-3+\sqrt{11},-3-\sqrt{11}\right\}$

 e) $\left\{-\dfrac{1}{2},-1\right\}$

2)   a) No roots
        b) Two real roots
        c) One real roots
        d) No roots
        e) One real roots.

3)   36

4)   a) $\{2, -2\}$

        b) $\frac{9}{20}$

        c) $\frac{27}{4}$

        d) $\pm\sqrt{8}$

5)   a) k>1 or k<-1

        b) $k = \sqrt{6}$ or $k = -\sqrt{6}$

        c) $k > \frac{1}{15}$

        d) $k < \frac{9}{16}$

EXERCISE 2.10

a) figure

x^2-2x+1

b) figure

x^2-3x+6

The graph touches the x-axis at a point 1, so the solution set is: {1}

The graph doesn't crosses or touches the x-axis, so the solution set is: { }

EXERCISE 2.11

1a) figure

$y=x^2+6x+8$

a) The x-intercepts are -2 and -4
b) The y-intercept is 8
c) The axis of symmetry is x=-3
d) V(-3, -1)
e) y≥-1

1b) figure

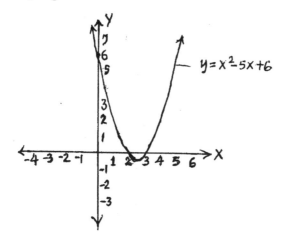

a) The x-intercepts are 2 and 3.
b) The y-intercept is 6.
c) The axis of symmetry is $\dfrac{5}{2}$.
d) The vertex of the parabola is $V\left(\dfrac{5}{2}, \dfrac{-1}{4}\right)$
e) The range is $y \geq -\dfrac{1}{4}$

EXERCISE 2.12

1) a) -4
    b) 4
    c) $9\!\!/\!\!_2$
    d) -7
    e) $-2\!\!/\!\!_3$

2) a) 6
    b) $\dfrac{-10}{3}$
    c) $\dfrac{1}{10}$
    d) -4
    e) -18

3) $3\!\!/\!\!_2$

4) a) 8 and -2
    b) -4 and 3
    c) -11 and -7
    d) $1\!\!/\!\!_9$ and $\dfrac{1}{2}$
    e) -9 and -5

5) $\left\{\dfrac{8}{3}, -1\right\}$

6) k=1 and $r_2 = \dfrac{5}{2}$

7) k=3

8) a) 13
    b) $\dfrac{5}{36}$
    c) $\dfrac{5}{6}$

9) k=0

EXERCISE 2.13

1) 1, 6, 15, 20, 15, 6, and 1
2) 1, 9, 36, 84, 126, 126, 84, 36, 9, 1
3) 128
4) -1080
5) a) $(x-4)^4 = x^4 - 16x^3 + 96x^2 - 256x + 256$
    b) $(4-y)^5 = ((-y)+4)^5 = -y^5 + 20y^4 - 160y^3 + 640y^2 - 1280y + 1024$
    c) $(3x-2)^3 = 27x^3 - 54x^2 + 36x - 8$
    d) $(2x+6)^4 = 16x^4 + 192x^3 + 864x^2 + 1728x + 1296$
    e) $(4x+2)^4 = 256x^4 + 512x^3 + 384x^2 + 128x + 16$
6) -4,320
7) -405

CHAPTER - 3

EXERCISE 3-1

1) a) odd
    b) neither odd nor even
    c) neither odd nor even
    d) even
    e) even
    f) neither odd nor even
    g) odd
    h) odd

2) figure

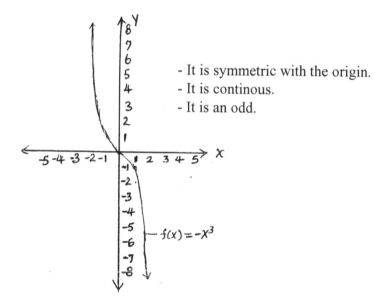

- It is symmetric with the origin.
- It is continous.
- It is an odd.

$f(x) = -x^3$

3) a) symmetric with y-axis.
    b) symmetric with origin.
    c) symmetric with origin
    d) symmetric with y-axis.
    e) symmetric with origin.

EXERCISE 3.2

a)   Neither increasing nor decreasing
b)   Decreasing
c)   Increasing
d)   Decreasing
e)   Increasing
f)   Decreasing
g)   Neither increasing nor decreasing

EXERCISE 3.3

(1)   a) $(x+3)^3$
       b) $x(x+7)^3$
       c) $-4(x+3)^3$
       d) $3x(x-2)^3$
       e) $(x+6)^3$
       f) $-2(x-3)^3$

2)   a) $-4(ab+4)^3$
       b) $(7xy+z)^3$
       c) $(rst+5)^3$
       d) $3(pq-6)^3$

3)   a) 12
       b) 144

4)   900

EXERCISE 3.4

a)　　　figure

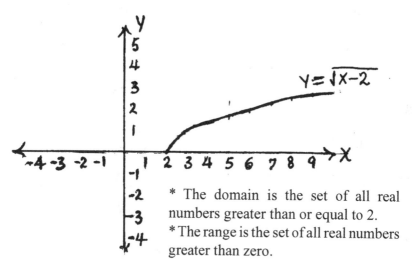

* The domain is the set of all real numbers greater than or equal to 2.
* The range is the set of all real numbers greater than zero.

b)　　　figure

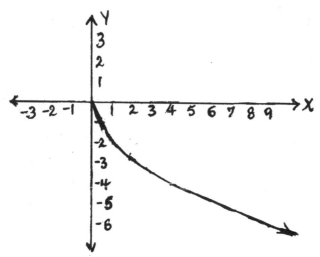

* The domain is the set of real numbers greater than or equal to zero.
* The range is the set of all real numbers less than or equal to zero.

EXERCISE 3.5

a)   32
b)   54
c) $30\sqrt{10}$
d) $18\sqrt{6}$
e) $7\sqrt{2}$
f) $3\sqrt{14}$
g)   24

EXERCISE 3.6

1) $|x+3|$
2) $|y|\,|y-2|$
3) $(x+1)$
4) $|x+5|$
5) $-2xy$
6) $3xy$
7) $11|x-5|$

8) $|xyzw|\sqrt{zw}$

9) $48\sqrt{3}$
10) $2\sqrt[3]{2}$
11) $(a+b)\sqrt[3]{(a+b)}$

12)   59049

13) $\dfrac{2}{|y|}\sqrt{\dfrac{3}{y}}$

14)   5
15)   25

16) $\dfrac{1}{5}|x|\sqrt{x}$

17) $\dfrac{8x}{y}$

18) $2\sqrt{5}\left(5-16\sqrt{2}\right)$

19) $\sqrt[3]{2xy}\,(12y+7)$

20) $-\left(9\sqrt{6}+11\right)$

21) $\sqrt[3]{3/25}$

22) $\sqrt{\dfrac{3x}{7y}}$

23) $\dfrac{2y}{x}\sqrt{\dfrac{2}{3}}$

EXERCISE 3.7

1)   a) 6

b) $\sqrt{39}$

c) $x^2 y^2 zw\sqrt[3]{z^2}$

d) $0.01x^2 y^3 |z|\sqrt{0.1}$

e) $x^9 y^{12} z^{-4}$

f) $\sqrt{7}\left(1-6\sqrt{2}\right)$

g) $|x-3|$

h) $|2c-3|$

i) $|z|\sqrt{xy}$

j) -6912

490

k) $|pqz|$

l) $\dfrac{|x^3y^3|}{z^2}\sqrt{\dfrac{y}{2}}$

m) $|p+6|$

n) $|x-3|$

o) $\dfrac{3}{2}\sqrt[3]{3}-32|xy|+4\sqrt{xy}$

p) $-9-4\sqrt{5}$

q) $x\sqrt[3]{3x^2y}$

r) $8|rsk|\sqrt{3}+2|rs|-\sqrt{rst}$

s) $25|mnr|\sqrt{2}$

2)   a) 1

b) $\dfrac{3x-25}{3\sqrt{3x}-15}$

c) $\dfrac{4xy}{\sqrt{2x}\square\sqrt[3]{4^2xy^2}}$

d) $\dfrac{y-3}{y+2\sqrt{3y}+3}$

e) $\dfrac{-(2+y)}{\sqrt{6+y}(y+3)+2y+6}$

3)   a) $\dfrac{\sqrt[3]{20y}}{2}$

b) $\dfrac{\sqrt[3]{a^2b^2}}{ab}$

c) $\dfrac{63-7\sqrt{10}}{71}$

d) $\dfrac{3\sqrt{6}+\sqrt{10}}{2}$

e) $\dfrac{\sqrt{2}\left(8-3\sqrt{5}\right)+2\sqrt{7}\left(8-3\sqrt{5}\right)}{-26}$

f) $\dfrac{a+b-2\sqrt{ab}}{a-b}$

g) $-\left(\sqrt{x-1}-\sqrt{x+2}\right)$

h) $\dfrac{x\left(2-\sqrt{x}\right)}{4-x}$

i) $\dfrac{2\sqrt[3]{5^2}}{5}$

j) $\sqrt{x+1}+2$

k) $\dfrac{5\sqrt{3}-9\sqrt{2}}{12}$

l) $-\left(\sqrt{3}-6\right)$

m) $\sqrt{x+3}-2$

n) $\dfrac{x+h+x\sqrt{x}+\sqrt{x+h}\left(x+\sqrt{x}\right)}{x+h-x^2}$

EXERCISE 3.8

1) { }

2) $\left\{\dfrac{1}{4}\right\}$

3) { }

4) $\left\{\dfrac{1}{4}\right\}$

5) {1294}

6) { }

7) {5}

8) { }

9) $\dfrac{1}{2}$

10) {-4}

11) -3 is an extraneous solution.
12) { }
13) {6}

EXERCISE 3.9

a) x=2
b) x=2 or -2
c) No vertical asymptotes

EXERCISE 3.10

a) No horizontal asymptote
b) y=0 or the x-axis.
c) y=0 or the x-axis.
d) y=3

e) y=0
f) y=2
g) y=0
h) Has no horizontal asymptote

EXERCISE 3.11

a) $x = \dfrac{3}{5}$ (V.A.)

 y=0 (H.A)

b) x=-5 (V.A.)
 y=7 (H.A.)

c) No vertical asymptote
 y=5 (H.A.)

d) x=2 or x=½ (V.A.)
 y=1 (H.A.)

e) No vertical asymptote
 y=0 (H.A.)

f) x=1 or x=-1 (V.A.)
 y=1 (H.A.)

g) x=1 or $x = {}^{-5}\!/_3$ (V.A.)
 y=0 (H.A.)

h) $x = {}^{5}\!/_4$ (V.A.)

 $y = {}^{3}\!/_4$ (H.A.)

i) x=-1 or x=2 (V.A.)
 y=3 (H.A.)

j) x=2 or x=-3 (V.A.)
 y=7 (H.A.)

2)   a) x=2 (V.A.)
        y=2x-1 (O.A)

        b) x=2 (V.A.)
        y=3x-1 (O.A.)

        c) x=2 or x=-2 (V.A.)
        y=5x (O.A.)

        d) x=5 (V.A)
        y=2x+3 (O.A.)

        e) x=1 (V.A)
        y=3x+3 (O.A.)

EXERCISE 3.12

1) $q(x)=3, r(x)=-4x+7$
2) $q(x)=x^2+3x+9, r(x)=26$
3) $q(x)=x^2+3x+1, r(x)=0$
4) $q(x)=x^2+x+1, r(x)=x^2-3$
5) $q(x)=2x-1, r(x)=-6x+6$

EXERCISE 3.13

6) $q(x)=2x+5, r(x)=0$
7) $q(x)=3x-8, r(x)=20$
8) $q(x)=5x^2-16x+35, r(x)=-59$
9) $q(x)=5x^2-15x+48, r(x)=-146$
10) $q(x)=4x-3, r(x)=13$

EXERCISE 3.14

11)   24
12)   -11
13)   -1
14)   -3

15) $\dfrac{25}{8}$

16) 41

17) $8\!\!\not{}_{81}$

18) 167
19) -41
20) 190
21) -1

22) $K = \dfrac{98}{5}$

23) 4
24) -500
25 -3100

26) $-5\!\!\not{}_{4}$

27) 11

EXERCISE 3.15

1)   a) $\frac{2}{3}$

    b) No excluded value

    c) -1

    d) $3, -\frac{1}{2}$

    e) 0, 2, -2

    f) 3, -3

    g) $\frac{1}{2}, -\frac{5}{2}$

2)   a) $\dfrac{2x-1}{x}$

    b) $\dfrac{x-4}{x+4}$

    c) $\dfrac{-3x-2}{x+1}$

    d) $\dfrac{x+2}{x-2}$

    e) $\dfrac{x^2}{x+1}$

    f) $\dfrac{h^2}{h^2+6h+9}$

    g) $\dfrac{3x-1}{x+4}$

3)   a) $\dfrac{6x+16}{2x^2+13x+15}$

b) $\dfrac{7x+9}{6x^2-2x}$

4) M=x²-3x+2

EXERCISE 3.16

1) a) $\dfrac{x^2+2x+4}{3x}$

b) $\dfrac{3x^2-9x-30}{4x^2-22x+24}$

c) $\dfrac{x^3-3x^2+3x-1}{2x^2-8}$

d) $\dfrac{x^2-9}{x^2+4x}$

e) $\dfrac{x+3}{x-2}$

2) a) $\dfrac{28x^4+28x^3+24x^2-84x-42}{3x^3-15x^2-60x}$

b) $\dfrac{2x+6}{3}$

c) $\dfrac{x^2-2x}{x^2+x-2}$

d) 4x-4

e) $\dfrac{x^2+6x+8}{x-5}$

3)   a) $\dfrac{4x^2+x+2}{x^2-4}$

b) $\dfrac{x^3+x^2-3x+1}{(x-4)(x-1)}$

c) $\dfrac{8x^3+x^2+1}{20x^3+8x^2-5x-2}$

d) $\dfrac{12x+10}{6x+5}$

e) $\dfrac{5x^2+19x}{x^3+6x^2-x-30}$

4)   a) $\dfrac{x}{x-5}$

b) $\dfrac{x-6}{x+2}$

c) $\dfrac{-5x}{x^3-3x^2-22x+24}$

d) $\dfrac{-2x^2+11x+1}{x^2-16}$

e) $\dfrac{19x+10}{x^3-5x^2}$

5)   a) $\dfrac{-x+14}{7}$

b) $\dfrac{2x+4}{x+1}$

c) $\dfrac{x-4}{x-1}$

d) $\dfrac{1}{x+3}$

e) $\dfrac{1}{3}$

f) $\dfrac{x-2}{x+1}$

6) a) 1
    b) 8
    c) Undefined
    d) Undefined
    e) $\dfrac{23}{65}$

EXERCISE 3.17

1) a) { }
    b) { }
    c) {R |{0, 6}} or all real numbers except 0 and 6.

    d) $\left\{ 11\!\!\diagup\!\!_{30} \right\}$

    e) { }

    f) $\left\{ \dfrac{-1}{4}, 1 \right\}$

2) a) Not equivalent
    b) Not equivalent
    c) Equivalent
    d) Not equivalent
    e) Equivalent

EXERCISE 3.18

1) a) $15\!\!\diagup\!\!_{8}$

b) $-\frac{7}{8}$

2) a) $\frac{9}{2}$

b) -2

EXERCISE 3.19

1) a) 4, 13, 28, 49, 76, 109

b) $-\frac{1}{2}, -\frac{1}{5}, 0, \frac{1}{7}, \frac{1}{4}, \frac{1}{3}$

c) 4, 9, $\frac{512}{27}$, $\frac{625}{16}$, $\left(\frac{12}{5}\right)^5$, $\left(\frac{7}{3}\right)^6$

d) $\sqrt{3}$, $\sqrt{5}$, $\sqrt{7}$, 3, $\sqrt{11}$, $\sqrt{13}$

e) $\frac{1}{2}, \frac{1}{4}, \frac{1}{6}, \frac{1}{8}, \frac{1}{10}, \frac{1}{12}$

2) a) $a_n = 4n-1$

b) $a_n = -4n+14$

c) $a_n = 3\left(\frac{2}{3}\right)^{n-1}$

d) $\frac{n}{n+1}$

e) $0.3\left(\frac{1}{10}\right)^{n-1}$

3) 183

4) -198

5)   -7

6)   -602

7)   99

8)   26

9)   20.5

10)   -1, -4, -7, -10, -13, -16

11)   11, 15, 19, 23, 27

12)   a) $-\frac{1}{3}$

        b) 5
        c) -1
        d) 4

        e) $\dfrac{1}{5}$

13) $-2(4)^{33}$

14)   2

CHAPTER - 4

EXERCISE 4.1

1)   a) 10
        b) $7\sqrt{2}$
        c) $\sqrt{145}$

2)   a) $10\sqrt{3}$
        b) $\sqrt{119}$
        c) $2\sqrt{22}$

3)   a) $\sqrt{104}$
        b) $4\sqrt{2}$
        c) $2\sqrt{5}$
        d) $\sqrt{5}$
        e) $\sqrt{61}$

4)   5 or 3

5)   1 or 7

6)   a) 12.5 square miles

        b) $\left(10+5\sqrt{2}\right)$ miles

        c) Isosceles right angled triangle.

7)   a) (-2, -3)
        b) (5, -6.5)
        c) (4, 4.5)

        d) $\left(-\frac{1}{2},-1\right)$

        e) $\left(\frac{1}{2},5\right)$

f) $\left(\frac{3}{2}, \frac{7}{2}\right)$

8) a=4 and b=4

9) $x = \frac{-5}{2}$ and $y = \frac{8}{3}$

EXERCISE 4.2

1) 243

2) $\frac{1}{32}$

3)

4)

5) a) The number 0 is neither positive nor negative.
    b) $\sqrt{3}$ is irrational and $-\sqrt{3}$ is also irrational but

    $\sqrt{3} + \left(-\sqrt{3}\right) = 0$, which is rational number.

    c) The quotient of $\frac{x}{0}$, where x is any number is not defined.

    (Division by zero is undefined).
    d) The product of -2 and 3 is negative.
    e) The area of a rectangle having length 5 and width 9 is 45, which is an odd number.

EXERCISE 4.3

1)   a) Proposition
           b) Not proposition.
           c) Not proposition.
           d) Proposition
           e) proposition

2)   a) True
           b) False
           c) True
           d) True
           e) False

EXERCISE 4.4

1)   a) Compound proposition.
           b) Compound proposition.
           c) Compound proposition.
           d) Simple proposition
           e) Compound proposition.

2)   a) Disjunction
           b) Implication
           c) Negation
           d) Bi-Implication
           e) Disjunction

EXERCISE 4.5

1)   a) $r \wedge s$
           b) $\neg r \wedge s$
           c) $\neg r \vee \neg s$
           d) $r \Rightarrow s$
           e) $r \Leftrightarrow s$
           f) $r \Leftrightarrow \neg s$
           g) $\neg(r \wedge s)$
           h) $r \wedge \neg s$

2)   a) A rectangle doesn't have three sides.
        b) A rectangle has three sides and Tuesday is the day after
        Monday.
        c) A rectangle has three sides or Tuesday is the day after
        Monday
        d) If a rectangle has three sides then Tuesday is the day after
        Monday.
        e) A rectangle has three sides if and only if Tuesday is the day
        after Monday.
        f) If a rectangle doesn't have three sides, then Tuesday is the
        day after Monday.
        g) A rectangle doesn't have three sides if and only if Tuesday
        is not the day after Monday.
        h) If a rectangle doesn't have three sides, then Tuesday is not
        the day after Monday.
        i) Tuesday is not the day after Monday.

3)   a) T
        b) F
        c) T
        d) T
        e) F
        f) T
        g) F
        h) F
        i) F

4)   a) T
        b) F
        c) T
        d) T
        e) T
        f) T
        g) F
        h) F
        i) T
        j) F

5)   a) T
        b) F
        c) T
        d) T
        e) F
        f) T
        g) T
        h) T
        i) F
        j) T
        k) T

6) a)

Converse:- If Addis Ababa is a capital city of Ethiopia, then 2 is
        a prime.
Inverse:- If 2 is not a prime, then Addis Ababa is not a capital city
        of Ethiopia.
Contrapositive:- If Addis Ababa is not a capital city of Ethiopia,
        then 2 is not a prime.

   b)

Converse:- If all birds have three legs, then -5 is an integer.
Inverse:- If -5 is not an integer, then all birds do not have three
        legs.
Contrapositive:- If all birds do not have three legs, then -5 is not
        an integer.

   c)

Converse:- If 0 is a negative number, then $\sqrt{3}$ is rational.
Inverse:- If $\sqrt{3}$ is not rational, then 0 is not a negative number.
Contrapositive:- If 0 is not a negative number, then $\sqrt{3}$ is not
        rational.

7)   a) T
        Converse: T
        Inverse: T
        Contrapositive: T

        b) F
        Converse: T

Inverse: T
Contrapositive: F

c) T
Converse: T
Inverse: T
Contrapositive: T

8)   a) It is hot and not raining.
     b) It is not hot.
     c) It is raining.
     d) It is not raining if and only if it is hot.
     e) It is hot if and only if it is raining.
     f) It is not raining or it is not hot.
     g) If it is raining, then it is not hot.

EXERCISE 4.6

1)   a) ¬(P∧q)≡¬P∨¬q

P	q	¬P	¬q	P∧q	¬(P∧q)	¬P∨¬q
T	T	F	F	T	F	F
T	F	F	T	F	T	T
F	T	T	F	F	T	T
F	F	T	T	F	T	T

The last column are the same, showing that ¬(P∧q)≡¬P∨¬q

b) P∨(q∨r)≡(P∨q)∨r

P	q	r	q∨r	P∨(q∨r)	P∨q	(P∨q)∨r
T	T	T	T	T	T	T
T	T	F	T	T	T	T
T	F	T	T	T	T	T

T	F	F	F	T	T	T
F	T	T	T	T	T	T
F	T	F	T	T	T	T
F	F	T	T	T	F	T
F	F	F	F	F	F	F

Column (5) and (7) are the same showing that: $P\lor(q\lor r)\equiv(P\lor q)\lor r$

c) $P\land q\equiv q\land P$

P	q	P∧q	q∧P
T	T	T	T
T	F	F	F
F	T	F	F
F	F	F	F

Column (3) and (4) are the same showing $(P\land q)\equiv(q\land P)$.

d) $\neg(P\lor q)\equiv\neg P\land\neg q$

P	q	¬P	¬q	(P∨q)	¬(P∨q)	¬P∧¬q
T	T	F	F	T	F	F
T	F	F	T	T	F	F
F	T	T	F	T	F	F
F	F	T	T	F	T	T

Column (6) and (7) are the same showing $\neg(P\lor q)\equiv\neg P\land\neg q$

e) $P\Rightarrow q\equiv\neg(P\land\neg q)\equiv\neg P\lor q$

•

P	q	¬P	¬q	P⇒q	P∧¬q	¬(P∧¬q)	¬P∨q
T	T	F	F	T	F	T	T
T	F	F	T	F	T	F	F
F	T	T	F	T	F	T	T
F	F	T	T	T	F	T	T

Column (5), (7) and (8) are the same showing: P⇒q≡¬(P∧¬q)≡¬P∨q

2)   a) (P⇒q)⇔(¬P∨q)

P	q	¬P	P⇒q	¬P∨q	(P⇒q)⇔(¬P∨q)
T	T	F	T	T	T
T	F	F	F	F	T
F	T	T	T	T	T
F	F	T	T	T	T

All the truth value of the last column is true, thus, the complex proposition is a tautology.

b) (P∨q)⇔(q∨P)

P	q	P∨q	q∨P	(P∨q)⇔(q∨P)
T	T	T	T	T
T	F	T	T	T
F	T	T	T	T
F	F	F	F	T

All the truth value of the last column is true, showing that the complex proposition is a tautology.

c) q⇒(P⇒q)

P	q	(P∨q)	(q∨P)	(P∨q)⇔(q∨P)
T	T	T	T	T
T	F	T	T	T
F	T	T	T	T
F	F	F	F	T

All the truth value of the last column are true, showing, the complex proposition is a tautology.

3) a) P⇔¬P

P	¬P	P⇔¬P
T	F	F
F	T	F

All the last column has a truth value False, showing the complex proposition is contradiction.

b) (r∧s)∧¬(r∨s)

r	s	r∧s	r∨s	¬(r∨s)	(r∧s)∧¬(r∨s)
T	T	T	T	F	F
T	F	F	T	F	F
F	T	F	T	F	F
F	F	F	F	T	F

All the last column have truth value False showing the complex proposition is contradiction.

4)   a) F
        b) T
        c) T
        d) F

EXERCISE 4-7

1)   24 inches
2)   20°
3)   a) m(∠a)=39°
        b) m(∠c)=39°
        c) m(∠a)≅m(∠c)
4)   a) x=18°
        b) m(∠P)=m(∠Q)=44°
5)   x=20°, y=20°
6)   a) x=10°, y=72°
        b) x=10°, y=30°
7)   a) x=21°
        b) 63°
        c) 27°

EXERCISE 4-8

1) $\dfrac{4\sqrt{10}}{5}$

2)   k=4.5

3) $k = \dfrac{7\sqrt{13}-14}{3}$ or $\dfrac{-7\sqrt{13}-14}{3}$

EXERCISE 4-9

1)   a) ∠S
        b) ∠T
        c) ∠U
        d) $\overline{SU}$

e) $\overline{TU}$
f) $\overline{ST}$
g) ΔTSU
h) ΔTUS

2)   a) $\overline{FH}$
b) $\overline{FG}$
c) ΔGFH
d) ΔHFG

EXERCISE 4-10

1)   a) ΔEDF
b) ΔEDF

2)   a) ΔAEB
b) ΔOQP

3)   a) ΔSRQ
b) ΔRQP

4)   a) ΔCDP
b) ΔCAD

5)   a) ΔDAC
b) ΔQPR

6)   a) ΔCDB
b) ΔNPM

7)   a) ΔORA
b) ΔAOC

8)   ΔXBZ≅ΔYAZ and $\overline{XB} \cong \overline{YA}$

9)   a) x=4
b) x=8

CHAPTER-5

EXERCISE 5-1

1)   a) 12ft

          b) 20cm

2)   17cm

3) $\overline{AC} = 28cm$ and $\overline{XY} = 8cm$

4)   4

5)   17cm

6)   69ft

7)   a) 6

          b) 29

8)   a) $\overline{PQ} = 10$

          b) $\overline{PS} = 11$

          c) $\overline{QR} = 10$

          d) $\overline{SR} = 11$

9)   88cm

10)   a) 5

          b) 3

11)   3

12)   6

13)   a) 12

          b) $4\sqrt{6}$

14)   a) $\dfrac{144}{13}$

          b) 6.4

15)   D(6.3, 4.6)

16)   a) x>2

          b) x>1

17)   a) cannot be, because the sum of any two sides of a triangle must be greater than the third side.

b) can be the sides of a triangle.
c) can be the sides of a triangle.
d) can be the sides of a triangle.
e) cannot be, the same reason as (a)
f) can be the sides of a triangle.

18)   a) The shortest side is the side opposite to 42°, which is $\overline{PR}$.
       - The longest side is the side opposite to 78°, which is $\overline{QR}$.
       b) The shortest side is the side opposite to 38°, which is $\overline{AC}$.
       - The longest side is the side opposite to 80°, which is $\overline{AB}$.

19)   a) The smallest angle is ∠R
       The largest angle is∠P
       b) The smallest angle is ∠E or ∠F
       The largest angle is ∠D
       c) The smallest angle is ∠C
       The largest angle is ∠A
       d) All angles are equal.

EXERCISE 5-2

1)   1080°

2)   7

3)   150°

4)   360°

5)   40°

6)   93°

7)   176cm

8)   156°

9)   30

10)   60°

11)   36°

12)   36°

13)   1080°

14)   18

15)   a) 30°
        b) 171°

EXERCISE 5-3

1)   a) 8
        b) 10
        c) 24°
        d) 10°

2)   a) 20°
        b) 30°
        c) 30°

3)   a) 36°
        b) 20°
        c) 10°

4) $\left(\dfrac{3}{2},5\right)$

5)   4 and 5

6)   a) 12cm
        b) 7cm
        c) 3cm

7)   a) True
        b) True
        c) True
        d) False
        e) True
        f) True

EXERCISE 5-4		EXERCISE 5-5	

EXERCISE 5-4   EXERCISE 5-5

1)   a) $2a\sqrt{2}$
      b) $3a\sqrt{2}$
      c) 5a

2)   31ft

3)   a) 46°
      b) 92°
      c) 46°
      d) 92°

4)   x=78° and y=54°

5)   36°

6) $\overline{UR}$

7)   a) x=12° and y=20°
      b) m($\angle$ADC)=60°
      c) 120°

8)   x=5, and the length of the sides of the rhombus is 16.

9)   a) True
      b) True
      c) False
      d) True
      e) True

EXERCISE 5-5

1)   21cm

2)   a) 114°
      b) 114°
      c) 66°

3)   a) 66°
      b) 90°
      c) 66°

4)   5

5)   a) y=4
      b) 11

CHAPTER - 6

EXERCISE: 6-1

1)   24

2)   94

3)   17

4)   6x8=48 ways

5)   10x9x8x7x6=30,240

6)   2x2x2=8 outcomes

7)   9x9x8x7x6=27,216

8)   a) 362,8800
        b) 720
        c) 1
        d) 120

9)   a) 2,860

        b) $\dfrac{1}{10}$

        c) $\dfrac{1}{336}$

        d) n^2-n

        e) n^3-3n^2+2n

10)   5

11)   10

12)   6

EXERCISE 6-2

1)   a) 1,680
        b) 120
        c) 15,120
        d) 42

2)   604,800

3)   132

4)   720

5)   2,520

6)   90

7)   420

8) $\dfrac{17!}{2!2!2!2!2!2!2!}$

EXERCISE 6-3

1)   a) 20
      b) 84
      c) 45
      d) 35

2)   a) 56
      b) 56
      c) 45
      d) 45

3)   a) a and b are equal
      b) c and d are equal

4)   924

5)   15

6)   15,504

7)   210

EXERCISE 6-4

1)   a) An event:- is one or more outcomes of an experiment.
      b) An outcome:- is the result of a single trial of an experiment.
      c) An experiment is any procedure that can be infinitely repeated
      and has a well-defined set of outcomes.

2)   {(h, h), (t, t), (h, t), (t, h)}
      There are 4 different possible outcomes in this experiment.

EXERCISE 6-5

1) $\dfrac{5}{13}$

2) $\dfrac{1}{4}$

3) a) $\dfrac{1}{20}$

   b) $\dfrac{19}{20}$

4) a) $\dfrac{1}{2}$

   b) $\dfrac{1}{2}$

   c) $\dfrac{5}{6}$

   d) $\dfrac{5}{6}$

   e) 1

   f) 0

5) $\dfrac{1}{3}$

EXERCISE 6-6

1) $\dfrac{2}{13}$

2) $\dfrac{2}{3}$

3) $\dfrac{1}{3}$

EXERCISE 6-7

1) $\dfrac{1}{6}$

2) a) $\dfrac{25}{169}$

   b) $\dfrac{1}{9}$

3) $\dfrac{1}{4}$

4) $\dfrac{2}{3}$

5) $\dfrac{1}{784}$

EXERCISE 6-8

1) $\dfrac{5}{6}$

2) $\dfrac{7}{12}$

3) $\dfrac{1}{2}$

EXERCISE 6-9

1) $\dfrac{17}{55}$

2) $\dfrac{47}{66}$

EXERCISE 6-10

1)   a) 76.9
       b) 78.5
       c) 85
       d) The new mean become 71.9 (The new mean decreases by 5)
       e) The new mean become 81.9. (The new mean increases by 5).

2)   52

3)   657

4)   sum=479, mean=79.8

5)   a) 77
       b) 67 and 94
       c) 78

6   a) 21.5
       b) 23
       c) 23

EXERCISE 6-11

1) $v = 28.8$

       $S.d. = \sqrt{28.8}$

2)   a) 9.5
       b) 25
       c) 15.4
       d) 20.4
       e) 20.6
       f) 28.7

4)   a) 9
       b) 2
       c) 8
       d) 17
       e) 38.3

f) $\sqrt{38.3}$
g) Average absolute mean deviation from mean=8.9
Average absolute mean deviation from mean=7
Average absolute mean deviation from mean=5.3

EXERCISE 6-12
1)   True

2)   False

3)   True

4)   True

5)   False

6)   True

7)   True

8)   True

9)   True

10)   a) 2.2
      b) 8
      c) 0
      d) 14
      e) 70.2%
      f) 29.8%

11)   a) -1
      b) -1
      c) 0
      d) $\sqrt{4.7}$
      e) 8
      f) 58.3
      g) 83.3